Springer Proceedings in Business and Economics

Springer Proceedings in Business and Economics brings the most current research presented at conferences and workshops to a global readership. The series features volumes (in electronic and print formats) of selected contributions from conferences in all areas of economics, business, management, and finance. In addition to an overall evaluation by the publisher of the topical interest, scientific quality, and timeliness of each volume, each contribution is refereed to standards comparable to those of leading journals, resulting in authoritative contributions to the respective fields. Springer's production and distribution infrastructure ensures rapid publication and wide circulation of the latest developments in the most compelling and promising areas of research today.

The editorial development of volumes may be managed using Springer Nature's innovative EquinOCS, a proven online conference proceedings submission, management and review system. This system is designed to ensure an efficient timeline for your publication, making Springer Proceedings in Business and Economics the premier series to publish your workshop or conference volume.

This book series is indexed in SCOPUS.

João Carlos de Oliveira Matias
Carina Maria Oliveira Pimentel
João Carlos Gonçalves dos Reis
Joana Maria Costa Martins das Dores
Gilberto Santos
Editors

Quality Innovation and Sustainability

3rd ICQIS, Aveiro University, Portugal, May 3-4, 2022

Springer

Editors
João Carlos de Oliveira Matias
Department of Economics, Management,
Industrial Engineering and Tourism
University of Aveiro
Aveiro, Portugal

Carina Maria Oliveira Pimentel
Department of Economics, Management,
Industrial Engineering and Tourism
University of Aveiro
Aveiro, Portugal

João Carlos Gonçalves dos Reis
Industrial Engineering and Management
Universidade Lusófona
Lisbon, Portugal

Joana Maria Costa Martins das Dores
Department of Economics, Management,
Industrial Engineering and Tourism
University of Aveiro
Aveiro, Portugal

Gilberto Santos
School of Superior Design
Polytechnic Institute of Cávado and Ave
Barcelos, Portugal

ISSN 2198-7246 ISSN 2198-7254 (electronic)
Springer Proceedings in Business and Economics

ISBN 978-3-031-12913-1 ISBN 978-3-031-12914-8 (eBook)
https://doi.org/10.1007/978-3-031-12914-8

This Springer imprint is published by the registered company Springer Nature Switzerland AG
The registered company address is: Gewerbestrasse 11, 6330 Cham, Switzerland

Preface

Introduction

The world is facing multiple and intertwined crises; thus, the present days are particularly challenging to policymakers, managers, academics and citizens in general. Human civilization is pushing the boundaries of political, social, economic and environmental equilibria, which may lead to unprecedented consequences. A welter of events arising from different grounds are forcing the redesign of the established economic models, demanding an overall societal re-organization. Energy transitions differ sharply from transitions in technology, industry and other sectors. Electricity decarbonization is an ongoing process, notwithstanding the decarbonization of industrial production, transport and heating is lagging further behind.

A collective effort towards the reduction in the use of fossil fuels, assuming the technological transition towards greener sources, will substantially raise economic efficiency, along with waste reduction while promoting sustainability and the control of geopolitical risks. An orchestrated effort in the promotion of innovation, quality improvement and sustainability will lead to environmental and climate justice. Hosting this tripod will enhance the societal transition to the communities, building vibrant ecosystems, which will be the seed to productive, dignified and ecologically sustainable livelihoods with fair-minded governance and ecological resilience.

Ground Setting

The pandemic crisis has unleashed an unprecedented uncertainty in financial and productive markets, causing further disruption in economic growth as well as sustainability. The present volume emerges in a particularly challenging context, around which the current macroeconomic scenario demands the re-shaping of the productive processes, the value chains and also the consumption habits. On top of this demanding shift re-organization, the clock of climate emergency continues ticking, giving firms and governments lesser time to encompass sustainability in all actions.

It is essential to tackle the climate crisis, accelerate the deployment of renewable energy and transform productive strategies in all economic activities [10]. Also, it urges to provide sustainable energy systems everywhere. Moreover, the healthcare emergency has collapsed the transcontinental value chains due to their vulnerability to critical conditions, evidencing malfunctioning and extreme fragility [14]. In this vein, world authorities are concentrating on political "re-shoring", re-nationalization and re-turning production and manufacturing back to the parent company's native country [8, 9], while promoting frugal and smarter ways of production [10, 11].

These crises have irreversibly affected the health sector, energy production and distribution and even the political order; as such, skepticism about free markets, international cooperation and open borders augmented [4, 13]. The emerging dynamics have multilateral implications for the international system of production and trade, which need to be carefully addressed. Therefore, the iron arm among globalization, deglobalization and re-globalization trends is not likely to be immediately settled and will irreversibly change the course of international affairs in the twenty-first century [7, 9].

These adjustments target increasing the reliability of the supplies, price controls and local ecosystems to gain endurance in facing future crises at the economic, environmental and sustainability dimensions [6]. The entrepreneurial does need to navigate and survive crises, disregarding its sector, dimension or capital structure [1, 5], minimizing adverse shocks and transforming the difficulties into entrepreneurial opportunities [2]. Leveraging agility and flexibility to react to adverse contexts will enhance resilience, which will further improve survivability in the local context [12, 14].

Producers are no longer allowed to use an excess of raw materials in the production of an asset if could be made with fewer inputs and, thus, at a lower cost.

Therefore, supplementary improvements need to be made. The change to a paradigm in which all agents develop smarter and responsible conducts will allow the emergence of a new societal order [3].

Crossing literature streams, at the level of organizational and technological activities, has been a growing interest towards the promotion of innovation and sustainability, which is indeed quite complex, multifaceted and multidisciplinary. Despite the previous efforts, there is plenty of room for adding research contributions in order to formalize and explore the fields of manufacturing, smarter solutions, logistics and sustainable use of resources, as such, the present volume aims to shed light on these fields.

Outline of the Volume

The volume encompasses miscellaneous contributions from the conference tripod: innovation, quality and sustainability. All these visions aim at widening the debate on the technological and societal changes to be taken in the recent future towards a better use of resources, while promoting smarter ecosystems to accommodate highly efficient productive processes which must be human-centric.

It starts with the insights of "Cost and Learning Factories". Then, the energetic shift is nailed in "Agile 8D Problem-Solving Framework in a Renewable Energy Sector". Knowledge transfer practicalities are explored through "The Transfer Process of Lean Approach Within Multinational Companies' Network: The Schnell S.p.A. Case Study". Innovation methods in health sectors are appraised in "Lean healthcare: A Critical Analysis". Also, the challenges in logistics are highlighted in "Road Freight Transport in Europe: Alternatives for Increasing Capacity". New technological frameworks are presented in "Knowledge Management: An Overview of Roadmaps for Additive Manufacturing" as well as "Life Cycle Inventory of Additive Manufacturing Processes: A Review".

In regards to quality management in the tertiary sector, several aspects are raised in regards to its determinants "Critical Success Factors of TQM for Sustainability in Higher Education Institutions: A theoretical contribution" as well as consumer loyalty "Customer Loyalty in Hospitality: Can the Quality Perception Be a Key Factor?".

Considering the digital transition and industry 4.0, the debate is also open in "The Role of Enabling Technologies from Industry 4.0 in the Formulation of Public Policies for Smart Cities", but also on the "Impact of TQM and Industry 4.0 on Sustainable Performance: An Empirical Study on the Bangladeshi Garment Industry" and "Academic Community Perceptions of Open Innovation: An Exploratory Study".

Greener transitions and innovation in the production processes are debated in "Green Human Resource Management: The Performance of Women Researchers Based on Bibliometric Indicators" and "Innovative Responses to the Covid-19 Pandemic in Primary Health Care: The Case of the Arte Nova Family Health Unit" and still, "Competitive and Business Intelligence: A Bibliometric Analysis". Additionally, new systems and methods were explored in "Integrated Management System Role-Play Simulation: Training and Development Tool" and in "Towards a Conceptual Framework for Agroforestry Residual Biomass Sustainable Business Models", and also in "Environmental and Economic Sustainability of Electric Vehicles vs Combustion Engine Vehicles Fueled with B15 and B30 Blends of Biodiesel", and lastly appraising the consumer perspective "The Influence of Consumer Optimism and Pessimism on Purchasing Intention of Eco-friendly Clothing by Generation Z: Model Proposal".

In regards to innovative and sustainable strategies to encompass smarter productive methods and resource-saving strategies, contributions like "Persistence in Innovation: Do Low-Tech Sectors Differ Much from the High-Tech?", "Towards a

More Sustainable Use of the Portuguese Road Network: The A25 and IP5 Case Study", "Improvement of a Porcelain Baseboard Production Line Capacity Using Simulation: A Case Study", "Multifunctional Furniture for Tiny Houses: Design, Quality, Innovation and Sustainability in Advanced Materials" and "Angolan Cement Industry: Marketing Channel and Distribution Channel Strategies" analysed in different angles relevant generalist and sectoral solutions.

Once more, the importance of responsible innovations and behaviours along the value chain towards a sustainable framework was discussed in "How Can Customer Experience Improve Retail Operations Sustainability?", "Operational Planning of Integrated Urban Freight Logistics Combining Passenger and Freight Flows Through Mathematical Programming", "Towards Digital Transformation: A Case Study to Identify and Mitigate COVID-19 in the Retail Industry", "Technician Routing and Scheduling Problem: A Case Study" and "Application of Fuzzy Methodologies in Navy Systems Maintenance".

The plethora of heterogeneous approaches with the red thread of promoting innovative and sustainable ecosystems will boost the societal transitions taking place. As such, their diversity must be addressed as transdisciplinary rapport to the construction of a larger venture, society 5.0.

Research Agenda

The world has no alternative but to keep pushing its boundaries and strengthening efforts towards a more just and sustainable economic model. The financial and environmental burden of resource depletion along with the negative ecological impacts of overproduction are beginning to stall economic growth, jeopardizing social sustainability. All over the world, there is an urgent need to find alternative ways to save and manage finite endowments of resources, while meeting the increasing needs of a growing and progressively urban world population. The immediate integration of sustainable consumption and production patterns in the agendas of all economic agents is a path to be walked if we want to give humanity a future. As such, policymakers need to design instruments, which lead to more efficient and responsible production processes while disciplining consumption habits. These strategies will generate increased utility to both producers and consumers with positive externalities to communities and the environment.

At the macroeconomic level, new policies need to provide factual proactive intervention promoting quantitative and objective achievements, as well as deadlines for each target. Given the fragility of the economic context, monetary injections from central banks will provide entrepreneurs the liquidity to reinforce their endeavours, as well as an expansionary industrial policy promoting sectoral-structural transitions and energetic shifts.

The global economy, heavily impacted by the multiple crises, forced the reshaping of the multilateral order in both the productive and commercial domains. The downsizing is taking place leading to an uprising of the regional ecosystems, and also the unprecedented changes occurring in the multilateral equilibria made the policy design harder and demanding for constant monitoring and shorter time frames. Continuous feedback loop is required demanding high flexibility and resilience. However, putting the focus on purely economic terms may lead to civilizational throwbacks postponing the sustainability path, something that humanity cannot afford.

The future demands for a policy-making strategy with a holistic approach and enrolling all agents in the communities. Strict collaboration between multiple institutions, areas and layers in both the private and the public domains is necessary to design, develop, implement, monitor and evaluate successful policies which promote socio-economic and energetic justice. Hence, it is just not a matter of improving the productive systems, but also changing the mindsets towards societal responsibility. Production and consumption choices must become frugal; otherwise, there will be nothing to leave to future generations. This human-centred paradigm calls for collective action in the promotion of the transition, from governments to universities as well as user communities and the civil society. The following chapters highlight several angles in which this future can be addressed.

Aveiro, Portugal — João Carlos de Oliveira Matias
Aveiro, Portugal — Carina Maria Oliveira Pimentel
Lisbon, Portugal — Joao Carlos Gonçalves dos Reis
Aveiro, Portugal — Joana Maria Costa Martins das Dores
Barcelos, Portugal — Gilberto Santos

References

1. Calabrò, A., Frank, H., Minichilli, A., & Suess-Reyes, J. (2021). Business families in times of crises: The backbone of family firm resilience and continuity. *Journal of Family Business Strategy, 12*(2), 100442.
2. Campopiano, G., De Massis, A., & Kotlar, J. (2019). Environmental jolts, family centered noneconomic goals and innovation: Toward a framework of family firm resilience. In E. Memili, & C. Dibrell (Eds.), *The Palgrave handbook of heterogeneity among family firms* (pp. 773–789). Palgrave Macmillan.
3. Carayannis, E. G., & Morawska-Jancelewicz, J. (2022). The futures of Europe: Society 5.0 and Industry 5.0 as driving forces of future universities. *Journal of the Knowledge Economy*. https://doi.org/10.1007/s13132-021-00854-2
4. Casson, M. (2021). International business policy in an age of political turbulence. *Journal of World Business, 56*(6), 101263.
5. Czakon, W., Hajdas, M., & Radomska, J. (2022). Playing the wild cards: Antecedents of family firm resilience. *Journal of Family Business Strategy*, 100484. https://doi.org/10.1016/j.jfbs.2022.100484
6. Espitia, A., Mattoo, A., Rocha, N., Ruta, M., & Winkler D. (2021). Pandemic trade: COVID-19, remote work and global value chains. *Policy Research Working Paper* 9508, World Bank.
7. Kornprobst, M., & Paul, T. (2021). Globalization, deglobalization and the liberal international order. *International Affairs, 97*(5), 1305–1316.

8. Morgan, J. P. (2021). *2022 market outlook: More upside for stocks, economic growth to rebound*. Retrieved from: https://www.jpmorgan.com/insights/research/market-outlook-2022. Accessed in 28 Apr 2022.
9. Raza, W., Grumiller, J., Grohs, H., Essletzbichler, J., & Pintar, N. (2021). *Post Covid-19 value chains: Options for reshoring production back to Europe in a globalised economy*. European Parliament, ISBN: 978-92-846-7831-0.
10. United Nations. (2015a). *Population 2030: Demographic challenges and opportunities for sustainable development planning*. Report. Retrieved from: http://www.un.org/en/development/desa/population/publications/pdf/trends/Population2030.pdf. Accessed in 29 Apr 2022.
11. United Nations. (2015b). *Sustainable consumption and production. A handbook for policymakers*. Global edition. Retrieved from: https://sustainabledevelopment.un.org/content/documents/1951Sustainable%20Consumption.pdf. Accessed in 2 May 2022.
12. United Nations Development Programme. (2014). *Breaking through: Inclusive business and the business call to action today*, United Nations Development Programme. Report. Retrieved from: http://www.businesscalltoaction.org/wpcontent/uploads/2014/09/BCtAImpactReport_BreakingThrough.pdf. Accessed in 29 Apr 2022.
13. van Bergeijk, P. (2019). Deglobalization 2.0: trade and openness during the Great Depression and the Great Recession. Edward Elgar
14. World Bank Group. (2021, June). *Global economic prospects*. Retrieved from https://openknowledge.worldbank.org/bitstream/handle/10986/35647/9781464816659.pdf. Accessed 28 Apr 2022.

Contents

Costs and Learning Factories

Alfonso Redondo, Imoleayo Abraham Awodele, and Angel M. Gento

Abstract Learning factories have been a tool used by companies and universities to be training students and employees in a simulated environment. In Valladolid, we have developed a learning factory jointly between the University of Valladolid and Renault Nissan Consulting for training in lean manufacturing tools (and in the future of industry 4.0) by students of the university and Renault employees and supplier. The flagship product used is a toy car made up of nearly 100 components of different sizes, manufactured in several workstations divided into independent manufacturing cells. In the first learning phase, we produced a few versions that can be increased to more than 1500 if we continue learning up to a fifth phase (only for advanced groups). Although costs (labor costs, material costs, warehouse costs, manufactured products costs, etc.) are one of the key indicators of the performance of a production process in any industry, they are not usually considered directly as one of the key indicators of progress in the different learning stages in a learning factory. This chapter shows how costs represent a tactical decision variable in our Lean School, just as it happens in real life. Costs allow us to represent through a single measure the gains obtained by the application of lean tools at different learning phases that go from traditional production to lean manufacturing. The cost estimation is based on the activity-based costing (ABC) method so that all participants in the training identify where the main sources of costs are located and focus on how to reduce them.

Keywords Activity-based costing · KPI · Lean School · Learning factories

A. Redondo · A. M. Gento (✉)
Dpto. Organización de Empresas y CIM, Valladolid, Spain
e-mail: gento@eii.uva.es

I. A. Awodele
Construction Management and Quantity Surveying Dpt., DUT, Durban, South Africa

J. C. de Oliveira Matias et al. (eds.), *Quality Innovation and Sustainability*, Springer Proceedings in Business and Economics,
https://doi.org/10.1007/978-3-031-12914-8_1

1 Introduction

The constant changes taking place in the world today have caused us to move from an "industrial" economy to a "knowledge" economy, which makes education, life-long learning, a necessity and a differential aspect of society [1, 2]. Therefore, teachers' efforts should be focused on developing learning skills to apply knowledge, rather than on teaching content [3].

It is in this field where the collaboration between Renault-Nissan Consulting and the University of Valladolid arises in order to apply the principles of lean management in a physical environment and close to reality for the development of a learning factory [4–6], where costs are a fundamental tool for decision-making [7].

This chapter first presents the basic concepts of activity-based costing (ABC) and learning factories, then show the learning factory we have at the University of Valladolid (Lean School) and how participants in the trainings use ABC to focus where to make improvements in the production system.

2 Integration of the Production and Financial Areas Through Costs: Activity-Based Costing (ABC)

Multidisciplinary is increasingly necessary for all areas: health, industry, and services. It is a culture that must be inculcated from the training centers. By way of example, nowadays we do not understand a production system that is not synchronized with logistics/warehouses, finances, resources, etc. The alignment and/or integration of all functional areas, in a common objective, are one of the key factors for the success of the organizations. And the integrated element is the "activity" [8]. One of the sub-objectives of this work is precisely to carry out training that integrates cost calculation (financial area) with the improvements that students propose and implement in the Lean School (production area).

The first problem is to choose the cost system to be used: (a) traditional cost system (full costing, direct costing, standard cost, cost per process, etc.); (b) activity-based costing/activity-based management (ABC/ABM): and (c) modern systems such as value stream costing (VSC). The answer is not white or black; different approaches to management accounting [9] may be appropriate depending on the circumstances and the environment [10]. Traditional cost systems "emphasize financial management, not operational management or cost management" [11], and some authors highlight the easy integration of traditional cost systems with ABC [12] and how ABC is best integrated with MRP and JIT systems. Arbulo, Fortuny, and Cuatrecasas [13] state that "due to the problems of traditional costing and the complexity of ABC, another alternative has been developed, which is called 'lean accounting' (VSC)," although "VSC can only be implemented after a company has achieved the maturity stage of lean manufacturing." The ABC/ABM system is therefore the best-cost model according to its quality/cost ratio [14, 15]. What is an ABC/ABM system vs. a traditional cost system is summarized in Fig. 1.

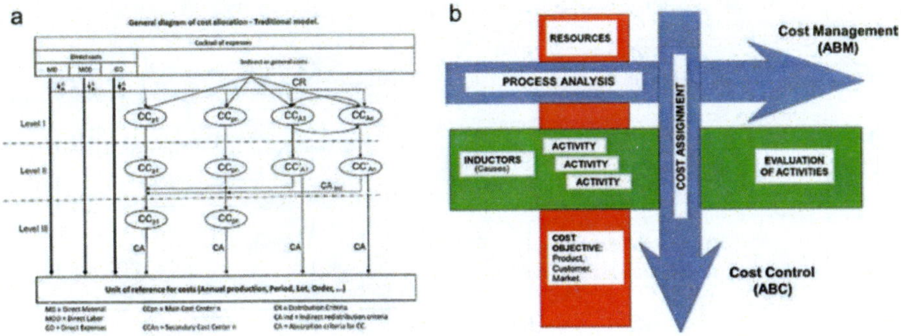

Fig. 1 (**a**) Traditional cost model vs. (**b**) cost ABC/ABM [16]

Traditional cost models are still valid, it is not an obsolete model, but it is outdated with the tools that are used where everything is pre-established and boxed in. In contrast, ABC/ABM systems are totally dynamic and flexible. For example, in a traditional costing system, time is divided according to accounting and/or planning cycles, and each function in each phase of the cycle is optimized separately and in isolation, which is aggravated by the fact that different control tools are used in each phase. However, ABC/ABM systems base their entire operation on the management of the activities performed by the organization, together with their natural evolution throughout their life cycle, using cause/effect chaining, and identifying performance drivers.

Brimson [17] perfectly synthesizes the steps/stages needed to implement an ABC costing system (see Fig. 2).

The critical point when designing an ABC/ABM system is step 1 "Analysis of activities" and its subsequent integration with its life cycle (step 2) in order to achieve a dynamic character [18]. This analysis of activities requires three phases: identification (using top-down and bottom-up approaches); collection of information (based on material and information flow diagrams, Delphi method, etc.); and its structuring by means of processes.

With the information already structured and characterized (name, description, location, unit of measurement, capacity, inputs, outputs, control variable, efficiency range, etc. the so-called "activity map" is drawn up and its management is carried out: identifying the effective activities (e.g., those of the critical path); and establishing the cause–effect relationships and the cost drivers (Fig. 2b).

3 Learning Factories

The new learning environment derived from the guidelines of the European Higher Education Area (EHEA) obliges university teachers to adapt their teaching methodologies so that students, in addition to acquiring knowledge, are able to manage

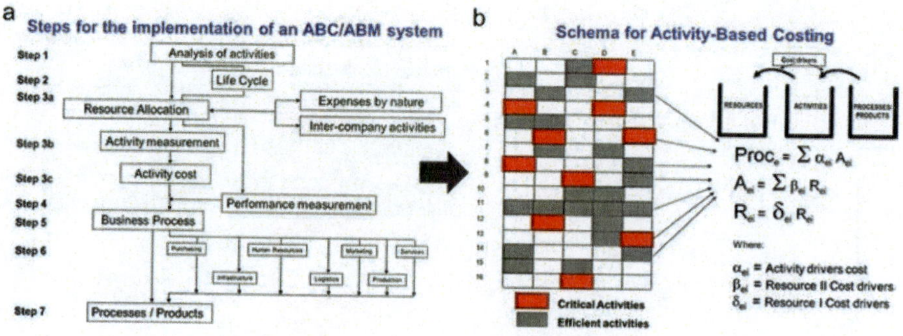

Fig. 2 (**a**) Steps [14] and (**b**) schema for implantation of an ABC/ABM system

their learning process and develop the skills and abilities necessary for the development of their professional future [19]. In view of the ever-changing environment in which we are immersed, the performance of training is particularly important in the teaching of engineers. The technique, closely related to engineers, is evolving so rapidly that we cannot expect today's teachings to be the same as 10 years ago or as we will have in 10 years. As a result, many universities (and companies) around the world (though especially European) [20] have developed learning factories where students (and workers) "learn by doing" based on the idea initially proposed in 1994 by Lamancusa, Zayas, Soyster, Morell, and Jorgensen [4].

A unified definition of a learning factory can be found in [21]: "A learning factory in a narrow sense is a learning environment specified by processes that are authentic, include multiple stations, and comprise technical as well as organizational aspects, a setting that is changeable and resembles a real value chain, a physical product being manufactured, and a didactic concept that comprises formal, informal and nonformal learning, enabled by own actions of the trainees in an on-site learning approach. In a broader sense, learning environments meeting the definition above but with a setting that resembles a virtual instead of a physical value chain, or a service product instead of a physical product, or a didactic concept based on remote learning instead of on-site learning can also be considered as learning factories."

Most existing learning factories cover one or more business processes in the value chain (e.g., planning, sourcing, manufacturing, delivery, and return), and typically include the planning and design of products (or services) and processes, as well as the improvement of existing products (or services) and processes. In literature, you can find many references to learning factories focused on product development, in the application of lean tools and recently, in industry 4.0, or to improve energy and resource efficiency [22].

4 Lean School

Renault-Nissan Consulting (RNC) (consulting company of the Renault Group) has been collaborating with the University of Valladolid for more than 25 years. The relationship began because several Renault workers had been students of the University of Valladolid and therefore known to the university professors. They began collaborating selflessly by giving a practical session based on their professional experience. These first informal contacts were increased with internships, factory visits for students, and the use of some laboratories for testing.

In 2015, this collaboration resulted in the development of a laboratory for teaching lean tools to students at the University of Valladolid following the example of other European universities [20]. This has allowed the University of Valladolid to be the first Spanish university to have a learning factory for teaching lean tools: Lean School (LS) [23]. With this purpose, an old chemistry laboratory has been transformed into a modern space consisting of different workstations, warehouses, shelves, transport elements, etc., which allow students to "learn by doing" rather than through a master class.

It is important to highlight that this learning factory is used (in a coordinated way) for the training of students at the University of Valladolid and for the training of Renault employees and suppliers and customers of RNC. Sometimes joint training session is carried out with students and employees to exchange experiences and also participating university professors and RNC consultants, thus ensuring a balance between theoretical concepts and their applicability in different real situations.

Figure 3 provides a 3D image of the learning factory showing the most important areas: machining and stamping (continuous rectangle), assembly (dashed rectangle), and storage (dotted rectangles).

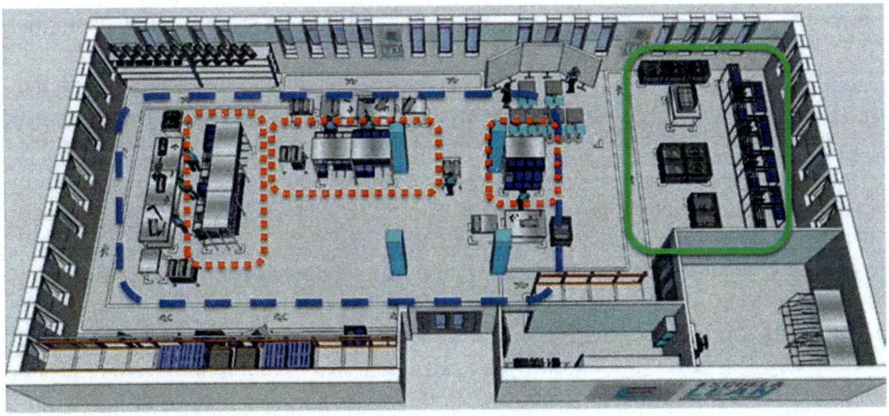

Fig. 3 Initial layout of the Lean School to manufacture educational cars (L34N)

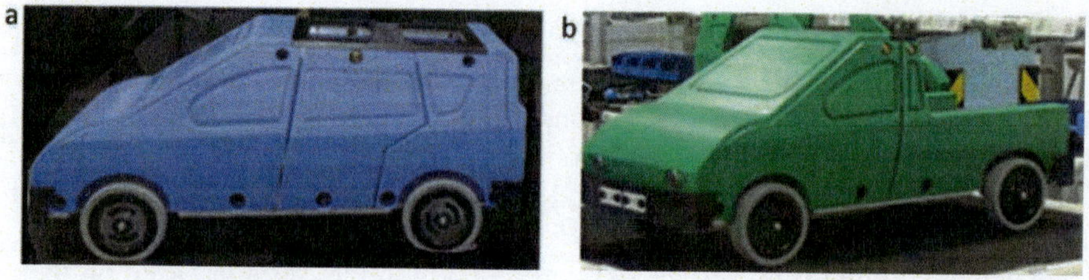

Fig. 4 (**a**) Minivan; (**b**) pickup

Our main industrial partner in the Lean School is Renault-Nissan Consulting, so it is quite logical that the initial product in the learning factory is a car (Fig. 4). This car weighs about 25 kilos and is built by assembling 30 pieces of different sizes and about 70 connectors (screws and plastic rivets). Although there are two basic models (minivan and pickup), it can manufacture more than 1500 different versions depending on the options chosen by the client: color (blue or green), headlights (halogen or xenon), wheels (normal or off-road), sunroof, seats (fabric or leather), and dashboard (air conditioning, radio, navigator, etc.). Depending on the initial training and experience of the students or workers who come to the Lean School, we start from a basic configuration with two, four, or eight different versions to which different options are incorporated according to the objective of the training.

5 Costs in the Lean School

The key performance indicators (KPIs) most commonly used to evaluate the efficiency of any production system are those shown grouped in a scorecard (Fig. 5): quality, time, surface area, human resources, and costs. The KPIs used are according to the objectives to be achieved in the different learning processes depending on the experience and initial training of the participants in the Lean School. On this board, participants must reflect the evolution of the most important indicators based on the results they obtain in the different productions of the L34N product. In this way, they can make decisions and see the results of their actions.

How have we integrated the "lean improvements" proposed by the students with the costs of their implementation? Providing them with an Excel application, totally parameterized for the initial situation, in which they must introduce the improvements they consider should be applied (taking into account customer requirements and the objectives set by the plant manager) (Fig. 6). Zone 2 shows the actual times consumed by the activities carried out in the defined processes, and zone 3 shows the breakdown of costs (broken down by processes and main components), as well as the % under-utilization of resources, both for the target situation (zone 1, not shown in the figure) and the actual situation.

Therefore, training participants can see the progression in the different key indicators (Fig. 5) from the data they update in the spreadsheet (Fig. 6).

SCORECARD

		PROD. 1	PROD. 2
	Tack Time	2	2
		Result	Result
Q	CUSTOMER DEFECTS BY PRODUCT	[illegible]	[illegible]
Q	Internal Defects by Product	[illegible]	[illegible]
Q	Rework (Retouched products on line)	56	59
Q	STRAIGHT TROUGH RATIO (STR)	0,8	0,9
T	% ON-TIME DELIVERIES TO DISTRIBUTION (OTD)	0,8	0,9
T	Production cycle time Assembly	5,25	3,88
T	Direct Labor Assembly [Manufac. / Logistics] (1 shift)	9	9
T	Direct Labor Machining (1 Shift assembly)	1,75	1,75
T	Indirect Labor Factory (1 shift assembly)	2	2
T	Assembly DL Time per product [Manufac. + Logistics]	47,25	34,92
T	Saturation rate Assembly (sum(tcx) / DL time Assembly)	41,01	46,86
T	DSTR ASSEMBLY [Manufacturing + Logistics]	8,59	6,34
T	OEE MACHINING	0,85	0,85
T	LEAD TIME [Assembly / Factory]		
S	Travelled distance by the product [Assembly / Factory]		
S	Logistics Transport (steps)		
S	SURFACE AREA M2	250	250
S	Warehouse Stock (raw material + WIP Machining)		
S	Stocks in workstations (product quantity)		
S	WIP + Finished products (product quantity)		
HR	Nº WORK ACCIDENTS	0	0
HR	Absences	0	0
HR	Nº Faults [identified / resolved]		
C	Salary cost per product [Factory] (1)	33,46875	24,735
C	Internal defects per product (not reworked in line)	[illegible]	[illegible]
C	Customer Defects per Product	[illegible]	[illegible]
C	Cost of Non-Quality per product (2)	15,625	4,25
C	Cost of Equipment and Tools per product (3)	0,015	0,029
C	Stock Cost (withou raw material) per product (4)		
C	Surface cost per product (5)		
C	TOTAL COST= 1 + 2 + 3 + 4 + 5	49,10875	29,014

Fig. 5 Scorecard with the most important KPIs for a typical learning process

ZONE 2

	ACTUAL TIME RECORDING OF PRODUCTION =		1
	Working shifts:		2
	Daily production per shift:		368
	Useful working day (min.):		202.100
	8 h/day	8	
	50 min. Break	50	
	5 days/week	5	
	47 weeks per year	47	
	Annual production:		172.735
	Batch size	6	
	Time-taking of activities:	Ø min.	
Logistics Processes	Logistic 1	471	
	Procurement for assembly	359	
	Moving Finished Product to Customer	112	
	Logistic 2:	558	
	Remove Recycled Materials	381	
	Recibir Cubitrones del Proveedor	106	
	Place Products to be Recycled in R0	71	
	Logistic 3:	558	
	Handle Recycled Products	302	
	Supply materials to assembly	256	
Assembly Process	Activities Assembly Operator 1	495	
	Activities Assembly Operator 2	702	
	Activities Assembly Operator 3	505	
	Activities Assembly Operator 4	698	
	Activities Assembly Operator 5 - Quality	232	
Recycling Process	Activities Recycling Operator 1	385	
	Activities Recycling Operator 2	603	
	Activities Recycling Operator 3	381	
	Activities Recycling Operator 4	595	
		% Under-utilisation	
Control Process	Assembly Team Leader	15,00%	
	Recycling Team Leader	20,00%	

ZONE 3

Breakdown of COSTS and % INFRAUTILISATION of Resources				
	Target		True	
RESOURCES:	% under-utilisation	Value in €	% under-utilisation	Value in €
Surface	0,00%	0,00 €	0,00%	0,00 €
Lighting	0,00%	0,00 €	0,00%	0,00 €
Recycling facilities	10,12%	156,85 €	13,96%	216,38 €
Machining facilities	10,71%	321,43 €	14,53%	435,90 €
Logistics Operator 1	34,82%	13.232,14 €	32,91%	12.504,27 €
Logistics Operator 2	23,81%	8.571,43 €	20,51%	7.384,62 €
Logistics Operator 3	20,21%	7.802,86 €	20,61%	6.974,36 €
Assembly Operator 1	26,79%	11.464,29 €	29,49%	12.620,51 €
Assembly Operator 2	0,00%	0,00 €	0,00%	0,00 €
Assembly Operator 3	26,79%	11.410,71 €	28,06%	11.954,70 €
Assembly Operator 4	0,00%	0,00 €	0,57%	245,58 €
Assembly Operator 5	66,82%	28.396,58 €	66,95%	28.454,42 €
Recycling Operator 1	44,64%	17.857,14 €	45,16%	18.062,68 €
Recycling Operator 2	13,69%	5.613,10 €	14,10%	5.782,05 €
Recycling Operator 3	44,64%	18.035,71 €	45,73%	18.473,50 €
Recycling Operator 4	13,69%	5.640,48 €	15,24%	6.279,77 €
Assembly Team Leader	20,00%	12.800,00 €	15,00%	9.600,00 €
Recycling Team Leader	25,00%	15.000,00 €	20,00%	12.000,00 €
UNIT COST OF L3C4		Target		True
TOTAL unit cost =		26,77		26,91
Supply Logistics Process		0,19818966		0,2154083
Assembly Process		25,47		25,53
Direct Material		24,28		24,28
Direct Labour		0,90456599		0,9333489
Indirect Labour		0,28374072		0,3149332
Recycling Process		0,91155346		0,9618216
Recycling Process Logistics		0,19294635		0,2066162
External Logistics Process				
Unit Cost of Production		26,38		26,49
Unit Logistics Cost		0,39113601		0,4220246

Fig. 6 Spreadsheet used for activity-based costing

6 Conclusions

Lifelong learning is a necessity in the society in which we live where changes in both technology and customer needs are happening faster and faster. In this respect, universities and businesses must learn to work more closely together to facilitate this learning and recycling of knowledge so that scarce resources are better used. The collaboration between the University of Valladolid and Renault is a good example of this collaboration, starting with the delivery of conferences and courses, through student internships and up to the shared use of laboratories (Lean School).

On the other hand, and taking into account the training carried out in the Lean School, we have detected how costs become a fundamental tool for decision-making, both from a strategic and operational point of view, where costs based on activities make it easier for us to see the plant as a whole, identifying the drivers with the greatest impact. The most important limitation we encounter is the quantification of costs and the imputation of some of the activities carried out in the Lean School.

The main problem of ABC is the initial configuration of our system because if we want to have a high control of our costs it is necessary to feed it with a large amount of details and its subsequent updating. Nowadays, this data feed can be done automatically in many cases, so the biggest handicap is the initial system configuration stage. However, the result is worth the effort.

Acknowledgments Special thanks to Renault-Nissan Consulting for its commitment to the University of Valladolid that facilitates the training of students in the first learning factory in Spain.

References

1. OECD. (2008). *21st century learning: Research, innovation and policy directions from recent OECD analyses.*
2. OECD. (2013). *OECD skills outlook 2013: First results from the survey of adult skills.* OECD Publishing.
3. Kuh, G. (2008). *High-impact educational practices: What they are, who has access to them, and why they matter.* Association of American Colleges and Universities.
4. Lamancusa, J. S., Zayas, J. L., Soyster, A. L., Morell, L., & Jorgensen, J. (2008). The learning factory: Industry-partnered active learning. *Journal of Engineering Education, 97*(1), 5–11.
5. Helleno, A. L., Simon, A. T., Papa, M. C. O., Ceglio, W. E., Rossa Neto, A. S., & Mourad, R. B. A. (2013). Integration university–industry: Laboratory model for learning lean manufacturing concepts in the academic and industrial environments. *International Journal of Engineering Education, 29*(6), 1387–1399.
6. De Vin, L. J., & Jacobsson, L. (2017). Karlstad lean factory: An instructional factory for game-based lean manufacturing training. *Production & Manufacturing Research, 5*(1), 268–283.
7. Cockins, G. (2019). Lean accounting and activity-based costing - a choice or a blend? *Cost Management*, January-February, 5–15.
8. Sanz-Angulo, P., Benito-Martín, J. J., Gento-Municio, A. M., Redondo-Castán, A., Pascual-Ruano, J. A., Galindo-Melero, J., & Javato-Martín, M. (2016). A review of the innovative

teaching activities carried out by the School of Industrial Engineering of the University of Valladolid in the field of Business Organization. In C. Hernández (Ed.), *Advances in management engineering* (Lecture notes in management and industrial engineering) (pp. 221–231). Springer International Publishing.
9. Ahrens, T., & Chapman, C. S. (2007). Theorizing practice in management accounting research. In C. S. Chapman, A. G. Hopwood, & M. D. Shields (Eds.), *Handbook of management accounting research* (Vol. 1, pp. 9–112). Elsevier.
10. Hancock, D. R., & Algozzine, B. (2011). *Doing case study research*. Teachers College Press.
11. Johnson, H. T., & Kaplan, R. S. (1987). *Relevance lost: The rise and fall of management accounting*. Harvard Business School Press.
12. Meade, D. J., Kumar, S., & Houshyar, A. (2006). Financial analysis of a theoretical lean manufacturing implementation using hybrid simulation modeling. *Journal of Manufacturing Systems, 25*(2), 137–152.
13. de Arbulo, R., Lopez, P., Fortuny-Santos, J., & Cuatrecasas Arbós, L. (2013). Lean manufacturing: Costing the value stream. *Industrial Management & Data Systems, 113*(5), 647–668.
14. Abdulla. (2021). Optimizing the value of activity based costing system: The role of successful implementation. *Management Science Letters, 11*, 179–186.
15. Quesado, P., & Silva, R. (2021). Activity-based costing (ABC) and its implication for open innovation. *Journal of Open Innovation: Technology, Market and Complexity, 7*, 41.
16. Castelló, E., & Lizcano, J. (1994). *El sistema de gestión y de costes basado en las actividades*. Instituto de Estudios Económicos.
17. Brimson, J. A. (1991). *Activity accounting. An activity-based accounting approach*. Wiley.
18. Jourdaine, M., Loubet, P., Sonnemann, G., & Trébucq, S. (2021). The ABC-LCA method for the integration of activity-based costing and life cycle assessment. *Business Strategy and the Environment, 30*, 1735–1750.
19. Sheppard, S., Macatangay, K., Colby, A., Sullivan, W., & Shulman, L. (2008). *Educating engineers: Designing for the future of the field*. Jossey-Bass.
20. Abele, E., Metternich, J., Tisch, M., Chryssolouris, G., Sihn, W., ElMaraghy, H., Hummel, V., & Ranz, F. (2015). Learning factories for research, education, and training. *Procedia CIRP, 32*, 1–6.
21. Abele, E. (2016). Learning factory. In S. Chatti & T. Tolio (Eds.), *The international academy for production engineering, CIRP encyclopedia of production engineering*. Heidelberg Springer.
22. Abele, E., Metternich, J., & Tisch, M. (2019). *Learning factories: Concepts, guidelines, best-practice examples*. Springer International Publishing.
23. Pascual, J., Hoyuelos, I., Mateo, M., & Gento, A. M. (2019). Lean school: A learning factory for training lean manufacturing in a physical simulation environment. *Management and Production Engineering Review, 10*(1), 4–13.

Agile 8D Problem-Solving Framework in a Renewable Energy Sector

Cláudia Sousa Silva and Ana Luísa Pereira

Abstract Quality management (QM) has evolved due to the current market needs, where innovation and flexibility are essential. In this sense, there is a tendency to combine and integrate different methodologies with QM, namely, the integration of agile approaches given their synergies. This work is supported by the proposition that QM can benefit from agile methods, improving the organization management facing current challenges. Supported by an action research strategy, it describes a case study in a renewable energy organization aiming to understand how agile practices can be adapted to QM, specifically in problem-solving methods. This research also presents theoretical contributions as it provides a framework named "Agile 8D problem-solving," resulting from integrating the Scrum framework with the 8D methodology. However, it also has practical implications for organizations and QM professionals once it hands over a framework composed of guidelines for implementing more collaborative and efficient problem-solving methods.

Keywords Quality management · Agile · Problem-solving 8D · Renewable energy

1 Introduction

Given the technological and turbulent current context, it is essential to explore the development of quality management (QM) from this perspective [1]. The dynamic expectations of customers, global competition, and the growing complexity of

C. S. Silva (✉)
Research Unit on Governance, Competitiveness and Public Policies (GOVCOPP), Department of Economics, Management, Industrial Engineering and Tourism, University of Aveiro, Aveiro, Portugal
e-mail: claudia.margarida@ua.pt

A. L. Pereira
Mechanical Engineering Department, University of Coimbra, Coimbra, Portugal

J. C. de Oliveira Matias et al. (eds.), *Quality Innovation and Sustainability*, Springer Proceedings in Business and Economics,
https://doi.org/10.1007/978-3-031-12914-8_2

products are the main reasons for integrating QM into operational activities, as well as into organizations' strategies, objectives, and policies [2]. For such, it is necessary to rethink QM, namely, through the development of flexible and adaptive QM models.

Since the beginning of the twentieth century, the QM area has been widely disseminated and implemented in organizations. However, the companies' processes to plan, define, obtain, control, continuously improve, and demonstrate quality have undergone a remarkable evolution over the last few years. There has been a tendency to combine methodologies with various QM tools [3]. Also, the interest in agile methods has increased due to market dynamics, hence organizations need to respond quickly to changing customer needs while producing high-quality products at an ever-increasing pace. In the same line, Krehbiel and Miller [4] argue that a QM System (QMS) can benefit from the integration of the agile approach given the synergies of both, namely, a focus on a clear purpose, bet on short inspection and evaluation cycles, focused on processes, not just the final product, and focus on the customer.

Thus, this work, supported by a collaborative action research strategy, presents a singular case study in the renewable energy sector. The main objective of this research is to develop QM methods suited to market dynamics, therefore more adaptive and agile. Furthermore, the authors defend that this path should be modular and evolutionary by combining agile knowledge (principles, values, or frameworks) with specific QM tools. So, the main research question is:

- How to adapt agile practices (principles and values, methodologies, frameworks) to QM tools, specifically problem-solving processes?

To this end, the literature review portrays the state of the art regarding problem-solving methodologies, as well as the theoretical foundations of agile management. Therefore, this section ends with a discussion of the integration of agile into QM. Subsequently, the methodological procedures for developing a case study are described, and, finally, a synthesis of the main contributions, limitations, and future contributions.

2 Literature Review

As a vital part of QM, continual improvement should be the primary goal of any organization, forcing it to continually seek the effective resolution of problems [5]. However, problem-solving usually requires effective teamwork, ensuring that all elements are focused on the same goal. Thus, one of the most valued skills in an organization is teamwork.

Agile is an iterative approach that helps teams deliver value to their customers where the results are evaluated continuously so that teams have a natural mechanism for responding to change quickly. One of the main benefits of agile methodologies is directly related to the teams and the excellent interaction between them, such as improved communication and increased collaboration, expanding the team's motivation to achieve the proposed goals.

Thus, the symbiosis between QM and agile has been explored, and its deepening is encouraged. Aligned with this, the literature review presented here introduces the importance of effectively solving problems, the fundamental principles of agile management, and the interconnection of both areas.

2.1 Problem-Solving Methods: Tools

When noncompliance occurs, including external complaints, the organization must react to control, correct effectively, and deal with its consequences. It must also assess the actions to eliminate the causes to avoid its repetition. Therefore, it is essential to review and analyze the noncompliance; determine the problem causes; check if there are similar noncompliances; or if they could occur. According to the ISO 9001:2015, the problem-solving process encompasses the definition of actions, reviewing the effectiveness of corrective actions, updating the risks and opportunities, and updating QMS [6].

Although there are several problem-solving techniques, Manaica [7] presents a comparative study between some of them (MASP, DMAIC, FTA, RCA, 8D), proving that all are supported by the PDCA cycle (Deming in 1986). This methodology is structured in four cyclical steps, leading to the continuous improvement that should be part of the organizational culture:

- Plan – study the current situation and plan improvement actions.
- Do – after planning, transfer the actions to practice achieving the objectives.
- Check – control the effects of the actions, it was expected, or there were deviations.
- Act – standardize in case the results are the expected ones, otherwise, new actions are formulated.

Some authors reveal that the main cause of the limited use of these methodologies is the lack of teams' availability and training [8]. The 8D tool is considered one of the most used problem-solving methodologies in the industry, whose objective is to reach the root cause of the problem and eliminate it at its source [9].

2.1.1 8D Framework

The 8D methodology was initially developed by the Ford Motor Company, combining various elements of other problem-solving techniques to shape the eight disciplines described in the Ford "Team-Oriented Problem Solving" manual [11]. It is structured into eight disciplines and emphasizes the synergy within a team. Considering Pradeep, Kumar, Karthik, Suraj, and Kishore [10], 8D is applied within an organization whenever serious nonconformities or recurring problems are detected to identify the root cause of the problem.

The eight steps or disciplines that must be followed sequentially without exceeding or ignoring any of them are: D1 – multidisciplinary team training >> D2 – problem description >> D3 – containment actions >> D4 – root-cause analysis >> D5 – corrective actions >> D6 – implementation and validation of corrective actions >> D7 – preventive actions >> D8 – conclusion, evaluation, and congratulation of the team.

Organizations currently implement this problem-solving methodology to gain a competitive advantage since 8D is responsible for quickly identifying the root cause and implementing corrective actions, preventing the reoccurrence of problems, ensuring customer satisfaction, and learning through the lessons learned [12]. Thus, whenever this tool is applied correctly, benefits are expected, such as

- Improved team problem-solving skills,
- Greater familiarity with a problem-solving framework
- Creation of a database with past failures
- Lessons learned to avoid problems in the future
- Better understanding of how to use basic tools to solve problems
- Improved effectiveness and efficiency in problem-solving
- A practical understanding of root-cause analysis
- Improved skills to implement corrective actions
- Open communication in the discussion of problem-solving, increasing efficiency
- Improved understanding of effective problem-solving

However, Rambaud [13] defends that methodology can be time-consuming and challenging to develop and the team involved must receive appropriate training to have a good performance. Still, some organizations do not promote the involvement of 8D teams, which does not motivate them to find the root cause of the problems. Therefore, making 8D an efficient process that favors constant communication becomes necessary. Thus, intellectual capital is a key factor in achieving the tool's basic objectives.

Chlpeková et al., [9] refer to a study in Slovakia, showing that most respondents say the tool is inefficient. This perception may come from the tool being mainly used for customer complaints and not continuous production process improvement. Also, according to this study, three main 8D weaknesses were identified: insufficient and superficial research and intuitive conclusions that are not based on facts; the application of the methodology to random errors in a nonsystematic way; and finally, the omission of the motivation factor. These weaknesses come from the lack of time to pay attention to this tool, and sometimes the pilot is not the best person for this task. In addition, lack of motivation, knowledge, good team relationship, and support from the management are the other problems for the correct implementation of the 8D methodology.

2.2 Agile Management and Scrum

Nowadays, agile methods are adopted widely to face business dynamics challenges as agile responses tend to provide deliverables with good quality at a faster rate [14]. The Agile Manifesto was created by 17 professionals in 2001 (https://agilemanifesto.org/principles.html), being a declaration of essential values and principles, summarized as follows:

- Persons and interaction between them more than processes and tools
- Develop prototypes at the expense of excessive documentation
- Collaboration with the client more than contract negotiation
- Responding to change rather than following a plan

The concept has expanded to product development projects and team management, being implemented in several areas [15].

Within the agile management approaches, Scrum and Kanban stand out. Scrum is not a methodology, much less a technique, but rather a framework focused on solving complex and adaptive problems while productively and creatively delivering products with the most significant value added by integrating the customer into the process. Agile bets flexibility, both in terms of deadlines and results; it is intuitively formed by small groups and requires frequent collaboration and review. Its knowledge base is empiricism, everything can be acquired through experience. The main benefits are directly related to the teams and the good interaction between them, such as improved communication, increased collaboration between those involved in the projects, increased team motivation and productivity, and decreased project duration [16].

As in Rugby sport, Scrum is characterized by the team's union in acting together, where each member plays a specific role, and everyone helps each other toward a common goal. Thus, the Scrum team comprises the Product Owner, Development Team, and Scrum Master. Each of them has its specific and indispensable role.

The Product Owner is responsible for describing and reporting all requirements and changes, having all the information about the product development. The other elements must respect his decisions. The Scrum Master is the person that plays the leading role in the group and follows all the steps until the final delivery of the product. He must ensure that the entire team respects and follows the values and practices established in Scrum. Finally, the Development Team is responsible for analyzing, building, developing, and testing the product. This team manages their work autonomously. For this, the tasks are planned in short sprints with deadlines and results to meet.

Every day, during the duration of the Sprint, there is a 15-minute meeting called the Daily Scrum, where the Scrum Master asks three essential questions to the Development Team: "What did you do yesterday to help achieve the goals?" "What are you going to do today to help achieve your goals?" and "What is the obstacle not achieving the goals?" These meetings improve communication, eliminate other meetings, identify impediments to development, highlight and promote rapid decision-making, and improve the Development Team's level of knowledge. It takes place next to a board where teams also use visual task- tracking tool. Each task is assigned an estimated effort that must be reduced over the weeks, whose main objective is to reach the goals in the shortest time possible. When the Sprint ends, a review is carried out to see if it is necessary to make any changes or adjustments, and the process is repeated [16, 17].

2.3 *Agile Management and Quality Management*

Agile approaches are already beginning to be used in QM, integrating the activities into collaborative and self-organized teams. In this way, promote personal contact between members (face-to-face), and a culture of adaptability and transparency. The planning of activities is supported by the "Timeboxed" concept determined by the team, where the results (deliverables) are perfectly defined. This symbiosis between QM and agile approaches can be further reinforced by the active participation of stakeholders and the more frequent assessment of customer satisfaction, and finally, promoting a regular review of the results achieved [18]. The most used domain area in agile is adaptive planning, continuous improvement, and value-driven delivery [19].

Krehbiel and Miller [4] argue that there is considerable alignment between 14 Deming's points and agile management, namely, "constancy of purpose," "cease dependence on mass inspection," and "dive out fear." Agile practices promote short iterations and frequent customer feedback reducing final mass inspection. At the same time, responding to change over following a plan and using empirical methods can help mitigate fear. The same authors point out that an effective QMS should include multiple tools and mindsets, incorporating the agile approach if organizations are prepared. Thus, developing an emerging QM approach more suited to the current dynamic context and focused on a preventive perspective and continuous improvement [18].

However, most QMS are developed and certified according to normative references, most commonly by ISO 9001. Although, at first analysis, it is supposed that ISO and agile approaches are not compatible, there is still the idea that ISO can make processes more bureaucratic and "heavy." Contrary to this perception, some authors state that ISO 9001 is not only compatible with agile, but it also promotes the necessary structure to help agile processes be implemented and followed [18].

3 Methodology

An intrinsic case study adopted the action research strategy since one of the researchers participated in problem-solving [20]. A mixed research approach (qualitative and quantitative) supports this research by developing a single case study. As a research methodology, case studies allow research development while preserving the events of the organizational context, namely, structure, culture, processes, among others [21].

A case study was developed with the main objective of contributing to the integration of agile in QM, more specifically, integrating the Scrum framework into the problem-solving methodology, handling a new framework that can lead the organization to develop an agile problem-solving model.

Thus, the planning of this case study followed five steps:

1. Company selection criteria
This case study was developed in a company that belongs to the renewable energy sector, with its QMS certified by ISO 9001:2015. This is an international company with about 1000 employers, being wind turbines, its main product is developed by the following process: initial design, construction, assembly, and maintenance. The convenience criterion was also followed since one of the researchers was already integrated into the organization. The internal researcher started an internship for five months to improve the implementation of the 8D methodology in the company.
2. Initial diagnosis – the problem
Since the manufacturing process is still very manual, there is high variability in the production process, leading to more errors and defects. Sometimes, the objectivity of the control becomes difficult because attributes like OK and NOK classify most defects. All defects incur high nonquality costs as the unit cost of each product is also very high. The 8D tool was already implemented in this organization whenever unpredictable defects occurred. The responsibility for opening an 8D report lies with supervisors, section leaders process, maintenance, or quality engineers, depending on the type of deviation. However, it is not always effective and neither is well done leading.
3. Procedures for data collection
To understand the main causes of the problems with 8D methodology in this organization, we collected amount information, such as:

    Database of 8D reports opened during 2018
    Survey of employees who work with the tool, with responses obtained a week and disclosed in the 8D tool training session
    Direct observation in six 8D meetings

    The survey was sent to 65 employees; however, only 32 answered, with 16 supervisors, 7 engineering workers, 5 from the quality department, 1 from the production director, 1 from health and safety, 1 from human resources, and two from logistics. The majority (about 20) have at least a degree, seven have completed secondary education, and the remaining have technical courses.
4. Data analysis
Section 4 describes the main problems identified in the 8D methodology in this organization.
5. Discussion
Section 5 presents a proposal of an agile problem-solving framework.

4 Results

A quantitative analysis was carried out through descriptive statistics using Excel® and a qualitative approach by the content analysis of the 8D reports.

4.1 *Database of 8D Reports (Year 2018)*

At the first level, the main problem of the 8D implementation by the number of errors identified was checked. Thus, the main deviations are summarized in the Pareto chart represented (Fig. 1), the incorrectly filled 5 Why's tools; the no reopening report when problems reoccur; and the poor plan for corrective actions. These three problems account for 80% of the total issues that avoid a more effective use of the 8D methodology.

4.2 *Direct Observations*

On the other hand, more conclusions can be drawn by participating in the six 8D meetings. The little importance of the tool was perceived through the participants' behavior, such as not arriving at the meeting on time, inhibiting some opinions and suggestions, trying to carry out the 8D meeting in the minimum time possible, and constant interruptions due to external factors deviating from the 8D focus.

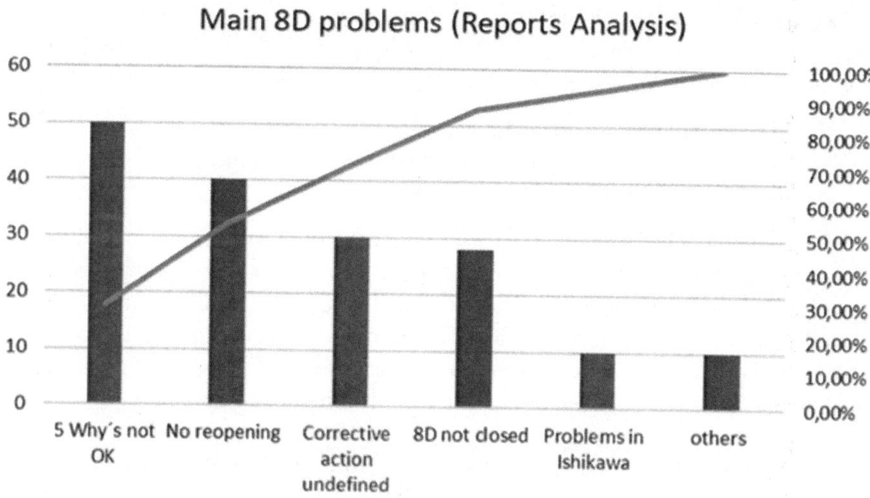

Fig. 1 Main 8D problems (reports analysis)

Due to the lack of some elements, poor knowledge, and ineffective leadership, it was impossible to find the root cause of the problem. Also, some discomfort appears when the cause was related to their work area, increasing the discussions, and deviating from the main 8D objective. Most of the meetings are not held at the place defined. In summary, the main problem that emerges is the reduced ability to solve problems as a team.

4.3 *Employees' Survey*

All participants claimed to know the 8D tool; however, only 15 had training (47%). Even so, 90% affirmed knowing the Ishikawa diagram and 5 Why's tools. Nevertheless, 22% of respondents said they do not know how to reach the root cause of the problem. Seven do not know when to open an 8D, 44% were unknown about the entire procedure. The main steps less dominated are closing and validation. About 13 left some stages unfilled, and the justification given was the lack of knowledge of the entire tool, not knowing how to complete it, and the difficulty in closing all the actions.

In short, concerning Pareto's results survey, 80% of the main problems with 8D methodology are identifying how many 8D are open and reopened; lack of 8D training; and lack of knowledge procedure and difficulties in reaching the root cause of the problem (Fig. 2).

Considering only the employees who have already opened 8Ds, the main problems are completing all stages, eliminating the difficulties felt during their completion, internalizing the congratulation after closing, helping to reach the root cause of

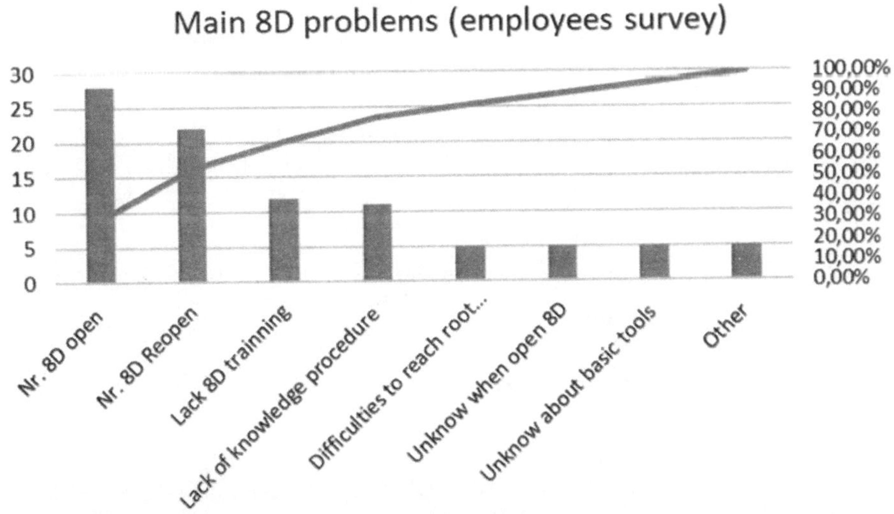

Fig. 2 Main 8D problems (reports analysis)

the problem, implementing corrective actions within the defined time, providing training, and knowing the entire procedure of the tool.

5 Discussion

The Pareto charts resulting from the survey and the report's content analysis enhance the importance of acting by the 8D teams to facilitate the interaction between elements and increase their involvement. One of the main problems faced through direct observation of 8D meetings was the relationship between departments to solve problems. Asset in the literature review and good team relationship was a problem identified for the correct implementation of the 8D methodology [9]. These issues could impact most of 8D's problems, such as not being completed, not being well filled, difficulties in follow-up, and consequently not being closed or validated. This research proposes combining some agile elements with problem-solving methods to overcome this difficulty.

As identified in the literature review, a good relationship team can be encouraged to solve quality problems through agile methodologies. The team must be organized and follow the agile principles, such as prioritizing customer satisfaction, being open to changes and accepting them, adopting continuous feedback systems, and privileging multidisciplinary teams to solve problems [18]. This philosophy helps the team increase autonomy and avoid bureaucracies to solve a problem, suggesting brief informal meetings based on feedback. On the other hand, rushing and exhausting teams do not increase agility and often lead to demotivation, impacting the quality of work. The self-organized teams let the elements decide how to distribute the work to avoid conflicts between them. This approach will enhance autonomy and responsibility, promoting motivation, and, consequently, better results at work.

Different methods stand out within this agile methodology, such as Lean, Kaizen, and Scrum, among others. The Scrum framework inspires our suggestion to restructure the problem-solving team. Thus, the proposed Agile 8D problem-solving framework consists of the following steps as illustrated in Fig. 3

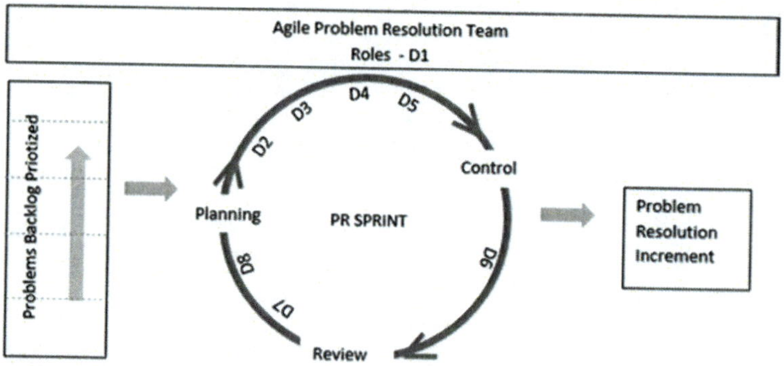

Fig. 3 Agile 8D problem-solving

5.1 Step 1: Agile Problem Resolution Team (APRT)

Different roles should be assigned to the elements involved in problem-solving, as such:

- "Problem Owner" – 8D pilots have a role like the "Scrum Product owner." They must be the most interested in the 8D resolution and committed to solving the problem as soon as possible and effectively.
- "Problem Master" – He/she is a facilitator element in parallel to "Scrum Master," that is, the person who helps to understand the principles, values, and practices to perform each action planned.
- "Team Problem Resolution" (TPR) – All elements responsible for the actions to solve the problem, so it must be an "Agile" team: self-organized, collaborative, and multidisciplinary.

5.2 Step 2: Problem Resolution Backlog

The APRT, supported by the disciplines defined in 8D, must define the added value actions (containment, correction) to solve the current problem and prevent its recurrences effectively:

- Actions that do not significantly impact should be eliminated, thus avoiding waste in actions that are not focused on the main problem. Subsequently, the team should prioritize the 8D actions based on their contribution to solving the problem.
- The "Problem Resolution Backlog" is built with only the value-added actions, sorted by priority.
- The problem-solving team should estimate the effort needed to implement the actions, considering this priority.
- The "Problem Resolution Backlog" should be a dynamic document, which must be updated considering the results feedback.

5.3 Step 3: Problem Resolution Sprint

Considering the team structure and the complexity of the problems, the organization should define a timebox for the duration of its sprints to solve the problems, recommending between 1 and 4 weeks. In each sprint, the team estimates and plans the actions to be implemented and completed, being committed to this plan.

5.4 Step 4: Control and Monitoring

A meeting should be held weekly related to the open 8Ds, with the main objective of monitoring the 8D report and evaluating its results. This meeting would start with the open 8D listed in "Problem Backlog." Then, a visual Kanban Board could be used, in which the first column is reserved for actions that are not yet performed or whose implementation has not been efficient to eliminate the problem, followed by the actions "to do" and then the progress actions. This board visually monitors activities that have already been carried out, later verified by the 8D pilot.

This practice is important to commit the team to 8D goals and identify impediments that should be solved as soon as possible to proceed with the action.

5.5 Step 5: Problem Resolution Review and Retrospective

This last step focuses on steps D7 – preventive actions and D8 – conclusion, evaluation, and congratulation of the team's 8D methodology. The main objective is to reflect on the implemented actions' results and the lessons learned about the problem-solving process to enhance teamwork.

6 Conclusions

Teamwork is appointed as a strategy to promote new work visions to achieve a common goal, prioritizing communication. For that reason, agile teamwork is essential in QM, namely, in multidisciplinary teams to solve problems. Thus, this research presents a framework that combines some Scrum elements with the 8D problem-solving methodology, described as the "Agile 8D problem-solving framework."

Furthermore, this framework intends to eliminate the main obstacles in the problem-solving process, namely, relationships between people, incomplete 8D reports, and difficulties in follow-up, and closing, or validating the results. The proposed model provides a methodology focused on the priority problems, on the involvement of employees, attributing different roles necessary for the systematization of the 8D methodology. In this way, it is expected to enhance the importance of problem-solving and 8D methodology in the company's culture.

The main limitation of this work is related to a single case study, suggesting that future studies can replicate the concept in other organizations and then assess the improvements with this new framework.

The authors propose that the adaptation of agile to QM should follow a modular and evolutionary integration strategy through the combination of agile knowledge (principles, values, or frameworks) with specific QM tools, building this path step by step. Thus, it is planned to extend this integration to other QM tools and methodologies as a future research line.

References

1. Gunasekaran, A., Subramanian, N., & Ngai, W. T. E. (2019). Quality management in the 21st century enterprises: Research pathway towards industry 4.0. *International Journal of Production Economics, 207*, 125–129., Jan.
2. Foidl, H., & Felderer, M. (2016). *Research challenges of industry 4.0 for quality management* (pp. 121–137).
3. P. D. Morris "The perfect process storm: Integration of CMMI, agile, and lean six sigma.,"Dec. 2012.
4. Krehbiel, T., & Miller, D. (2018). Should agile be part of your quality management system? *Journal of International Technology and Information Management, 27*(1) Jan.
5. R. Cook & A. Jenkins, (2014). Building a problem-solving culture that lasts. In *The lean management enterprise* (pp. 113–118).
6. ISO. (2015). *ISO 9001:2015 – Quality management systems — Requirements*. ISO.
7. C. R. Manaica, "KAIZEN – Desmistificando o PDCA, 8D, A3, DMAIC – SEIS SIGMA, FTA, MASP E RCA," *Télios*, 2019. [Online]. Available: https://www.telios.eng.br/site/kaizen-desmistificando-metodos-para-melhoria-continua.
8. Łuczak, J., & Wolniak, R. (2015). Problem-solving and developing quality management methods and techniques on the example of automotive industry. *Manager, 22*(1), 237–250.
9. Chlpeková, A., Večeřa, P., & Šurinová, Y. (Jan. 2014). Enhancing the effectiveness of problem-solving processes through employee motivation and involvement. *International Journal of Engineering Business Management, 6*(Godište 2014), 6–31.
10. Pradeep, S., Kumar, S., Karthik, R., Suraj, A., & Kishore, S. (2019). Application of 8D methodology for productivity improvement in assembly line related papers application of 8D methodology for productivity improvement in assembly line. *International Journal Of Advance Research, Ideas And Innovations In Technology, 5*(2).
11. Broday, E. E., & Andrade Júnior, P. P. (2013). Application of a quality management tool (8d) for solving industrial problems. *Independent Journal of Management & Production, 4*, 377–390.
12. Banduka, N., Veža, I., & Bilić, B. (2016). An integrated lean approach to Process Failure Mode and Effect Analysis (PFMEA): A case study from automotive industry. *Advances in Production Engineering And Management, 11*(4), 355–365., Dec.
13. Rambaud, L. (2006). *8D structured problem solving: A guide to creating high quality 8D reports*. Phred Solutions.
14. Boehm, B., & Turner, R. (2003). Using risk to balance agile and plan-driven methods. *Computer (Long. Beach. Calif), 36*(6), 57–66., Jun.
15. J. Verret, "Implementing agile methodology: Challenges and best practices," 2018.
16. Sutherland, J. (2014). *Scrum: The art of doing twice the work in half the time*. Currency.
17. Schwaber, K., & Sutherland, J. (2020). *The scrum guide*. Creative Commons.
18. Sakurai, R., & Zuchi, J. D. (2018). As Revoluções Industriais Até A Industria 4.0. *Revista Interface Tecnológica, 15*(2), 480–491., Dec.
19. S. Abdalhamid, A. O. M. Mohammed, & A. Mishra, "Agile and quality: A systematic mapping study," *2019 International Conference of Computer Science and Renewable Energies (ICCSRE)*, Jul. 2019.
20. Amado, J., & Vieira, C. C. (2014). Apresentação dos Dados: Interpretação e Teorização. In J. Amado (Ed.), *Manual de Investigação Qualitativa em Educação* (2nd ed.). Imprensa da Universidade de Coimbra.
21. Yin, R. K. (2015). *Estudo de Caso: Planejamento e Métodos*. Bookman editora.

The Transfer Process of Lean Approach Within Multinational Companies' Network: The Schnell S.p.A. Case Study

Giada Pierli, Federica Murmura, and Laura Bravi

Abstract In a highly dynamic and competitive market environment, a crucial factor for companies is the ability to pursue objectives strongly oriented to continuous improvement, especially in terms of flexibility, quality, and efficiency of business processes. This aspect becomes particularly important in the current scenario, where competition for high levels of quality and innovation and cost containment no longer takes place locally, but globally. In this sense, the adoption of Lean Thinking could significantly support businesses, extending the concept of continuous improvement beyond the boundaries of their corporate system and encouraging integration between different value chains. From this perspective, the chapter aims to investigate how Lean initiatives are inserted and disseminated in an international context, supporting businesses in their value creation process. To this end, a single case study research based on the empirical method of the semi-structured interview has been developed. In particular, the study focuses on the analysis of the Schnell company, which has been successfully implementing Lean Manufacturing and Six Sigma practices for years. The project has thus enabled the Schnell Group to achieve important results, further increasing its efficiency and capacity to respond to market needs. Limited to the involvement of the subsidiary, the main results achieved are the opening of forms of dialogue and collaboration, the increase in the efficiency of its processes, a greater sense of belonging to the group and the acquisition of new skills, in addition to the full achievement of the operational objectives set by headquarter.

Keywords Lean Manufacturing · Six Sigma · Improvement project · Lean production

---

G. Pierli · F. Murmura (✉) · L. Bravi
Department of Economics, Society, Politics, Carlo Bo University of Urbino, Urbino, Italy
e-mail: federica.murmura@uniurb.it

J. C. de Oliveira Matias et al. (eds.), *Quality Innovation and Sustainability*, Springer Proceedings in Business and Economics,
https://doi.org/10.1007/978-3-031-12914-8_3

1 Introduction

Since the 1990s, the expression of Lean Production has started to extend worldwide as a new management paradigm oriented to increase operational performance and value created for customers through waste elimination and continuous improvement of business processes [1–3]. Over the years, the Lean concept has deeply evolved and applied in numerous companies in different sectors, from manufacturing to services [4, 5]. Although the implementation of Lean practices begins within the company with a focus on the efficiency of its production processes, Lean principles and techniques could be successfully extended not only to all different activities and departments of an organization, but also to external partners [6]. According to Netland and Powell [7], firms that exclusively exploit Lean approach internally are missing out on external opportunities.

This aspect becomes particularly important in the current scenario, where competition for high levels of quality and innovation and cost containment no longer takes place locally, but globally. In this sense, the adoption of Lean Thinking could significantly support businesses, extending the concept of continuous improvement beyond the boundaries of their corporate system. In addition, lean transfer process has become an important strategic issue as it can considerably improve efficiency at the network level.

Nevertheless, as stated in the literature by Demeter and Losonci [19] the lean transfer process is particularly difficult in the relationships between the companies of a same network and the knowledge in this field is limited. Therefore, the chapter aims to investigate how lean approach is transferred within the multinational companies' network by developing a single case study research based on the empirical method of a semi-structured interview. In particular, the study focuses on the analysis of the Italian multinational company Schnell S.p.A., operating in the manufacturing sector, which has been successfully implementing Lean Manufacturing and Six Sigma practices for years. The chapter is structured as follows: Sect. 2 develops the theoretical background, Sect. 3 describes the research methodology used, Sect. 4 illustrates and discusses the case study, and Sect. 5 presents the main conclusions and limitations.

2 Literature Review

The expression lean production identifies a production system that aims to reduce waste and variability in processes, adding more value to customers and improving operational performance [8]. Historically, the term lean production was first coined by John Krafcik in 1988 in the article *Triumph of the Lean Production System*, but it is the three authors James P. Womack, Daniel T. Jones, and Daniel Roos [9] in the book *The Machine That Changed the World* who deepened and analyzed in detail the topic, making it known worldwide the productive and organizational revolution started at the beginning of the 1950s by Toyota Motor Company, a Japanese company operating in the automotive sector. This revolution started with the Toyota Production System (TPS), that is, the production system created by the engineer Taiichi Ohno (1912–1990), who undertook a continuous fight against waste at all levels to bring out clearly the problems to be solved, always focusing on the total satisfaction of the end customer. From this perspective, it is possible to state that the intention of lean implementation is to identify and eliminate activities that do not bring value [10–12] to the final product or customers. Specifically, lean production or lean manufacturing identifies seven different kinds of waste (in Japanese, muda), such as "overproduction, waiting, transport, defects, inappropriate processing, unnecessary inventory and unnecessary motion" [11]. Recently, an eighth muda has been identified, namely, the waste of staff capabilities [13]. In addition to muda, which are the physical waste and therefore the most visible, there are also two other types of waste, muri and mura. The first are waste due to irregularities in workloads, while the mura are waste of overload or excess [14].

Over time, this approach has become so effective that the Lean concept does not only concern business operations and production systems, but it becomes a Lean Thinking able to involve the entire company and guide it toward continuous improvement. Womack and Jones [15] defined five key principles of Lean Thinking: identify value, map the value stream, create flow, establish pull, and seek perfection.

It is therefore evident that lean approach, strongly directed to a constant search for perfection, allows companies to reach the highest quality of their products or services and increases their operational efficiency, consequently maximizing customer satisfaction and business performance. Over the years, the lean concept has deeply evolved and applied in numerous companies in different sectors from manufacturing to services [4, 5], demonstrating to significantly improve their competitiveness [5]. Many scholars have focused their attention on the application of lean practices and operational performance, confirming the existence of a positive correlation [16–18].

Although it is widely recognized that Lean practices are primarily aimed at eliminating waste and inefficiencies in internal production processes, Lean principles and techniques could be successfully extended not only to all different activities and departments of an organization, but also to external partners [6]. According to Netland and Powell [7], firms that exclusively exploit Lean approach internally are missing out on external opportunities. In this sense, coordinating and managing the transfer of lean approach in the company's network becomes an important strategic issue, especially within the scope of multinational companies, which allows to significantly improve efficiency and value at the network level [19]. In particular, transferring lean-related knowledge represents a challenging and crucial process that involves a complex and interrelated system of practices spread across multiple dimensions within the corporate organization [20, 21]. Studies that describe mechanisms to transfer knowledge in multinational companies can be classified into two different categories. The first concerns the external sources (buyers or suppliers) of lean knowledge [22–25] and the second considers the lean knowledge transfer within internal networks [20, 26–28]. In this chapter, the focus is on the latter.

In recent decades, there are an increasing number of companies experiencing international growth through establishing production subsidiaries in different countries. Driven by the success achieved by headquarters, several multinational companies have started to transfer lean knowledge to foreign subsidiaries [29]. Some examples are Toyota, Mercedes, Caterpillar, Bosch, Siemens, and Volvo [30].

However, managers must cope with geographically dispersed and heterogeneous units that further complicate the lean transfer process [31], as well as other important elements must be considered. It is widely acknowledged that many contextual factors can influence the implementation of the lean approach within the network, such as technological and geographic distance, information technology, and cultural values [32–34]. Therefore, it is important to properly consider and manage the different possible critical factors as a conflict between contextual conditions and knowledge transferred could generate a resistance to change [35, 36] and prevent the sharing of lean practices.

In addition, lean production is characterized by explicit and tacit knowledge and both transfers are a crucial element in the lean context [26]. Scholars agree that technical and analytical practices are more codifiable and easy to transfer, using manual, training material, or shared database. On the other side, concepts and practices regarding people and relationships are difficult to codify and share. Several studies highlight the importance of using social mechanisms, such as interactions among lean experts, rotation of personnel and knowledge brokers, to effectively share and improve tacit knowledge in lean transfer [26, 37].

Another major factor is represented by organizational context that supports lean knowledge transfer in multinational corporations and roles of different hubs in the transfer process. According to Demeter and Losonci [19], global and local teams play an essential role in successful transfer. In particular, global offices are involved in coordinating and organizing lean transfer process, while the local office is primarily responsible for exploiting lean knowledge at the plant level [19]. Furthermore, it is necessary to consider the strategic role of internal stakeholders. For instance, Netland and Ferdows [38] underlined corporate-level improvement offices' personnel as the key players in the development and transfer of lean knowledge and program-related information to and from plants. Similarly, Inkpen [27] highlights the chief executive officer's role in his study about the knowledge transfer between General Motors plants and NUMMI, a joint venture of General Motors and Toyota.

Businesses are therefore faced with a huge challenge, which requires the development of specific skills and the definition of policies and strategies oriented to an integrated, coordinated, and efficient management of the complex lean transfer process. From this perspective, this chapter aims to investigate how lean approach is transferred within multinational networks, with a particular focus on headquarter and its foreign subsidiary.

3 Research Methodology

The research has been developed with a qualitative approach. In order to investigate the mechanisms that characterize the transfer of the lean approach in a multinational company, a qualitative single case study research based on the empirical method of the semi-structured interview has been developed [39]. The analysis focused on Schnell S.p.A. as it is a leading company within the Italian industrial context in implementing successful Lean Manufacturing and Six Sigma techniques, becoming an important model of best practices for companies in the sector. The interviews were conducted periodically, following the main development phases of the project, that is, the sharing phase, the implementing and maintaining ones, with in addition a final interview to draw the conclusions of the project. These were addressed to the Head of Group Purchasing since he was the figure responsible for the development of the entire project. The Lean Six Sigma Project started on May 1, 2020, and ended on November 30, 2020. Table 1 shows the dates, duration, and topics covered during the interviews with the Head of Group Purchasing.

Subsequently, the interviews were triangulated with secondary data [40], such as company reports and website, which were used respectively for the analysis and presentation of the quantitative data of the Schnell S.p.A. project and for the description of the company profile. Given the complexity of the subject under investigation, a single case study allows developing an in-depth understanding of the phenomenon, together with the context conditions in which it occurs.

Table 1 Schnell S.p.A. interview

Date	Duration (h)	Topic
May 31, 2020	1	Description of the sharing phase of the project with the subsidiary (objectives set, critical issues to be resolved, lean techniques used in the development of the project)
July 22, 2020	2	Description of the implementation phase of the project with the subsidiary (state of advancement of set objectives, indicators for analysis, monitoring activities implemented, problems detected)
September 24, 2020	1	Description of the maintaining phase of the project with the subsidiary (verification of achievement of set objectives, closure of ongoing issues, implementation of continuous improvement)
December 15, 2020	1.5	Final analysis of the project with evaluation of the benefits obtained and the difficulties encountered

4 Case Study

4.1 Company Profile

Schnell S.p.A. is an Italian company founded in 1962 that has been operating for almost 60 years in the manufacturing sector of automatic machines and plants for processing iron for reinforced concrete. Currently, the Schnell Group is characterized by a staff of over 700 employees worldwide and is made up of five production plants. It operates in over 150 countries through its 11 subsidiaries, over 50 agents and resellers, and a dense network of service centers. Since 2018, the introduction of Lean Manufacturing principles and methodologies has significantly contributed to the maintenance and strengthening of its competitive advantage in the market. Currently, the company has 440 initiatives for continuous improvement and 216 initiatives are previewed for 2022.

In 2020, Schnell started the first project of Lean Six Sigma Manufacturing characterized by an international momentum through the involvement of the Chinese subsidiary, Schnell Machinery Tianjin. In particular, Schnell Tianjin was involved as a supplier to the Italian headquarters in Monteprandone.

In line with the objectives of the broader corporate project Group Purchasing Excellence, the project allowed to start an important empowerment process with Schnell Machinery Tianjin, with significant results in both directions. In 2021, the implemented system has allowed producing 59% more than the estimates, maintaining the required performance. The line of Monteprandone has gone from a production of 24 to 38 machines, always considering the same time frame of 12 months.

4.2 Lean Six Sigma Project at Schnell

To understand how the Lean approach was extended to the Chinese subsidiary, it is possible to subdivide the Schnell lean manufacturing project into three phases:

1. *Sharing*. In this first stage, Schnell shared with the Chinese subsidiary the improvement project defined by the headquarters in order to start a first phase of dialogue and identification of critical components in the processes implemented by the company. As indicated by the Head of Group Purchasing, every week brainstorming activities were carried out and the Chinese teamwork participated actively and constantly, proving to be particularly motivated and interested in increasing their performance. Due to the pandemic crisis, these meetings took place using video conference tools and online sharing of documentation and files useful for the implementation of the project.
2. *Implementing*. During the implementation phase, the central work team has consistently supported the Chinese subsidiary in achieving the agreed goals in terms of Transit Time (50 days) and Estimation Time Delivery (10 days). In particular,

the Head of Group Purchasing noted that the main support activity concerned the supply chain management with respect to which the Chinese company showed difficulties and problems, for example, the sampling of suppliers was completely absent. In this phase, monitoring activities were also agreed and carried out once every 2 weeks through the brainstorming technique. An online sharing folder was specially designed for uploading files and documents.

3. *Maintaining*. For continuous improvement, meetings with the Chinese subsidiary continue to be held every month, although the project objective has been fully achieved. As the Head of Group Purchasing stated, Schnell China today has a Lead Time of about 45 days.

In the following sections, the quantitative data extracted from the company reports provided by the purchasing manager will be reported to complete the information obtained during the interviews.

The Project Chart Document of Table 2 illustrates a synthesis of the improvement project that has been carried out in the period May–November 2020.

"As Is" Analysis

From a preliminary analysis, on a sample of 542 deliveries, there is an average of 17.618 days with a minimum of 7.035 days in advance and a max value of 68 days. From an ulterior analysis, it can be affirmed that 75% of the deliveries fall back with values until 21.137 days, while advanced to 75% there are values until 68 days of Delta Delivery. Without carrying out any analysis and improvement actions, it can be affirmed to have values that go from 16.66 days to 18.569 days of delay (see Fig. 1 and Table 3).

"GAP" Analysis

From an analysis carried out after 3 months from the start of the activities based on Lean Manufacturing techniques, on a sample of 542 deliveries, there is an average of 7.03 days with a minimum of 2.998 days in advance and a max value of 12.31 days. From an ulterior analysis, it can be asserted that 75% of the deliveries fall back with values until 8.05 days, while advanced to 75% there are values until 12.31 days of Delta Delivery. Without carrying out any analysis and improvement actions, it can be affirmed to have values that go from 6.91 to 7.16 days of delay. The distribution is a possible approximation to a Gaussian model (see Fig. 2 and Table 4).

"To Be" Analysis

From an analysis made after 12 months from the start of activities based on Lean Manufacturing techniques, on a sample of 542 deliveries, there is an average of 5.03 days with a minimum of 2.34 days in advance and a max value of 8.7935 days. From an ulterior analysis, it can be affirmed that 75% of the deliveries fall back with values until 5.66 days, while advanced to 75% there are values until 8 or 79 or 35 days of Delta Delivery. Without carrying out any analysis and improvement actions, it can be affirmed to have values ranging from 4.97 to 5.19 days of delay. The distribution is a possible approximation to a Gaussian model (see Fig. 3 and Table 5).

Table 2 Schnell S.p.A. project chart

Project chart Schnell Group S.p.A.			
Title	Schnell Monteprandone turns color and renews its production!		
Business case	First project developed on a line of machines produced to order, subject to strong customization and with standard forecasting methods in the Schnell Monteprandone (AP) plant. By analyzing the market requests, the project requirements were generated and then all the Lean Manufacturing approaches were applied in order to reach the estimated target (time to market 20 days Data Customer order received). Once the process maps (Value stream map and Spaghetti Chart) have been generated, the production times As Is (Tackt Time), the special and common causes, the process variability measured (Lean 6Sigma-DMAIC) and finally the supplier Lead Time, we have generated initiatives for the optimization and stabilization of the machine production process. All methods were used in a bottom-up approach involving and collaborating with all team participants.		
Project objectives	*Start of model machine Prima 13 SMART* I. Definition of product/process requirements II. Definition of KanBan elements III. Stabilization of the balanced production line at a Demand of Nr. 2 Machines-month with a supply Lead Time of 3 months IV. Increasing efficiency and accuracy in delivery V. Stabilization of time to market		
Ring (scope)	*In scope*	*Out of scope*	
	All suppliers who participated in the Schnell HQ portfolio from January 1, 2020, to December 31, 2020	All suppliers who do not accept the centralization	
Project team	*Project team*	*Name of leader*	
	Project leader	Group purchasing director	
	Sponsor	Plant director	
	Team member	Local strategic buyer 1	
	Team member	IT information team	
	Team member	Senior electric worker	
	Team member	Senior mechanical worker	
	Team member	Production plant planner	
	Team member	Quality department	
	Team member	Industrial controller	
	MBB/coach	Mirko Bartolucci	
Milestone	Start	End	Status (%)
Define	May 1, 2020	May 31, 2020	100
Measure	June 3, 2020	June 30, 2020	100
Analyze	July 1, 2020	August 30, 2020	100
Improve	September 1, 2020	September 30, 2020	100
Control	October 1, 2020	November 30, 2020	100

(continued)

Table 2 (continued)

Project chart Schnell Group S.p.A.		
Voice of customer (V.o.C.)	Customer	Needs – "stomach ace"
	Market	Time to market reduction
	Central purchasing	Obtaining targeted involvement at the central purchasing organization
	Central procurement	Reduction of needs management activities
	Warehouse	Elimination of stock and reduction of freight movements
	Procurement (China)	Be sure of the incoming material as confirmed by the managing supplier without intermediate processes
	Production (China)	Reduction of the set-up for lack of semifinished coming from the suppliers
	Testing	Certainty of the date of presence of the machine to have no overlap on the test
Critical to quality (CTQs)	*Primary CTQ*	
	Punctual delivery (deliveries ≤5gg confirmed date) On time delivery = 1 (punctual deliveries) On time delivery = 0 (late deliveries)	
"As Is" value	0.0% performance (Process Sigma 0)	
"To Be" value	80% performance (Process Sigma 2.5)	

Concluding Evidence

Summarizing the results of the project at the end of the "To Be" phase, efficiency is passed from 0.30% to 84.50%, passing from the "As Is" phase with a range of delivery times ranging from 7 to 68 days to the "To Be" phase where the range has been reduced to 2–8 days (see Fig. 4).

5 Conclusions and Limitations

This study was carried out with the main objective to describe the Lean approach transfer process in the internal network of a multinational company, with a particular focus on the foreign subsidiary. Although coordination and implementation

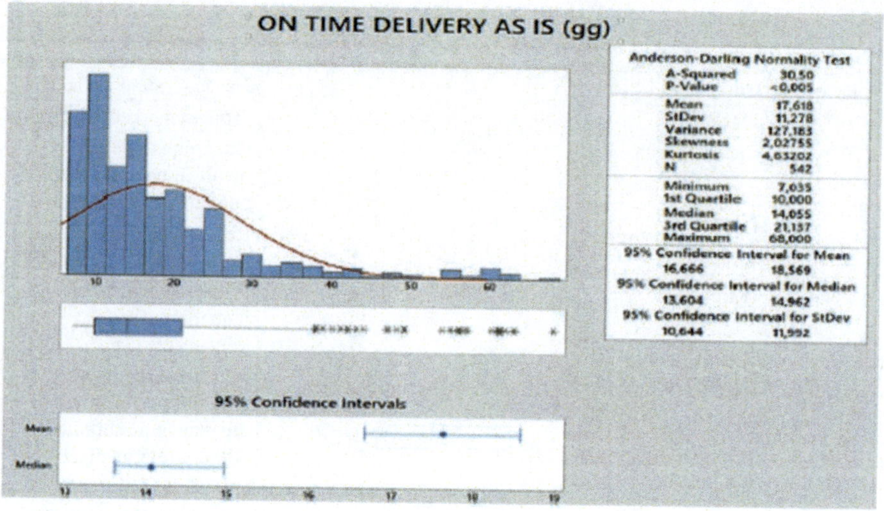

Fig. 1 On-time delivery "As Is"

Table 3 Process Sigma calculator "As Is"

Opportunity defect (O)	1
Sample number (U)	542
Defect (D)	540
DPMO (defect per million opportunity)	996,310
Efficiency	0.37%
Process Sigma	−1.18

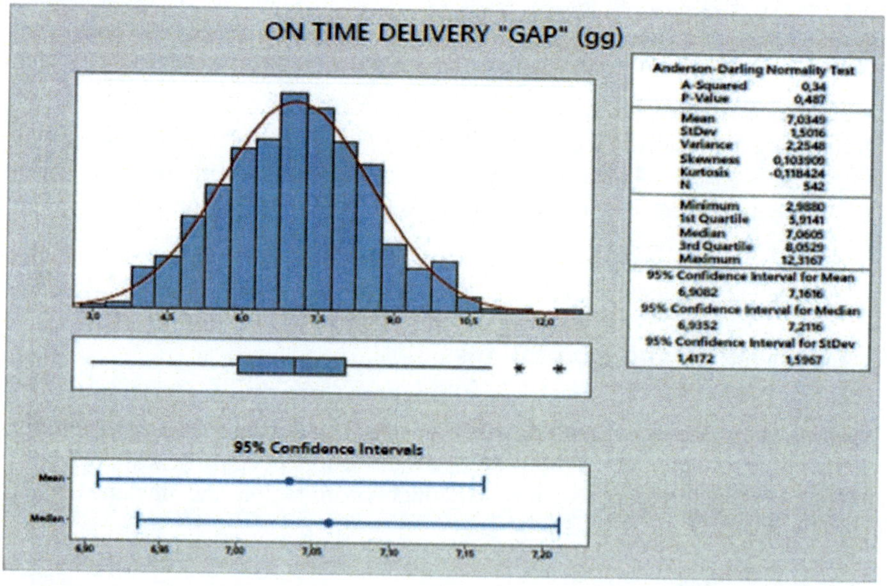

Fig. 2 On-time delivery "GAP"

Table 4 Process Sigma calculator "GAP"

Opportunity defect (O)	1
Sample number (U)	542
Defect (D)	398
DPMO (defect per million opportunity)	734,317
Efficiency	26.57%
Process Sigma	0.87

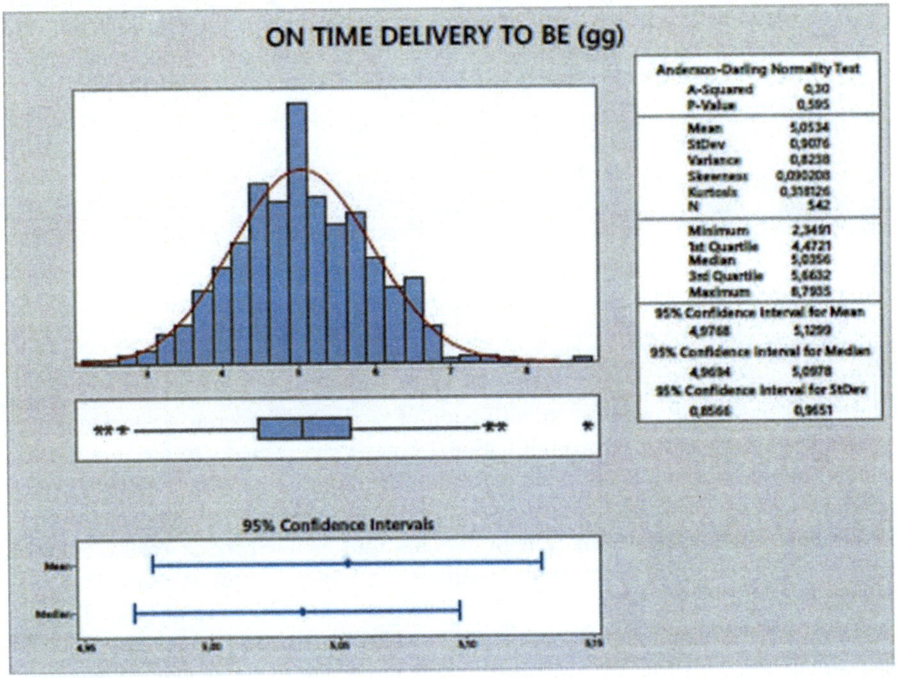

Fig. 3 On-time delivery "To Be"

Table 5 Process Sigma calculator "To Be"

Opportunity defect (O)	1
Sample number (U)	542
Defect (D)	84
DPMO (defect per million opportunity)	154,982
Efficiency	84.50%
Process Sigma	2.52

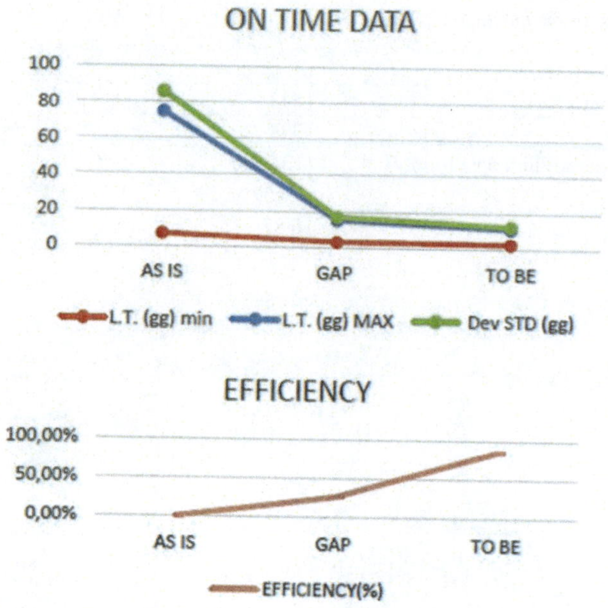

Fig. 4 Summarizing results of the entire project

efforts are significant, creation and diffusion of projects oriented to Lean Thinking represent a strategic issue for multinational companies. Thanks to the qualitative analysis carried out, it is possible to highlight the important results achieved by the company under investigation that allowed to extend the Lean Thinking even beyond the borders of the headquarters. The project has thus enabled the Schnell Group to achieve important results such as the opening of forms of dialogue and collaboration, the increase in the efficiency of its processes, a greater sense of belonging to the group and the acquisition of new skills, in addition to the full achievement of the operational objectives set by headquarter. It is important to stress that the involvement of the foreign subsidiary results from an internal improvement project. Although the involvement has been indirect, it has brought many benefits and has allowed the Chinese subsidiary to increase its performance and empowerment, approaching lean philosophy. In this sense, the study contributes to enriching the literature on lean transfer, also showing situations of implicit transfer.

Despite the success of this first international lean project, Schnell S.p.A. is fully aware of the difficulties encountered and the need to consider these aspects in future projects. In particular, the obstacles mainly encountered concern the linguistic and cultural aspects and a different level of maturity in the processes management (above all, supply chain management), in addition to the geographic distance increased by the COVID-19 pandemic.

The main limitation of this research derives from the qualitative methodology adopted. In spite of allowing the developing of in-depth insights into a phenomenon and its context, a single case study makes it difficult to generalize the results achieved. Further research should investigate a large sample of companies that have developed similar projects to be more generalizable. Another limitation is related to the analysis of a Lean project from a headquarters' perspective. In the future, it could be interesting to combine headquarter with subsidiary perspective to obtain an overall analysis framework.

References

1. Liker, J. (2004). *The Toyota way: 14 management principles from the world's greatest manufacturer*. McGraw-Hill.
2. Shah, R., & Ward, P. T. (2007). Defining and developing measures of lean production. *Journal of Operations Management, 25*(4), 785–805.
3. Bortolotti, T., Boscari, S., & Danese, P. (2015). Successful lean implementation: Organizational culture and soft lean practices. *International Journal of Production Economics, 160*, 182–201.
4. Hines, P., Holweg, M., & Rich, N. (2004). Learning to evolve: A review of contemporary lean thinking. *International Journal of Operations and Production Management, 24*(10), 994–1011.
5. Moyano-Fuentes, J., & Sacristan-Díaz, M. (2012). Learning on lean: A review of thinking and research. *International Journal of Operations & Production Management, 32*(5), 551–582.
6. Dos Santos, L. D. B., Holmen, E., & Pedersen, A. C. (2020). Viewing lean supply from the IMP perspective. *Journal of Business & Industrial Marketing, 35*(1), 172–182.
7. Netland, T., & Powell, D. (2017). A lean world. In T. H. Netland & D. J. Powell (Eds.), *The Routledge companion to lean management* (pp. 465–473). Taylor and Francis.
8. Shah, R., & Ward, P. T. (2003). Lean manufacturing: Context, practice bundles, and performance. *Journal of Operations Management, 21*(2), 129–149.
9. Womack, J. P., Jones, D. T., & Roos, D. (1990). *The machine that changed the world*. Harper Perennial.
10. Rodrigues, J., Sá, J. C. V. D., Ferreira, L. P., Silva, F., & Santos, G. (2019). Lean management "quick-wins": Results of implementation. A case study. *Quality Innovation Prosperity, 23*, 3–21.
11. Verrier, B., Rose, B., Caillaud, E., & Remita, H. (2014). Combining organizational performance with sustainable development issues: The Lean and Green project benchmarking repository. *Journal of Cleaner Production, 85*, 83–93.
12. Santos, G., Gomes, S., Braga, V., Braga, A., Lima, V., & Teixeira, P. (2019). Value creation through quality and innovation - A case study on Portugal. *TQM Journal, 31*(6), 928–947.
13. Brito, M., Ramos, A. L., Carneiro, P., & Gonçalves, M. (2019). The eight waste: Non-utilized talent. In *Lean Manufacturing: Implementation, opportunities and challenges*. Nova Science Publishers, Inc.
14. Slack, N., Brandon-Jones, A., & Johnston, R. (2016). *Operations management*. 8th edition Edinburgh Gate, pp. 429–451. Financial Times Prentice Hall.
15. Womack, J. P., & Jones, D. T. (1996). *Lean thinking: Banish waste and create wealth in your organisation*. Rawson Associates.
16. Marodin, G. A., & Saurin, T. A. (2013). Implementing lean production systems: Research areas and opportunities for future studies. *International Journal of Production Research, 51*, 6663–6680.
17. Rosa, C., Silva, F. J. G., Ferreira, L. P., & Sá, J. C. (2019). *Lean manufacturing applied to the production and assembly lines of complex automotive parts in Lean Manufacturing: Implementation, opportunities and challenges* (pp. 189–224). Nova Science Publishers, Inc.
18. Rodrigues, J., Sá, J. C., Silva, F. J. G., Ferreira, L. P., Jimenez, G., & Santos, G. (2020). A rapid improvement process through "quick-win" lean tools: A case study. *Systems, 8*(4), 1–19.

19. Demeter, K., & Losonci, D. (2019). Transferring lean knowledge within multinational networks. *Production Planning & Control, 30*(2–3), 211–224.
20. Maritan, C. A., & Brush, T. H. (2003). Heterogeneity and transferring practices: Implementing flow manufacturing in multiple plants. *Strategic Management Journal, 24*(10), 945–959.
21. Inkpen, A. C. (2008). Knowledge transfer and international joint ventures: The case of NUMMI and general motors. *Strategic Management Journal, 29*(4), 447–453.
22. Boyle, T. A., Scherrer-Rathje, M., & Stuart, I. (2011). Learning to be lean: The influence of external information sources in lean improvements. *Journal of Manufacturing Technology Management, 22*(5), 587–603.
23. Dyer, J. H., & Nobeoka, K. (2000). Creating and managing a high performance knowledge-sharing network: The Toyota case. *Strategic Management Journal, 21*(3), 345–367.
24. MacDuffie, J. P., & Helper, S. (1997). Creating lean suppliers: Diffusing lean production through the supply chain. *California Management Review, 39*(4), 118–151.
25. Sako, M. (2004). Supplier development at Honda, Nissan and Toyota: Comparative case studies of organizational capability enhancement. *Industrial and Corporate Change, 13*(2), 281–308.
26. Henriksen, B., & Rolstadås, A. (2010). Knowledge and manufacturing strategy – How different manufacturing paradigms have different requirements to knowledge. Examples from the Automotive Industry. *International Journal of Production Research, 48*(8), 2413–2430.
27. Inkpen, A. C. (2007). Learning through alliances: General motors and NUMMI. *California Management Review, 47*(4), 114–136.
28. Secchi, R., & Camuffo, A. (2016). Rolling out lean production systems: A knowledge-based perspective. *International Journal of Operations and Production Management, 36*(1), 61–85.
29. Boscari, S., Danese, P., & Romano, P. (2016). Implementation of lean production in multinational corporations: A case study of the transfer process from headquarters to subsidiaries. *International Journal of Production Economies, 176*, 53–68.
30. Netland, T. H., & Aspelund, A. (2013). Company-specific production systems and competitive advantage: A resource-based view on the Volvo production system. *International Journal of Operations & Production Management, 33*(11/12), 1511–1531.
31. Danese, P., Romano, P., & Boscari, S. (2017). The transfer process of lean practices in multi-plant companies. *International Journal of Operations & Production Management, 37*(4), 468–488.
32. Bortolotti, T., & Boscari, S. (2016). Internationalisation of Lean Manufacturing: The influence of environmental conditions. In *Understanding the lean enterprise* (pp. 185–201). Springer.
33. Rich, N., & Bateman, N. (2003). Companies' perceptions of inhibitors and enablers for process improvement activities. *International Journal of Operations & Production Management, 23*(2), 185–199.
34. Kull, T. J., Yan, T., Liu, Z., & Wacker, J. G. (2014). The moderation of lean manufacturing effectiveness by dimensions of national culture: Testing practice-culture congruence hypotheses. *International Journal of Production Economics, 153*, 1–12.
35. Lozeau, D., Langley, A., & Denis, J. L. (2002). The corruption of managerial techniques by organizations. *Human Relations, 55*(5), 537–564.
36. Ansari, S. M., Fiss, P. C., & Zajac, E. J. (2010). Made to fit: How practices vary as they diffuse. *Acadademy of Management Review, 35*(1), 67–92.
37. Ferdows, K. (2006). POM forum: Transfer of changing production know-how. *Production and Operations Management, 15*(1), 1–9.
38. Netland, T. H., & Ferdows, K. (2014). What to expect from corporate lean programs. *MIT Sloan Management Review, 55*(4), 83–89.
39. Yin, R. K. (1994). *Case study research: Design and methods* (2nd ed.). Sage Publishing.
40. Flick, U. (2007). Concepts of triangulation. In *Managing quality in qualitative research*, pp. 38–54. Sage.

Lean Healthcare: A Critical Analysis

Yesica Pino, José A. Pascual, and Ángel M. Gento

Abstract The current pandemic situation caused by COVID-19, also known as coronavirus, has put the health sector in the spotlight. It has led to the emergence of new plans and strategies that can help society control and stop the spread of COVID-19. These strategies include the Strategic Preparedness and Response Plan, led by the World Health Organization (WHO), which highlights and outlines the public health measures that countries should adopt to prepare for and respond effectively to the spread of the coronavirus.

The pandemic is much more than a health crisis. It requires a response, matching the main problems detected in health and emergency centers. Some of them are quick and exponential spread of the disease, a spike in the number of COVID-19 cases, the successive waves, an increase in the number of patients, and insufficient response healthcare capacity. These problems become more important when there is also an economic downturn, which has made it impossible to expand budgeted public spending on healthcare.

For this reason, research for alternative methods that can improve the quality of healthcare management, but without generating increases in budgetary costs, is timely. Implementing Lean Methodology in healthcare can enhance the quality of service, process improvement, and improve patient satisfaction.

This chapter analyzes the application of lean manufacturing in the healthcare sector: from specific departments of a hospital (emergency, surgery, etc. to the healthcare management of a region. It identifies also the most commonly used lean tools and their benefits in order to achieve sustainable competitive advantages.

Keywords Lean manufacturing · Healthcare · Lean management

Y. Pino (✉) · J. A. Pascual · Á. M. Gento
Dpto. Organización de Empresas y CIM, U. Valladolid, Valladolid, Spain
e-mail: yesica.pino@uva.es

J. C. de Oliveira Matias et al. (eds.), *Quality Innovation and Sustainability*, Springer Proceedings in Business and Economics,
https://doi.org/10.1007/978-3-031-12914-8_4

1 Introduction

The current pandemic situation caused by COVID-19, also known as coronavirus, has put the spotlight on the health sector. It has led to the emergence of new plans and strategies linked to the need to incorporate more efforts to help society to control and stop the spread of COVID-19. These include the Strategic Preparedness and Response Plan, led by the World Health Organization (WHO), which highlights and outlines the public health measures that countries should take to prepare for and respond effectively to the spread of the pandemic [1].

The 2030 Agenda for Sustainable Development was adopted by all UN Member States in 2015, which set out a plan to enhance national implementation and strengthen institutions to achieve the 17 Sustainable Development Goals, leaving no one behind [2]. The Sustainable Development Goal number 3 aspires to ensure health and well-being of all ages.

The pandemic is much more than a health crisis [2]. It requires a response, matching the main problems detected in health and emergency centers. Some of them are the quick and exponential spread of the disease, a spike in the number of COVID-19 cases or hospitalizations, the successive waves, an increase in the number of patients, and insufficient response healthcare capacity (increased waiting time, cancellations of scheduled surgeries, stress levels at work, long working hours and massive patient presence, etc.) [3]. These problems become more important when there is also an economic downturn, which has made it impossible to expand budgeted public spending on healthcare.

For this reason, the research for alternatives or study methods that improve the quality of healthcare management, but without generating increases in budgetary costs, is timely. Implementing lean methodology in healthcare and reviewing processes and systems can enhance the quality of the service, decrease inefficiencies to ensure that all work performed adds value, reduce costs, minimize waste with ongoing process improvement, increase productivity without stress, and improve patient satisfaction [4].

2 Lean Manufacturing for Service Activities

2.1 Basic Concepts

Although Lean Manufacturing was originally applied particularly to industry production processes, such as chains of production or supply, increasingly, lean methodology was being implemented in other sectors, including service activities [5–8].

The main goal of the Lean manufacturing philosophy is to support companies to add value to client deliverables as efficiently as possible. It is based on the principles that seek to optimize processes, reduce waste, and pursue continuous improvement to make the process more efficient. Even though it is difficult to identify and detect higher quality waste because of intangible assets, the main principles of lean are the same: achieving efficiency of the value chain with the fewest resources [9].

For any business in the service industry, the priority is to provide excellent customer service [10]. Business success depends on its ability to deliver its service on time, with high quality and according to customer requirements. Currently, it is becoming more common due to increased competition and customer demand. The service activities, although they are different from industrial production, have also unnecessary tasks without value that could be eliminated or reduced [9, 10].

There is a great scope of application of Lean tools in this context. The tools are used to analyze all existing processes and define the value proposition for the customer, define areas for improvement according to the waste areas identified, and define objective levels of optimization or showing measurement indicators [11, 12]. In addition, they provide solutions to reduce mistakes, enhance the quality of service, or achieve sustainable competitive advantages [13].

Many companies dedicated to the service activities have successfully implemented the model proposed by Lean Manufacturing some years ago [7]. In particular, those companies that are characterized by the variability of their processes and high quality of customers are those that have obtained more successful results [11]. Therefore, we can advance that, despite being a model designed for the manufacturing industry, the Lean methodology can succeed in any sector that is characterized by the complexity of its processes [13].

2.2 *Lean Healthcare*

Lean manufacturing is important from the point of view of health management since both industrial production and health services are based on similar key success factors, that is, customer/patient needs are the most important point. In fact, Lean Healthcare is a Lean methodology that puts the patient at the center of the system [14, 15].

The main objective of Lean Healthcare is to minimize waste in every process, procedure, and task through an ongoing system of improvement. All members of the organization, from clinicians to operations and administration staff using lean principles, identify areas of waste and eliminate anything that does not add value to patients.

The originator of lean principles, Taiichi Ohno [16] of Toyota, described seven areas of waste that occur in every industry. Later, Toyota identified the eight wastes. Implementation of lean ideas in healthcare and reviewing processes and systems through the lens of the eight wastes can potentially reduce waiting, minimize inventory, eradicate defects to improve quality of care and increase reimbursement, decrease the movement of patients, supplies, and equipment to improve patient flow, prevent injuries and save time by reducing motion, maximize resources by minimizing healthcare overproduction, remove waste from overprocessing, and understand how healthcare waste leads to untapped human potential. Table 1 shows the eight areas of waste and their main characteristics.

Table 1 Eight wastes of lean healthcare [17]

Lean waste	Main examples
Idle time	Patient sitting in waiting areas, meetings stalled for latecomers, large waiting lists, idle high-tech equipment
Overproduction	Preparing medications for a discharged patient, duplication of test, extending hospital stays
Transportation	Moving people, supplies, and medical equipment unnecessarily
Overprocessing	Unnecessary work goes into treating patients, needless test, filling out different forms with same information, performing data entry in more than one system
Inventory	Surplus supplies and medications, extraneous data, stockpiles of preprinted forms
Motion	Hospital workers perform movement within their workspace that does not add value to patients (stopping for frequently used supplies or increased motion due to poor building design, etc.
Defects	Process or system failures, medical mistakes, misdiagnosis, medication or surgical errors, avoidable readmissions, incomplete or erroneous medical records
Untapped human potential	Waste in healthcare decrease time that employees could use for education, relationships with patients or implementation of systems-based improvements

3 Methodology

Information access is no longer limited, but in order to carry out this study it is necessary to have reliable, real, and accurate data, so a fundamental requirement is the use of appropriate search criteria. Therefore, the study is based on the results obtained from a systematic literature review (SLR), which ensures structured and rigorous data and overcomes the weaknesses of other traditional narrative review methodologies [18].

The results aim to analyze the implementation models of Lean Manufacturing techniques in the healthcare system, collected in scientific journal articles found in the PubMed database. It contains more than 33 million citations and abstracts of biomedical literature.

In order to obtain the most relevant results, inclusion and exclusion criteria have been defined. Table 2 provides a summary of the results. The first was to select those articles that contained "Lean Healthcare" in title, abstract, and/or keywords (1.452 documents). The second criterion used were those articles that had a peer review process and were published in English (1.352 documents).

A third condition to obtain the highest quality in our results was to include the availability of abstracts to ensure relevance (1.281 articles). We also set the criteria for the period from 2009 to 2019 (965 articles). In order to select more specific papers, we chose the filter with the keywords "lean management" to exclude medical papers with no mention of lean philosophy (331 articles). Finally, those that did

Table 2 Results of systematic literature review (SLR)

First stage	Search within article title, abstract, and keywords Search documents: "Lean Healthcare"	1.422
Second stage	Limit to peer review process and English language	1.352
Third stage	Limit to abstract	1.281
Fourth stage	Limit to year (from 2010 to 2019)	965
Fifth stage	Limit to keywords "lean management"	331
Sixth stage	Eliminate articles that did not contain the desired parameters	298

not contain the desired parameters (not related to the context of the search) were eliminated.

The result of all stages, applying the exclusion criteria, was a total of 298 articles. In this way, we were able to analyze and classify the results obtained according to the hospital departments to the regions. In addition, the most commonly used tools and their benefits in achieving sustainable competitive advantages were identified. Finally, we use the articles defined as case studies (CS) in order to analyze how Lean Healthcare techniques are implemented. Thus, 136 articles have been analyzed in this study.

4 Results

The analysis of the results shows, according to the first of the variables, that the number of articles published on the object of study, Lean application in the health sector, is increasing over years. In 2009, there was only one publication. In 2010, there were 21 publications, and in the following year there were only 6. Subsequently, the number of publications increased with 39 publications in 2016 and 48 publications in 2019 (see Fig. 1).

From a geographic perspective, most of the publications come from North America, with the United States (127 papers) and Canada (29 papers) standing out. If we focus on Europe, the application of Lean methodology in the healthcare is also important, with significant contributions in the United Kingdom (27 papers), Sweden (18), the Netherlands (17), and Italy (14) (Table 3 shows the countries with three or more papers).

Of all the articles selected (298), only the "case study" articles (136 results) were analyzed in order to show the application models of the Lean Healthcare methodology. Thirty-four general hospital departments reported having implemented lean. Emergency departments (17 cases), surgery (13), and pharmacy (6) apply lean in their processes, as well as in specialized care areas such as in 10 oncology departments, 7 orthopedic departments, 10 oncology departments, and 7 orthopedic departments (see Fig. 2).

The results illustrate that the Lean methodology has been implemented in different healthcare areas and has served as a reference for other departments or for

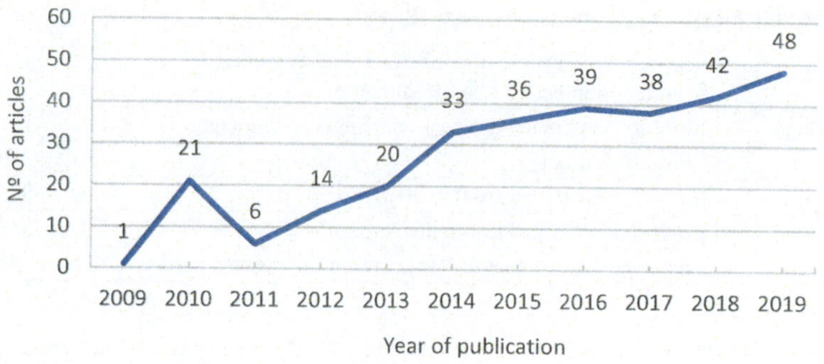

Fig. 1 Number of articles published on PubMed by year

implementation in other hospitals. These results prove that the Lean methodology is put into practice in numerous healthcare departments' processes with the aim of offering fast and quality care to customers/patients, where optimization is focused on reducing time and costs. Healthcare area considers the patient as a customer, and its priority is to provide fast and quality care. Patient care is focused on optimizing time with the human resources available.

The healthcare sector has as a priority the optimization of time to provide immediate medical attention to the patient; therefore, the most used Lean tool was value stream mapping (VSM), followed by define–measure–analyze–improve–control (DMAIC), standardized work, process mapping, and 5Ss (see Fig. 3).

The most usually used tools are those that allow us to analyze processes on a large scale (value stream mapping), at the workplace level (process mapping), and those related to the analysis and resolution of problems in a systematic way (DMAC, team approach to problem-solving and PDCA).

The list of the most commonly used Lean tools and their application in different departments provides evidence that the application of this methodology has reduced waiting times (41 cases), improved patient satisfaction (37 cases), increased productivity (31 cases), and reduced costs (29 cases) and response time (25 cases), depending on the objectives of each area.

5 Conclusions

This study has allowed us to learn about the usefulness of the Lean methodology in the service sector and specifically in the healthcare sector.

The analysis of current problems in the healthcare sector and the research on 298 articles showing the application of this methodology have shown us the great potential for improvement with the application of Lean methodology. It is proof that the trend in research and development of real cases is increasing (1 publication in 2009 and 48 in 2019).

Table 3 Number of articles published on PubMed by country

United States	Canada	United Kingdom	Sweden	Netherlands	Italy	Brazil	Others	Finland	Australia	Belgium	Ireland	Malaysia	Saudi Arabia	India	Jordan	Norway	New Zealand	Oman	South Africa
127	29	27	18	17	14	9	7	6	5	4	4	4	3	3	3	3	3	3	3

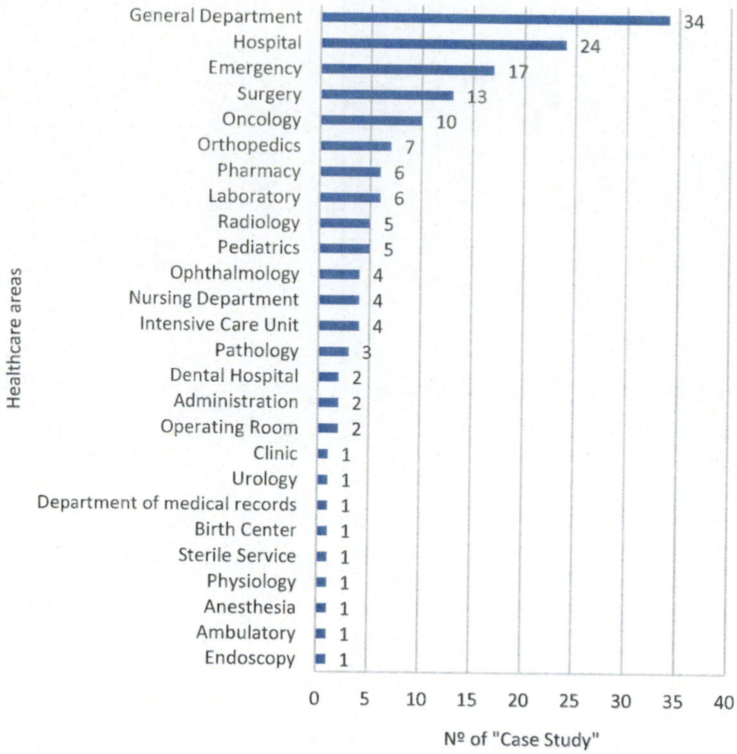

Fig. 2 Number of "case studies" by healthcare areas

The chapter demonstrates that the application of Lean methodology is not limited to the industrial area, as evidenced by the practical cases applied in different countries and departments of the healthcare system. In addition, the possibility of adapting the Lean methodology in the health sector, with optimal results, with the purpose of improving the efficiency of the service provided, achieving the goal of providing the patient with quick and quality care, and reducing medical errors and waste in processes of services without value.

The results of the healthcare areas that require immediate medical attention (emergency, surgery, and oncology) are the first where lean management has been implemented. This is due to the need to see quick results that facilitate the expansion to the other areas of the hospitals. All of this can be achieved with the application of the most commonly used tools: value stream mapping (VSM), define–measure–analyze–improve–control (DMAIC), standardized work, process mapping, and 5Ss.

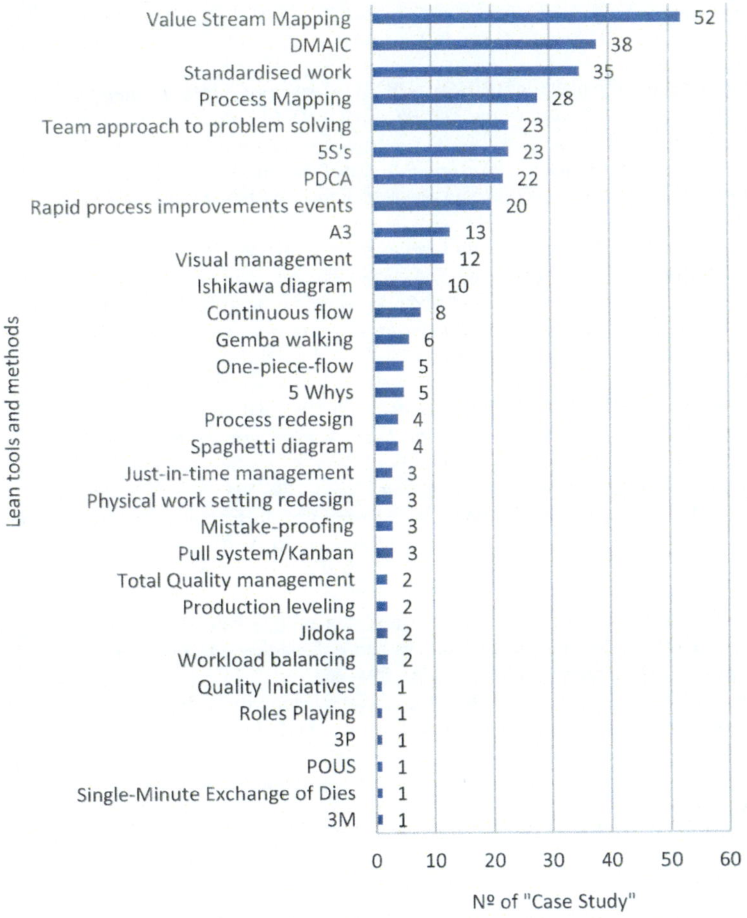

Fig. 3 Number of "case studies" by lean tools and methods

Lean tools enable quality improvement and cost reduction by improving and redesigning key processes or activities. For that reason, Lean tools implementation is highly profitable and helps to enhance patient care, professional satisfaction, and sustainability of the healthcare system. It seems appropriate that future areas of research should identify the improvements obtained after the application of Lean Manufacturing to the healthcare sector.

Acknowledgments The authors thank the students Daniela Guzmán Condori and Margarita Likhodei from the Faculty of Industrial Engineering for their contribution to the search for the papers analyzed for this literature review.

References

1. World Health Organization (WHO). (2020). *Prevention, identification and management of health worker infection in the context of COVID-19*. https://www.who.int/publications/i/item/10665-336265
2. United Nations (UN). (2021). *Sustainable Development Goals: 17 Goals to transform our world*. https://www.un.org/sustainabledevelopment/sustainable-development-goals. Last accessed 28 Feb 2021.
3. Ruiz, E., & Ortiz, N. (2015). Lean healthcare: una revision bibliográfica y futuras líneas de investigación. *Scientia et Technica, 20*(4), 358–365.
4. Delgado, N., Gisbert, V., Perez-Bernabéu, E., & Pérez, A. I. (2020). Lean Manufacturing as tools for improving hospital care. *Cuadernos de investigación aplicada*, 33–42.
5. Moyano-Fuentes, J., & Sacristán-Díaz, M. (2012). Learning on lean: A review of thinking and research. *International Journal of Operations & Production Management, 32*, 551–582.
6. Vignesh, V., Suresh, M., & Aramvalarthan, S. (2016). Lean in service industries: A literature review. *IOP Conference Series: Materials Science and Engineering, 149*, 012008.
7. Leite, H., & Vieira, G. (2015). Lean philosophy and its applications in the service industry: A review of the current knowledge. *Production, 25*(3), 529–541.
8. Liker, J. K., & Morgan, J. M. (2006). The Toyota way in services: The case of lean product development. *Academy of Management Perspectives, 20*, 5–20.
9. Womack, J., & Jones, D. (1996). *Lean thinking - Banish waste and create wealth in your corporation*. Simon and Schuster.
10. Simpson, S., & Cacioppe, R. (2001). Unwritten ground rules: Transforming organization culture to achieve key business objectives and outstanding customer service. *Leadership & Organization Development Journal, 22*(8), 394–401.
11. Song, W., Tan, K. H., & Baranek, A. (2009). Effective toolbox for lean service implementation. *International Journal of Services and Standards, 5*(1), 1–16.
12. Martinez Sanchez, A., & Perez Perez, M. (2004). The use of lean indicators for operations management in services. *International Journal of Services Technology and Management, 5*(5/6), 465–478.
13. Cuatrecasas, L. (2004). A lean management implementation method in service operation. *International Journal of services Technology and Management, 5*(5–6), 532–544.
14. Diane, M., Womach, J., Byrne, A. P., Flume, O. J., Kaplan, G. S., & Toussaint, J. (2005). Going lean in health care. *Innovation Series*, 1–20.
15. Kim, C. S., Spahlinger, D. A., Kin, J. M., & Billi, J. E. (2006). Lean health care: What can hospital learn from a world-class automaker? *Jorunal of Hospital Medicine, 1*(3), 191–199.
16. Ohno, T. (1988). *Toyota production system: Beyond large-scale production*. Productivity Press.
17. Zidel, T. G. (2006). *A lean guide to transforming healthcare: How to implement lean principles in hospitals, medical offices, clinics, and other healthcare organizations*. ASQ Quality Press.
18. Tranfield, D., Denyer, D., & Smart, P. (2003). Towards a methodology for developing evidence-informed management knowledge by means of systematic review. *British Journal of Management, 14*(3), 207–222.

Road Freight Transport in Europe: Alternatives for Increasing Capacity

Yesica Pino, José A. Pascual, and Imoleayo Abraham Awodele

Abstract Transport is a strategic sector of the European Union economy (about 5% of GDP) as it represents a significant source of jobs (5% of total employment), without forgetting its role in the proper implementation of the European single market. The volume of freight transport has grown in recent decades and is expected to continue to do so, with road freight transport contributing more than 75% of total inland freight transport in the European Union (EU) in 2020. Thus, more than three quarters (79%) of EU road freight transport in tonne-kilometres were carried by heavy goods vehicles with a maximum authorised mass above 30 tonnes. In addition, heavy-duty vehicles are responsible for about a quarter of the fuel consumption and greenhouse gas emissions of the transport sector in the EU. To reduce emissions, the EU has proposed an increase in transport efficiency which, among other strategies, includes better use of freight capacity and longer and heavier trucks. Currently, the dimensions and weights of trucks in international transport on European roads are regulated by Directive (EU) 2015/719. The European Commission allows EU members to test different dimensions than those proposed in the directive (without unfair competition) in order to study different alternatives for more efficient, greener and safer transport. The general situation of the EU and neighbouring countries in the field of road freight transport will be analysed, considering some alternatives such as those proposed in the Scandinavian countries, the United Kingdom and Italy, and their impact on costs and carbon footprint.

Keywords Heavy good vehicle · Large good vehicle · Sustainability

Y. Pino · J. A. Pascual (✉)
Dpto. Organización de Empresas y CIM, U. Valladolid, Valladolid, Spain
e-mail: pascual@eii.uva.es

I. A. Awodele
Construction Management and Quantity Surveying Department, DUT, Durban, South Africa

J. C. de Oliveira Matias et al. (eds.), *Quality Innovation and Sustainability*, Springer Proceedings in Business and Economics,
https://doi.org/10.1007/978-3-031-12914-8_5

1 Introduction

Nowadays organisations are facing economic, political, social, and technological changes increasingly faster. These changes have their origin in different aspects such as the globalisation of markets, uncertainty, high competitiveness, greater customer demand, concern for sustainability, and increased energy costs, among others [1]. Adapting to these changes has made both companies and supply chains increasingly competitive [2]. The competence in the logistics chain has revealed the importance of the logistics strategy as a generator of added value, a tool to satisfy customers and a facilitator of the achievement of business objectives. With the objective of generating value and obtaining competitive advantages over their rivals, companies seek greater efficiency and productivity in the different processes of the supply chain.

Therefore, the role and relevance of logistics in the strategic management of a company are increasing [3]. The generation of value and the search for competitive advantages make companies devote greater efforts to the design and organisation of logistics strategy. This has allowed in recent years a great development in logistics and transport, becoming one of the relevant areas of business management. Transport is an essential activity in the logistics strategy at the enterprise level, a competitive force in business, and an important contributor to the socio-economic and environmental development of countries [4–6]. Transport plays an essential role in development by providing people with accessibility to resources and markets, thus improving their quality of life, but doing so with the least possible environmental impact, which is sometimes difficult to implement [7, 8].

On the other hand, transport represents a relevant cost for most companies, around 10–20% of the total costs of a product [9, 10] and an important industry in all developed economies [11, 12]. Logistics accounts for 14% of the GDP of the European economy. Transport is a key sector of the economy: it accounts for more than 9% of the EU's gross value added (the EU's contribution to the economy). Transport services alone account for around €664 million and employ almost 11 million people. The costs associated with transport have increased significantly in recent years. The reasons are increase in the fuel price, greater distances travelled, reduction in delivery times, and lack of qualified personnel, among others. To reduce transport costs, many companies and institutions have promoted studies to improve utilisation and expand the available cargo capacity [13–15] in the different modes of transport.

The increase in load capacity (volume and weight) in the different modes of transport (road, rail, sea, and air) plays a fundamental role in the profitability of transport. The weight–volume ratio is very important in the planning of the occupation of the load capacity of the transport used since the volume is often exceeded before the weight that is transported [16].

In addition to the economic aspect of the more effective use of carrying capacity, there is a social facet related to the growing sensitivity towards sustainability and the climate impact of transport. Transport is the logistics operation with the greatest impact on the environment and the highest energy consumption, generating a large part of the world's greenhouse gas (GHG) emissions [17, 18]. According to the International Energy Agency [19], transport accounts for 28.9% of total energy consumption globally, and generating almost 25% of CO_2 emissions, for which road transport accounted for 72% in 2020. To improve the sustainability of transport, especially road freight transport, it is essential to study the efficient use of load capacity [16]. Operating with empty or partially loaded trucks implies an inefficient use of the company's resources and, consequently, operating in an unsustainable way [20–22].

The United Nations adopted in 2015 the 2030 Agenda for Sustainable Development [23], an opportunity for countries and their societies to embark on a new path to improve the lives of all. It has 17 Sustainable Development Goals (SDGs), including climate change and environmental advocacy. Precisely Goal 13 expresses how the levels of emissions of carbon dioxide (CO_2) and other GHG emissions in the atmosphere increased to record levels in 2019. Climate change is affecting all countries on all continents and alternating national economies [23].

Despite the fact that GHG emissions have fallen by around 6% in 2020 [24], due to movement restrictions and economic recessions resulting from the COVID-19 pandemic, this improvement is only temporary as global economy will recover from the pandemic and emissions will return to normal levels. In particular, the global transport sector is responsible for almost a quarter of GHG emissions, mainly CO_2, of which approximately 72% come from road transport [25]. In this sense, to realise a more optimal transport, it is essential to achieve a greater utilisation of the load capacity, which leads us to focus on the analysis of the percentage of volume and weight occupied as a system for measuring efficiency in the decision-making process [15, 16].

Cargo capacity concept has been studied from different perspectives and points of view (transport companies, customers, and government) which has led to a great diversity of opinions and results, without there being a standard model to calculate the most effective load capacity depending on the different situations of use [15, 16, 26]. One of the most used factors to determine efficiency of the freight transport is capacity utilisation (load factor), which relates the actual weight of goods to the maximum weight that can be transported [27]. This measure underestimates the actual use of the vehicle in sectors where utilisation is limited by volume and not by weight [28].

However, despite the importance of the concept of carrying capacity in the various logistics activities, most studies have focused on the analysis of the performance of the weight transported according to the type and age of the vehicles, without obtaining results on the economic and environmental consequences [29–34]. Based on data provided by Eurostat [35], the tonnes transported per vehicle are lower than the maximum allowed (EU average vehicle loads were 14.3 tonnes in 2020, with national loads of 13.5 and international loads of 15.9 tonnes) [35]: this is in many situations because lorries have a limited volume and cannot be filled more than the maximum permitted [28, 36].

Lumsden [37] states that long-distance road transport in Europe is very sensitive to load capacity measured on pallets as 2/3 of the total trips analysed were loaded to at least 90%, while less than 20% of the same trips were loaded above 90% of the weight. The volume capacity used, on average, was 82%, and half of the transport is charged to 90%. And the average weight capacity used was 57%. A recent study in Spain [38] estimated that only 50% of vehicles with a maximum authorised mass of 40 tonnes are loaded with 32 tonnes or more. This is usually because many products occupy much more volume than weight, so that the lorries are not saturated in weight but in volume.

The purpose of this study is to analyse the contribution of increased volume, particularly lorry length, on the efficiency and environmental impact of road transport, proposing a solution based on existing trailers (or adapting them in a cost-effective way) without increasing the maximum weight transported.

The chapter is organised as follows. In Sect. 2, a brief compilation of the dimensions and weights permitted in Europe is shown, and in Sect. 3, the most common configurations of articulated vehicles and road trains in Spain are analysed. Then, in Sect. 4, the proposed solution is presented and justified. The chapter concludes with some conclusions and future developments.

2 Current Lorry Size and Weight Regulations in Spain and Other European Countries

2.1 Europe

Although there is an EU directive to harmonise the weights and dimensions of heavy goods vehicles (HGVs) for reasons of road safety and avoid damaging infrastructure [39] (amending [40]), there are some differences between them and even more if we include non-EU countries. This directive ensures that Member States cannot restrict the circulation of vehicles complying with these limits to carry out international transport operations within their territories and also aims to prevent national operators from benefiting from undue advantages over their competitors from other Member States when carrying out national transport.

However, as there are non-EU countries in Europe, they do not have to comply with these limits and [39] itself provides for derogations on maximum lengths to make HGVs more environmentally friendly by improving their aerodynamic performance or safer by adding extra space in the driver's cab. And exemptions on weights are also allowed for vehicles powered by alternative fuels (due to the higher weight of batteries in the case of electric vehicles). Thus, in a country's internal traffic, the dimensions and weights of lorries in the different European countries vary according to a multitude of circumstances: goods transported (mainly cars, timber and agricultural products), intermodal transport (containers), types of roads (number of lanes, motorways), and of course certain construction characteristics of the lorries (number of axles, refrigerated lorries).

There are very few differences in width and height between countries. The maximum permissible height is 4.0 m in most countries, except for some countries that allow larger heights (4.2–4.5 m) for specialised lorries in transporting vehicles, cranes for removal of vehicles, or transporting containers approved for combined transport. And the maximum permissible width is 2.55 m except for some countries, which allow larger widths for refrigerated vehicles or for container transport. But although the length is broadly standard depending on whether we are talking about articulated vehicles (16.5 m) or road trains (18.75 m) in most European countries, there are significant differences when considering the transport of some goods, mainly cars. And there are some differences in the maximum authorised weight of articulated vehicles and road trains depending on the number of axles, the goods transported, the roads on which they travel, and above all whether the goods are transported in containers (Table 1).

Directive 97/27/EC [43] sets out provisions for the masses and dimensions of vehicles and also certain requirements relating to manoeuvrability. In the case of some semi-trailers, the manoeuvrability requirements will be deemed to be met by virtue of their dimensions.

2.2 *Typical HGV Configurations in Spain*

Articulated Vehicle The overall length of semitrailer tractor + semitrailer is 16.5 m in Spain (as in most European countries), so the semi-trailer can have a maximum length of 13.6 m (12 m from the kingpin) and a useful width of approximately 2.5 m. In order not to exceed the maximum overall height of 4 m, the semi-trailer is usually around 2.9–3 m (it can be slightly extended depending on tyre configurations). Therefore, we have approximately 33.7 m^2 of usable floor space and around 95–100 m^3 of volume for most models. This means that 33 euro pallets are usually transported (although it could be as many as 34).

Road Train The total length of a road train is 18.75 m, which allows an overall load dimension of 15.65 m for rigid drawbar trailers. Considering the same height and width dimensions, we obtain a usable area of approximately 39 m^2 and a volume of around 115 m^3. And a capacity of around 38 euro pallets.

Euro-Modular System (EMS) Since December 2015, road trains of up to 25.25 m and 60 tonnes have been allowed to operate in Spain, although their use is restricted to obtaining an authorisation to circulate. The EMS ('European Modular System') is a modular combination provided for in European legislation (Directive 2002/7/EC [44] amending Directive 96/53/EC [40]) which is made up of transport elements common in Europe, assembled in such a way as to form homogeneous units and optimise transport capacity: a 'carrier' lorry (7.82 m) pulling a semi-trailer across a platform (13.6 m) and a tractor unit with semi-trailer (13.6 m) plus trailer (7.82 m). Therefore, for the EMS, loading volume is approximately 156 m^3, loading area is about 53 m^2, and the maximum number of pallets (without re-assembly) is 52 pallets.

Table 1 Permissible maximum dimensions and weights of lorries in Europe

Country	Weights (in tonnes)			Length (in metres)	
	Road train		Articulated vehicle 5 axles and +	Road train[a]	Articulated vehicle
	4 axles	5 axles and +			
Albania	36	40	44	18.75	16.50
Armenia	36	36	36	20	20
Austria/Estonia /Germany	36	40/44	40/44	18.75	16.50
Azerbaijan	36	42	44	20	20
Belarus	38/40	40/42	42/44	20	24
Belgium	39	44	44	18.75	16.50
Bosnia-Herzegovina	36/38	40/42	42/44	18.75	16.50
Bulgaria /Liechtenstein /North Macedonia /Poland /Switzerland	36	40	40	18.75	16.50
Croatia	36	40	40/44	18.75	16.50
Czech Republic	32	48	48	18.75	16.50
Denmark[b]	38	44	44	18.75	16.50
Finland[c]	36	44	44	34.50	23
France	38	40/44	40/44	18.75	16.50
Georgia/Romania /Serbia	36	40	40/42	18.75	16.50
Greece	38	40/42	40/42/44	18.75	16.50
Hungary	36/38	40	40/42	18.75	16.50
Ireland	36	42/46	44/46	18.75	16.50
Italy	40	44	44	18.75	16.50
Latvia	36	40	40/42/44	18.75	16.50
Lithuania	36	40/42	40/44	18.75	16.50
Luxembourg	44	44	44	18.75	16.50
Malta /Moldova /Montenegro / Slovenia /Turkey	36	40	40/44	18.75	16.50
Netherlands	40	50	50	18.75	16.50
Norway	39	46–50	46–50	19.50	17.50
Portugal	37	44	44	18.75	16.50
Russia	36	40/44	40/44	20	20
Slovakia	40	40	40	18.75	16.50
Spain	36/38	40	42/44	18.75	16.50
Sweden	38	40[d]	44	25.25	24
Ukraine	38/44	40/44	40/44	22	22
United Kingdom	36/38	40/44	40/44	18.75	16.50

Adapted from International Transport Forum [41, 42]

[a]Road train specialised in the carriage of cars (loaded): 22 m (Ireland, Slovenia), 21.75 m (Romania), 21 m (Croatia, Montenegro, Serbia), 20.75 m (Czech Republic, Lithuania, Moldova), 20.55 m (Spain), 20.35 m (France); specialised road train: 20 m (Georgia); lorry with two trailers: 22/24 m (Hungary); road train with two trailers: 22.00 m (Turkey); heavy goods vehicle specially designed for the transport of timber: 24 m (Norway)

[b]Six-axle: 50 t; seven-axle or more: 56 t

[c]Five-axle: 44 t; six-axle: 56 t; seven-axle: 60 t; eight-axle: 64–68 t (restrictions for ADR), 69–76 t (not for ADR)

[d]On some roads, the permissible maximum weight is 74 t

2.3 Potential Benefits

Despite being a practical solution to reduce costs and emissions, longer and heavier vehicles (HLVs) often get a bad rap for being unsafe or requiring large infrastructure investments. HLVs [also known as toll doubles, eco-combis, high-capacity transportation (HCT), road trains, super trucks, or mega trucks] are trucks that are longer and heavier than the legal dimensions of a country.

By consolidating the load of many vehicles into a combination of high-capacity vehicles, HLVs can reduce carbon emissions and cost per unit of cargo transported: an individual LHV could reduce carbon emissions by 15–40%, a 33% reduction in its costs, and a 70% drop in fuel consumption and CO_2 emissions. There are also some positive effects on driver shortages; consolidating the load of many trucks into one allows transportation operators to operate with less manpower and potentially use the improved margins to pay better wages to skilled drivers.

3 Longer Articulated Vehicles: Similar Studies in Europe (Italy and the United Kingdom)

According to Article 4 of [40], a Member State may permit, through its national territory, the circulation of certain combinations of vehicles differing from maximum values of masses and dimensions set out in Annex I thereto, provided that this does not significantly affect international competition in the transport sector. This is enabled by a special permit, which is issued by the relevant state authorities.

The circulation of megatrucks is already allowed in European countries; however, a large part of these have applied this measure within the framework of pilot experiences to study and evaluate the impact of their implementation on traffic safety, infrastructure, emissions, costs, and transport. Sweden was the first country to implement gradually EMS systems, and subsequently the Netherlands. Also, Denmark, Finland, Belgium, Germany, Norway, and other European countries belong to the group of countries that have been gradually introducing this kind of transport. According to [45], the current studies are

- Sweden: Pioneer in the use of vehicles that exceed the limits of the regulations, it allows since 2009, in the north of the country, the use of trucks of 90 tonnes and 30 m long, for the transport of wood [46]. Up to 64 tonnes are currently allowed and 74 tonnes and 34 m long trucks (truck with two trailers) are being tested.
- The Netherlands is currently testing the use of 32 long truck trains with good results, which will soon allowed for usage.
- Denmark: Currently, the vehicle maximum authorised weight is 48 tonnes.
- Finland: It is similar to Sweden, but it is possible to use a complete trailer of 13.6 m long by train truck of 25.25 m, although it is not foreseen in the directive but cannot be used in international transport. In 2013, following a change in

legislation, the use of vehicles with the maximum authorised weight of 76 [47] was authorised. And they allow vehicles of 34.5 m since January 2019.
- Germany: Since 2017, trucks with a weight of 40 tonnes and a length of 25.25 m have been allowed to circulate, and the possibility of increasing the maximum gross weight to 44 t is being analysed [45].
- Norway: Norwegian regulations since January 2020 allow the circulation of vehicles 24 m long and up to 60 tonnes [45].

3.1 Progetto Diciotto (Italy)

With the publication in September 2021 of Decree Law number 121/2021 [48], the experiment known as Progetto Diciotto [49] (P18), developed by ANFIA (Associazione Italiana Filiera Industria Automobilistica) and several transport companies, has been put into operation. In 2009, the project began, which sought to increase the efficiency of road freight transport by increasing the length of articulated trucks from 16.50 m to 18 m (including towing devices), thus increasing their capacity in terms of volume (the maximum total mass remains at 44 tonnes).

P18 has shown that there are no problems in daily use and allows to reduce the number of commercial vehicles. In fact, an articulated truck of 18 metres, a metre and a half more than before (trailer of 15.1 m), can transport 37 euro pallets instead of the 33 of the 16.5 m (see Fig. 1). The experimental phase [50] after 10 million real kilometres, particularly on medium-length routes (200–600 km), proved that 18-m lorries have proven to be suitable for the transport of light and bulky goods loaded on pallets or in bulk, which can take advantage of the additional space available by staying within the legal limits of the maximum transportable weight (e.g. light food, packaging, toilet/household paper, household appliances, etc.).

In addition, allowing cargo saturation, saving on average 12% the number of trips, reducing fuel consumption per unit of goods transported, and consequently in CO_2 emissions has demonstrated to improve transport efficiency. In terms of driving comfort and manoeuvrability, for 74% of drivers there are no differences between the vehicles of the Progetto 18 and the traditional ones, except in some restricted parking lots, and they did not find manoeuvrability problems. In terms of safety, 83% perceived the same safety and 7% even better compared to 16.50-m trailers. When it comes to safety, only 6% of drivers believe that overtaking manoeuvres while driving a P18 vehicle is more difficult.

3.2 High-Volume Semi-Trailer Trial (the United Kingdom)

The Department for Transport [51] proposed an increase of 2.05 m in the length of semi-trailers and a maximum length of 18.75 m for articulated vehicles (allowing 30 standard UK pallets to be transported on one trailer instead of 26; see Fig. 2) within the existing weight limit of 44 tonnes gross vehicle weight (offering more efficiency

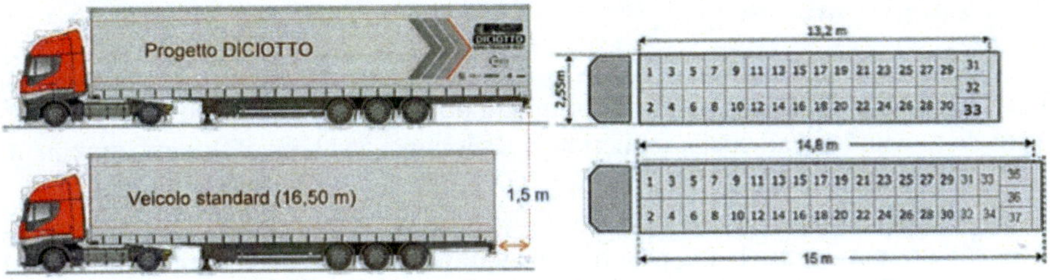

Fig. 1 Comparison between the P18 truck and the traditional one. (Taken from [50])

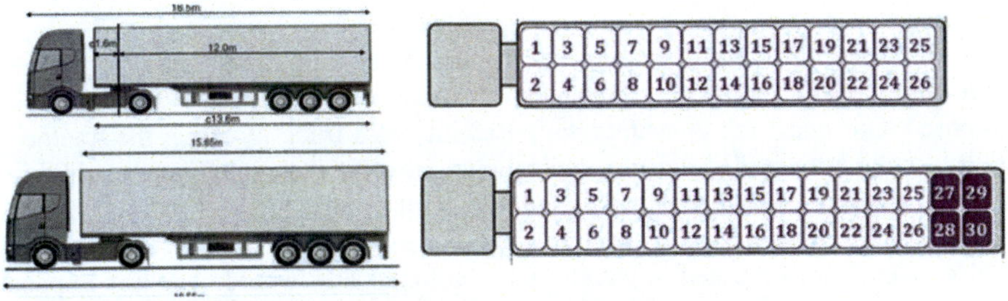

Fig. 2 Comparison between the LST truck and the traditional one. (Taken from [51, 52])

and fewer truck trips for goods limited by volume rather than weight). This will not involve vehicles longer than those on British roads as the current maximum allowable length for a rigid vehicle/trailer combination with drawbar is already 18.75 m.

The project began in 2009, with the main objective of analysing whether the introduction of these high-volume semi-trailers would generate general economic (more efficient freight transport, equal amount of cargo in less travels, etc.), environmental (local air pollution and carbon reduction), and land social benefits (positive benefit on congestion, reduction of accidents, noise, and infrastructure costs).

Testing with longer semi-trailers (LST) began in 2012, and until the end of 2019, the results of the trial indicated that [53, 54]

- It reduced, on average, 1 in 12 trips, saving between 54 and 60 million vehicle-kilometres.
- Showed reduction in emissions of 48,000 tonnes of CO_2 and 241 tonnes of NOx.
- LSTs have been involved in approximately 53% fewer collisions and personal injury victims than the average GB HGV.

Given these results, they have considered that they have enough data, and although initially it was planned that the tests would run until 2027, it is being considered to put an end to it and its use is approved, with certain limitations, this year; even a new variant of 48 tonnes is being thought.

4 Conclusions

The challenges faced by the transport and logistics sector worldwide include increase in demand and costs, reduction in emissions, and lack of drivers, making a series of measures necessary to improve the efficiency of the sector. One of these possible improvements is the efficient utilisation of truck capacity. In this sense, different alternatives are being proposed that allow to extend the length of the trucks especially if they are saturated in volume and not in weight.

An increase in the size of trucks means an increase in the capacity of the load and with it a decrease in the trucks that are necessary to transport the same amount of cargo, reducing congestion, accidents, consuming less fuel, and emitting less CO_2 than conventional vehicles per unit of cargo transported.

With respect to future developments, we are developing a pilot study, together with transport and industrial companies, in which a new alternative for trucks is proposed. Our approach considers the possibility of a truck in which the tractor–trailer combination is 20.5 m long, while maintaining the same maximum load limitation of 40 tonnes as at present. This solution, with a trailer of 18.75 m long, would allow 10 more euro pallets to be transported, which would be very efficient for routes that are not saturated in weight but in volume. This would have less impact on existing infrastructures and, given its manoeuvrability, would require little investment in infrastructure.

References

1. Bowersox, D. J., Closs, D., & Stank, T. P. (2000). Ten mega-trends that will revolutionize supply chain logistics. *Journal of Business Logistics, 21*, 1.
2. Atkins, R., & Gianiodis, P. (2021). An investigation at the intersection of the sharing economy and supply chain management: A strategic perspective. *International Journal of Logistics Research and Applications*. https://doi.org/10.1080/13675567.2021.1911970
3. Matwiejczuk, R. (2019). On the research concerning the relations between logistics and strategic management – Towards the strategic logistics concept in management. *Prace Naukowe Uniwersytetu Ekonomicznego we Wrocławiu.*
4. Chapter 1 Transportation in business and the economy definitive guide to transportation, The principles, strategies, and decisions for the effective flow of goods and services. By CSCMP, T, J. Goldsby, D. Iyengar, S. Rao. Published Jan 16, 2014 by Pearson FT Press. Part of the Council of Supply Chain Management Professionals series.
5. Sperling, D., & Salon, D. (2002). *Transportation in developing countries, an overview of greenhouse gas reduction strategies*. University of California.
6. Chatti, W. (2021). Moving towards environmental sustainability: Information and communication technology (ICT), freight transport, and CO_2 emissions. *Heliyon, 7*(10), e08190, ISSN 2405-8440. https://doi.org/10.1016/j.heliyon.2021.e08190. https://www.sciencedirect.com/science/article/pii/S2405844021022933
7. de Campos, R. S., Simon, A. T., & de Campos Martins, F. (2019). Assessing the impacts of road freight transport on sustainability: A case study in the sugar-energy sector. *Journal of Cleaner Production, 220*, 995–1004, ISSN 0959-6526. https://doi.org/10.1016/j.jclepro.2019.02.171

8. Ahmed, M. M. A. W., & El Monem, N. A. (2020). Sustainable and green transportation for better quality of life case study greater Cairo – Egypt. *HBRC Journal, 16*(1), 17–37. https://doi.org/10.1080/16874048.2020.1719340
9. Anderson, J. E., & Van Wincoop, E. (2004). Trade costs. *Journal of Economic Literature, 42*(3), 691–751.
10. Rodrigue, J. P., Comtois, C., & Slack, B. (2016). *The geography of transport systems* (4th ed.). Routledge.
11. Armstrong & Associates Inc Global 3PL Market Estimates. (2016). http://www.3plogistics.com/3pl-market-info-resources/3pl-market-information/global-3pl-market-size-estimates/ (16/10/2017).
12. Ballou, R. H. (2007). *Business logistics/supply chain management* (5th ed.). Pearson Education India.
13. Dorer, K., & Calisti, M. (2005). *And adaptative solution to dynamic transport optimization*. Whitestein technologies AG.
14. Wu, H. J., & Dunn, S. C. (1995). Environmentally responsible logistics systems. *International Journal of Physical Distribution and Logistics Management, 25*(2), 20–38.
15. Abate, M., & Kveiborg, O. (2013). Capacity utilisation of vehicles for road freight transport. In M. E. Ben-Akiva, H. Meersman, & E. van de Voorde (Eds.), *Freight transport modelling* (pp. 287–304). Emerald Group Publishing.
16. Abate, M. (2014). Determinants of capacity utilisation in road freight transportation. *Journal of Transport Economics and Policy, 48*(1), 137–152.
17. Rogers, M. M., & Weber, W. L. (2011). Evaluating CO_2 emissions and fatalities tradeoffs in truck transport. *International Journal of Physical Distribution and Logistics Management, 41*(8), 750–767.
18. Christopher, M. (1998). *Logistics and supply chain management: Strategies for reducing costs and improving service*. Pitman.
19. IEA. (2021a). *Tracking transport 2021*. IEA. https://www.iea.org/reports/tracking-transport-2021
20. Enarsson, L. (1998). Evaluation of Suppliers: How to consider the environment. *International Journal Physical Distribution & Logistics Management, 28*(1), 5–17.
21. Lieb, K., & Lieb, R. (2010). Environmental sustainability in the third-party logistics (3PL) industry. *International Journal of Physical Distribution & Logistics Management, 40*(7), 524–533.
22. McKinnon, A. (2003). Logistics and the environment. In D. A. Hensher & K. J. Button (Eds.), *Handbook of transport and the environment* (pp. 665–685). Pergamon.
23. United Nations. (2015). *Transforming our world: The 2030 agenda for sustainable development*. http://www.un.org/ga/search/view_doc.asp?symbol=A/RES/70/1&Lang=E
24. IEA. (2021b). *Global energy review: CO_2 emissions in 2020*. IEA. https://www.iea.org/articles/global-energy-review-co2-emissions-in-2020
25. *IPCC climate change 2014: Synthesis report* (Core Writing Team, Pachauri, R. K., & Meyer L. A., Eds.).
26. ITF. (2019). *High capacity transport: Towards efficient, safe and sustainable road freight* (International Transport Forum Policy Papers, No. 69). OECD Publishing.
27. Piecyk, M. I. (2010). *Analysis of long-term freight transport, logistics and related CO_2 trends on a business-as-usual basis* (Ph.D. Thesis). Heriot-Watt University.
28. McKinnon, A. C. (2009). Benchmarking road freight transport. Review of a government sponsored programme. *Benchmarking: An International Journal, 16*(5), 640–656.
29. Caplice, C., & Sheff, Y. (1995). A review and evaluation of logistics performance measurement systems. *The International Journal of Logistics Management, 6*(1), 61–74.
30. Contrell, W. D. (2008). *Performance metrics used by freight transport providers* [online]. Available at https://www.csusb.edu/leonard-transportation-center. Accessed 13 Dec 2021.
31. Harrinson, R., et al. (2006). *Freight performance measures guide*. Center of transportation research, University of Texas at Austin.

32. Lai, K., Ngai, E., & Cheng, T. (2004). An empirical study of supply chain performance in transport logistics. *International Journal of Production Economics, 87*, 321–331.
33. McMullen, S., & Monsere, C. (2010). *Freight performance measurements: Approach analysis*. Oregon Department of Transportation.
34. Mentzer, J. T., & Konrad, B. P. (1991). An efficiency/effectiveness aproach to logistic performance analysis. *Journal of Business Logistics, 12*(1), 33–61.
35. Eurostat. (2021). *Road freight transport by journey characteristic - Statistics Explained*. https://ec.europa.eu/eurostat/statistics-explained/index.php?title=Road_freight_transport_by_journey_characteristics&oldid=545754
36. Galos, J., Sutcliffe, M., Cebon, D., Piecyk, M., & Greening, P. (2015). Reducing the energy consumption of heavy goods vehicles through the application of lightweight trailers: Fleet case studies. *Transportation Research Part D: Transport and Environment, 41*, 40–49, ISSN 1361-9209, https://doi.org/10.1016/j.trd.2015.09.010
37. Lumsden, K. (2004). *Truck masses and dimensions - Impact on transport efficiency*. Department of Logistics and Transportation, Chalmers University of Technology.
38. Vassallo, J. M., Tarriño, J., & Pérez-Doval, J. M. (2020). *Impacto y Repercusión del Incremento del Peso Máximo de los Vehículos Pesados a 44 Toneladas en España*. Fundación Corell.
39. Directive (EU) 2015/719 of the European Parliament and of the Council of 29 April 2015 amending Council Directive 96/53/EC laying down for certain road vehicles circulating within the Community the maximum authorised dimensions in national and international traffic and the maximum authorised weights in international traffic (Text with EEA relevance) Official Journal L 115, 6.5.2015, pp. 1–10. http://data.europa.eu/eli/dir/2015/719/oj
40. Council Directive 96/53/EC laying down for certain road vehicles circulating within the Community the maximum authorized dimensions in national and international traffic and the maximum authorized weights in international traffic, Official Journal L 235, 17/09/1996, pp. 59–75. http://data.europa.eu/eli/dir/1996/53/oj
41. International Transport Forum. (2015a). *Permissible maximum dimensions of lorries in Europe*. http://www.internationaltransportforum.org/IntOrg/road/pdf/dimensions.pdf
42. International Transport Forum (2015b). *Permissible maximum weights of lorries in Europe*. http://www.internationaltransportforum.org/IntOrg/road/pdf/weights.pdf
43. Directive 97/27/EC of the European Parliament and of the Council of 22 July 1997 relating to the masses and dimensions of certain categories of motor vehicles and their trailers and amending Directive 70/156/EEC. Official Journal L 233, 25.8.1997, pp. 1–31.
44. Directive 2002/7/EC of the European Parliament and of the Council of 18 February 2002 amending Council Directive 96/53/EC laying down for certain road vehicles circulating within the Community the maximum authorised dimensions in national and international traffic and the maximum authorised weights in international traffic.
45. Dzioba, A., Markiewicz, M., & Gutsche, J. (2021). Analysis of modular transport systems functioning in selected European countries. *MATEC Web Conference, 332*, 01008. https://doi.org/10.1051/matecconf/202133201008
46. Muślewski, M., Lewalski, M., & Woropay, M. (2015). Analysis and evaluation of application of car modular systems in polish road transport. *Journal of Kones, 22*, 211–222.
47. Liimatainen, H., Pöllänen, M., & Nykänen, L. (2020). Impacts of increasing maximum truck weight – Case Finland. *European Transport Research Review, 12*(14), 1–12.
48. Disposizioni urgenti in materia di investimenti e sicurezza delle infrastrutture, dei trasporti e della circolazione stradale, per la funzionalita' del Ministero delle infrastrutture e della mobilita' sostenibili, del Consiglio superiore dei lavori pubblici e dell'Agenzia nazionale per la sicurezza delle infrastrutture stradali e autostradali. (21G00133) (GU Serie Generale n.217 del 10-09-2021)
49. ANFIA. Progetto DICIOTTO. Piano per la sperimentazione di complessi veicolari da 18 metri. http://palena.com/static/documentazione/progetto18.pdf

50. ANFIA. (2019). Progetto Diciotto – Long Vehicle Test. Monitoraggio della sperimentazione 2° report Fase II. Verona. https://www.anfia.it/allegati_contenuti/DOC/141_CONVEGNO_P18_PRESENTAZIONE_ANFIA.PDF
51. Department for Transport. (2010). *Longer semi-trailer feasibility study and impact assessment.* https://www.gov.uk/government/uploads/system/uploads/attachment_data/file/3820/report.pdf
52. Department for Transport. (2020). *GB Longer Semi-Trailer trial: 2019 Annual Report summary.* https://www.gov.uk/government/publications/longer-semi-trailer-trial-evaluation-annual-report-2019/gb-longer-semi-trailer-trial-2019-annual-report-summary
53. Department for Transport. (2021a). *Ending the longer semi-trailer trial: Consultation response.* https://assets.publishing.service.gov.uk/government/uploads/system/uploads/attachment_data/file/1012504/ending-the-longer-semi-trailer-trial-consultation-response.pdf
54. Department for Transport. (2021b). *Evaluation of the longer semi-trailer trial: Annual report 2020 update.* https://assets.publishing.service.gov.uk/government/uploads/system/uploads/attachment_data/file/1013304/evaluation-of-the-longer-semi-trailer-trial-annual-report-2020-update.pdf

Knowledge Management: An Overview of Roadmaps for Additive Manufacturing

V. Fernandes, F. Matos, and R. Godina

Abstract Additive manufacturing (AM), also recognized as 3D printing, has been winning more and more appreciation in recent times given its relevance to the emergence of Industry 4.0. AM technology has proven to have enormous potential, and the last decade has seen its rapid development and integration into the industry. Therefore, it becomes necessary to develop tools and support approaches to guide the implementation and adoption of AM technologies in organizations. The roadmap tool is one of the possible tools that can help in the innovation of an organization to support the integration and management of additive manufacturing. Although with the emergence of new materials and processes, it is predictable that AM becomes a trend in manufacturing processes and has more and more interest from both academia and industry, and the need to develop new studies and tools that support this technology is still required. The aim of this chapter is to study the current state of the use of the roadmap applied to additive manufacturing as a tool to map the technological knowledge.

Keywords Knowledge management · Roadmapping · Roadmap · Additive manufacturing · 3D printing

V. Fernandes
Department of Mechanical and Industrial Engineering, NOVA School of Science and Technology | FCT NOVA, NOVA University of Lisbon, Lisbon, Portugal

F. Matos
Lisbon University Institute – ISCTE – DINÂMIA'CET-IUL – Centre for Socioeconomic Change and Territorial Studies, Lisbon, Portugal

R. Godina (✉)
Department of Mechanical and Industrial Engineering, NOVA School of Science and Technology | FCT NOVA, NOVA University of Lisbon, Lisbon, Portugal

UNIDEMI – Department of Mechanical and Industrial Engineering, NOVA School of Science and Technology | FCT NOVA, NOVA University of Lisbon, Lisbon, Portugal
e-mail: r.godina@fct.unl.pt

J. C. de Oliveira Matias et al. (eds.), *Quality Innovation and Sustainability*, Springer Proceedings in Business and Economics,
https://doi.org/10.1007/978-3-031-12914-8_6

1 Introduction

Additive manufacturing (AM), or 3D printing, is seen as a technological trend of Industry 4.0 [1–3]. It is a manufacturing technique that is changing industry and manufacturing processes [4].

Recently, it became one of the most used technologies to manufacture and prototype different structures in different industries [5]. It is expected, with the development of new processes and materials, that AM would turn out to be a common and spread worldwide manufacturing process. However, there is a necessity to understand better how corporations could adopt AM technology to support innovation, improve their performance, and increase competitiveness [6].

The future of technology is uncertain, and companies need to support their planning to interact with upcoming future technological growth by defining a technology strategy [1]. Thus, the necessity arises to research and develop alternative methods and tools to regenerate and exploit knowledge in order to preserve or gain a competitive advantage [7].

Roadmapping emerged as a well-established method to support companies in getting involved in technology management and strategic planning [6]. A technology roadmap is essential and indispensable to ensure success in the digital transformation process required for Industry 4.0 [1].

The use of this kind of tool in the industry allows companies to manage their technology, showing great potential to support AM and industrial development; however, there are still hurdles that need to be overcome.

This chapter aims to study the current state of the use of the roadmap applied to AM as a tool to map the technological knowledge through a literature review of several scientific articles, as well as the identification of challenges and gaps that this tool might face in the near future.

2 State of the Art

2.1 Additive Manufacturing

AM has enabled a revolution in the way that products are designed and produced [8, 9]. In 1987, the first commercial AM machine, SLA-1, was introduced by 3D Systems, and since then, AM industry has been developing rapidly. As evidence of its growth, in 2016, a report reveals an impressive 26.2% annual growth rate of this technology over the past 27 years [10].

The advent of AM has attracted much attention from both academia and industry. This technology uses digital design and layer-by-layer (l-b-l) assembly to create complex 3D structures in an extensive range of scales from micrometer to meter and has a huge potential to transform the engineering, manufacture, and use of advanced materials [11].

Research has developed new AM processes and applied them in aerospace, automotive, biomedical, and other fields (e.g., digital art and architectural design) [3, 12].

Industries such as electronics, automotive, aerospace, and medical have shown an increased rhythm of implementation of AM. Their capacity for parts production that is functional and has a higher turnover contributed to an increased deployment of this technology. It also enables organizations to decrease the available occupied space and manufacturing time. Customers using AM technology can get prototypes faster and at a lower price compared to conventional rapid prototyping (RP) [10].

AM is a process that adds layer-upon-layer of material and allows parts to be made from 3D model data as opposed to subtractive and formative manufacturing methodologies. This technology presents a wide variety of processes that use an assortment of materials such as plastics, metals, ceramics, and composites, among others [6].

AM makes it possible from a CAD model to manufacture 3D objects through layer-by-layer [13]. ISO/ASTM52900-15 establishes seven different AM procedures: powder bed fusion, directed energy deposition, material jetting, material extrusion, sheet lamination, binder jetting, and vat photopolymerization [14]. Stereolithography, fused deposition modeling, inkjet printing, and selective laser sintering are the most common processes [6].

The Design for Additive Manufacturing (DfAM) has appeared in the literature. The researchers started to explore this concept with the goal of providing a basis to support professionals in the design practice. According to its main purpose and application, there are two types of DfAM. "DfAM in the strict sense" includes approaches that are directly associated to the design process itself, e.g., "AM design rules" to ensure AM-produced parts and "AM design potentials" to take advantage of AM capabilities. On the other hand, "DfAM in the broad sense" are the approaches indirectly associated; however, these are a part/application of the selection and manufacturability analysis [8].

There are several advantages to AM, such as design flexibility, the ability to produce complex shapes, simplicity of use, and product customization [4]. Also, its digital nature allows manufacturing to be performed directly from CAD files, eliminating inventory and costs associated, which turns out to be an advantage of this type of technology [6].

These technologies can be widely used for traditional manufacturing; however, it is still important to overcome certain challenges. Perhaps the most noteworthy is the lack of knowledge among designers on how to conceive compounds that make the most of these new technologies. In fact, codification of knowledge about design regulation, principles, standards, and best practices has been recognized as a key issue in the acceptance of this technology [8].

2.2 Knowledge Management

The application focusing on the "know-how" of the organization is the key to technology that differentiates it from more general kinds of knowledge. Technology is typically associated with science and engineering ("hard" technology), but the procedures that empower its application are also essential ("soft" aspects of technology). It is useful to address technology as a type of knowledge since knowledge management notions incorporate managing technology more effectively [15].

Knowledge management is the creation, organization, sharing, dissemination, application, and use of knowledge. It is a notable success factor for companies and a significant precursor to innovation. There are numerous models and approaches, and their effectiveness depends on the situations of organizations. Adapting to change can enable organizations to embrace technological change in order to remain competitive [7].

2.3 Roadmap

First used at Motorola in the 1970s, the technology roadmap has been widely used as an approach to technology management [16]. This tool is one of the most powerful tools for technology management that allows support in the development and implementation of strategic business and product plans as it makes available the processes, information, and tools necessary to create comprehensive plans [17].

Roadmapping consists of strategic planning and aims to support innovation by helping to develop products, services, or systems identifying the technology and other resources needed, in line with business aims and in response to market drivers [6].

Over the years, technology roadmapping (TRM) has become a frequently employed technique for technology management in companies and industries [5]. A survey of 2000 UK manufacturing companies revealed that the TRM method has been applied by 10% of companies and around 80% of these corporations are adopting the technique multiple times, or continuously [15].

To balance the various components presented in a roadmap is essential considering the time measurement, and its creation can be executed at the organizational, industrial, or national level. In order to have an even deeper understanding, the purpose of applying different layers in a roadmap allows organizations to explore where we are and where we hope to go [5].

Thus, the basic principle of a roadmap is to present the current state (e.g., where you are now) and desirable future state (e.g., where you want to go), and paths between the present and future (e.g., how to get there). This tool helps managers in all organizations to develop a common vision of a complex situation that usually gives rise to a high degree of uncertainty about the future [16]. An effective roadmap illustrates a company's short- and long-term vision and business opportunities, and can connect its technology resource to those opportunities [16], providing a clear visual format (but not always) to aid communication and cooperation [2, 6].

The graphic representations provided by a roadmap help companies understand the relationships between products, markets, and technologies in the course of time and help governments and industry to come to investment decisions efficiently and stay competitive by providing knowledge about the emerging technologies [17].

It can take many forms, and the roadmap development process is flexible as the approach can and should be customized to suit different organizational contexts and specific issues addressed [16]. There can be various types of technological roadmaps, depending on the intended purposes and graphic formats. Many classifications are defined by the scientific community, but the most common categorizes the roadmaps into four groups: product technology roadmaps, science technology roadmaps, product roadmaps, and industry roadmaps [17].

2.4 Limitations

The practice of the TRM method has many challenges for companies, although it is fairly straightforward in structure and concept. There is a lack of practical support available. Despite efforts to share experiences, these sources do not explain how to use this approach specifying step-by-step guidelines. The success of this tool is dependent on having the right people involved and commitment from high-end management. Overload of initiatives, interference from short-term tasks and information, data, or knowledge not being accessible are factors that compromise the application of this tool [15, 18, 19].

Furthermore, one of the major limitations of roadmapping is that there currently does not exist any "one-size-fits-all" approach that fits all companies, which means that for each company it is idiosyncratic and must be designed based on the business's fundamental principles, intent, capabilities, goals, motivations, priorities, and budgets [1].

A more specific limitation, already related to the application of roadmapping to AM technology, is the fact that future advances that have not been known may change part of the roadmap that has been done, also as future possibilities. Therefore, roadmap maintenance is extremely necessary [18, 20]. Usually, it is advisable to upgrade a roadmap every 2 years [21].

3 Methodology

The collection of articles was initiated by taking into account inclusion criteria and the chosen keywords. To carry out the search in a more detailed manner and proceed with the development of the review article, the relevance of each article was evaluated according to the following inclusion criteria:

- Articles published in scientific journals and written in English.
- Articles that address the development or possible development of a roadmap.
- Articles that address cases of application of this tool to Industry 4.0, in particular, to AM.

For the elaboration of this review article, a structured and exploratory search was conducted using the scientific journals *Science Direct*, *Emerald*, *Springer*, and *Taylor & Francis* and the academic database "Scopus," in order to obtain a broader search.

In the first phase, only the title and abstract of several articles found were analyzed in order to understand whether they were related to the theme. Twenty-seven articles were identified, and, therefore, some were rejected because they did not match the defined inclusion criteria and others were rejected because the access to the full article was not allowed.

After this screening, the 19 articles that meet the defined inclusion criteria and that presented themselves as the most significant for the literature review were analyzed in depth so that from then on these could be used in the preparation of the review article.

The flowchart in Fig. 1 summarizes the main steps of the methodology that was followed for the preparation of the review article.

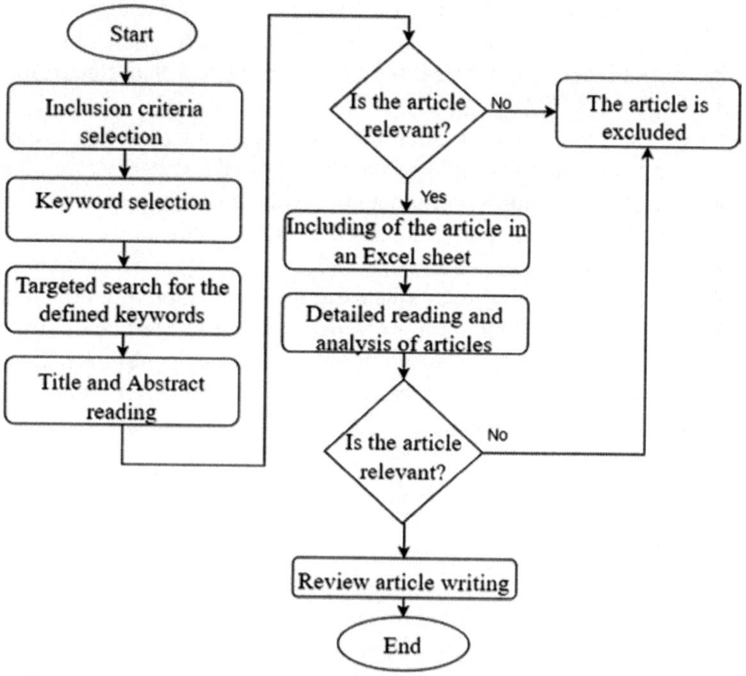

Fig. 1 Methodology flowchart

4 Analysis and Discussion of the Results

After the article selection process, an analysis of the type of article was performed. It can be seen that scientific articles represent 89% of all publications considered, of which one is a literature review article. Also, one book chapter and one conference paper were consulted.

In the literature, it was possible to find already some studies about the application of the roadmap tool to AM. In order to identify the different studies and researches that currently exist, which applied or addressed the roadmap for additive manufacturing technology, we considered only 10 articles in this initial analysis that contributed to the construction of Table 1. In this table, it is possible to find for each consulted article its year of publication, the application area, if applicable, the AM processes, and also the objective and the main conclusions (Table 1).

The articles analyzed showed a wide variety of application areas. In addition to those pointed out in the literature as industries where AM has been strongly adopted, it appears in the textile industry as a potential future path for a practical technology to create a real, new, complex, and diverse textile product [11]. It also emerges in the biomedical industry as the 3D bioprinting technology creates three-dimensional models, designing tissues or organs, through the prescribed deposition layer by layer of alive cells and biomaterials [5] and also in the plastic industry [22].

From the same table, it is possible to identify the different processes that are mentioned in the studies, which shows the great diversity of existing technologies of AM. The consulted literature mentions processes that are the most common to be used, such as selective laser sintering (SLS), fused deposition modeling (FDM), and also the laser beam powder bed fusion (PBF-LB process), also known as selective laser melting (SLM). However, less known processes such as 3D bioprinting were also mentioned.

Sustainability has been an increasingly current topic that has aroused great interest among researchers and organizations. AM technology and its potential advantages for driving environmental, economic, and social sustainability are causing research to begin to focus its attention on this topic. In three articles, studies related to sustainability and how AM can contribute to it are mentioned [6, 11, 22].

The articles [11, 22–24] mentioned developed studies and refer to the roadmap as a guide that aims to support innovation, identifying processes and other resources necessary to develop products, processes, and technologies, supporting future work. But no visual roadmap has been applied or developed, with a vision and opportunities in the short and long term.

The book chapter [10] reports the standards of AM that are essential to gather the raising needs of diverse industries, such as the consumer, commerce, and industrial areas, over a trustworthy verification of a product or market performance. The impact of the 2009 Roadmap for Additive Manufacturing Report (RAM) on advancing AM technology is assessed in the article [25].

The article [8] mentions the numerous problems and limitations of using AM, for each step from design to manufacturing and provides a framework to map current

Table 1 Characteristics of the consulted articles

Year	Application area	AM process	Objectives	Conclusions	Reference
2017	–	–	Build the sustainable value roadmapping tool (SVRT), combining the strategic roadmapping technique and the sustainable value analysis tool	There is still a need to refine and test what has been developed so that it can be a robust basis to assist companies in adopting AM	[6]
2018	–	–	Provide a basis for mapping current DfAM knowledge and identifying challenges and future guidelines for studies about DfAM	The study revealed that there are still several limitations affecting existing DfAM knowledge	[8]
2018	Plastic industry	Selective laser sintering (SLS) fused deposition modeling (FDM)	Demonstrate that a low-cost feed material (filament) for FDM can be fabricated from a mixture of residual SLS powder and tungsten carbide	It is shown that the filament can be made from a mixture of residual SLS powder and tungsten carbide, which will make the plastic AM more energy-efficient, self-sustaining, and contribute to the sustainability of the environment	[22]
2017	–	–	Introducing AM standards, their importance, historical context, and priority areas	–	[10]
2020	Textile industry	Direct ink writing (DIW)	Propose a potential for textile 3DP, naming the technique of printing with fibers to create textile structures	A potential path for 3D printing of textiles in the form of fiber printing has been outlined for further exploration	[11]

(continued)

Table 1 (continued)

Year	Application area	AM process	Objectives	Conclusions	Reference
2020	–	–	Provide an understanding of how AM and topology optimization can be used synergistically for design and manufacturing	Taking an aerospace support as an example, the procedure is performed and then the performance is verified where limitations and research directions for future studies are identified	[23]
2022	Biomedical industry	3D bioprinting	Develop and present a TRM for a startup to produce a roadmap for their product and business development	The roadmap developed allowed the identification of product and technology gaps, as well as knowledge gaps within the company, thus helping as a knowledge management tool	[5]
2021	–	Laser beam powder bed fusion (PBF-LB)	Present the roadmap for processing the H282 superalloy using PBF-LB technology	It has been proven that crack-free H282 can be processed by PBF-LB technology Selection of the appropriate parameters for this alloy was possible	[24]
2014	Biomedical applications, energy, and sustainability	–	Present the key findings of the 2009 roadmap for Additive Manufacturing Report (RAM) and some significant impacts the report has had on the advancement of AM technology	It is not possible to clarify whether in most cases the report was instrumental in triggering any AM activity; however, at least one case was used to create the Edison Welding Institute Additive Manufacturing Consortium	[25]
2017	Healthcare, aerospace, automotive, and manufacturing.	7 AM technologies	Identify the important aspects of the AM's evolutionary progress allows establishing trends in local markets, products, technologies, and capabilities	Determine possible advantages of the technology and identify opportunities for AM technology in the country, about its use, research, and development	[20]

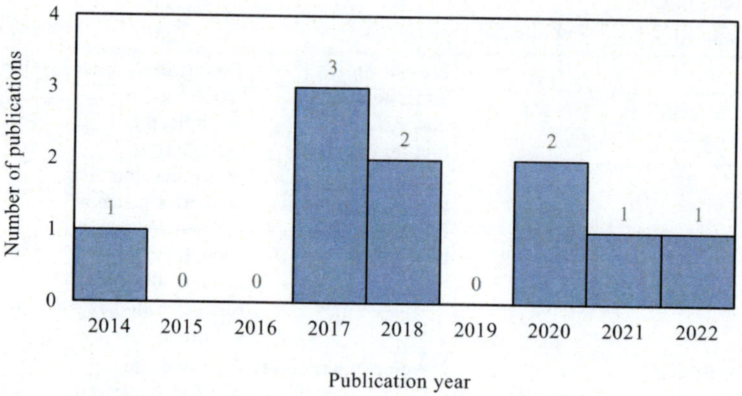

Fig. 2 Distribution of publications over time

DfAM knowledge and identify restrictions and future research directions for studies addressing DfAM. Related to sustainability, another article provides a structured methodology that consists of systematically considering the sustainability consequences of AM at all stages of the product life cycle [6].

Consequently, there are only a few articles that have developed and applied a visual roadmap tool to AM technology. At the national level, in Mexico, a roadmap was developed that permitted to create reference points in the upcoming progress of technology in terms of market, product, technology, and capacity in order to determine new opportunities for AM in the country [20].

On the other hand, in another article, a roadmap was developed at the organizational level for the specific case of a startup. The study showed that for a company at an early stage the use of a roadmap can be valuable for its development and, in addition to the application of the tool itself, the elaboration process allowed the identification of gaps in the organization [5].

Finally, through the year of publication of each article considered for the construction of the table, the graph in Fig. 2 presents the distribution by year of publication of scientific articles over the time interval considered.

After analyzing the graph, it is possible to conclude that, in addition to the year 2017 being the year with the highest number of publications, the real growth of interest in this topic has taken place since 2017. In recent times, there are a greater number of publications that are included in this article. It is possible to say that from 2017 onward the study of the roadmap for AM has stimulated a higher interest in research.

In a broader analysis, the application of roadmap for Industry 4.0 is also addressed by the research community and there are already studies available. The article [1] develops and introduces a strategic roadmap for the transition to Industry 4.0. This article mentions with more emphasis the best practices existing in the scope of strategic management, information system management, supply chain management, manufacturing technology management experience, and marketing. A strategic roadmap providing a basic framework for organizations to achieve Industry 4.0 implementation using Lean and Six Sigma principles also appears in the literature [26]. Finally, the article [2] developed an overview of European strategic roadmaps underlining the difference between "technology roadmap" and "strategic research agendas." Technology roadmaps normally are the inputs for the strategic research agendas.

Disruption of more traditional manufacturing practices and processes is unavoidable. Around the world, the rapid transformation through the implementation of Industry 4.0 appeared in industries. Besides AM, the high-performance cutting (HPC) technology is no exception and also in the literature it is already possible to find studies as in the article [27] that highlight some of the problems related to the development of a new roadmap for cutting processes. The roadmap tool shows to have a huge potential not only to be applied to AM but to any Industry 4.0 technology; however as verified, there is still a need for a great effort from both research and organizations for this tool to help organizations in its application.

5 Conclusion

With the continuous advancement and development of new technologies, Industry 4.0 is increasingly becoming present in organizations and starting to revolutionize industry. Additive manufacturing is currently recognized as one of the technologies of the 4th Industrial Revolution. This technology is transforming traditional manufacturing processes, so it is expected that in the near future, and as the industry adapts, its technological infrastructure becomes increasingly integrated into organizations. However, regardless of the growing interest in its use and its many advantages, there are still many difficulties that should be overcome. There is an evident need for more research to enable greater support and knowledge sharing to guide the adoption and implementation of AM technology by organizations. Proof of this is the difficulty in finding articles that applied roadmap to AM. Although it was possible to find a few studies, these were very specific, which ultimately highlights the gap in this topic, especially the lack of studies and developments in the application of this tool to AM.

The contribution of this chapter was to provide an improved understanding of the concept of a roadmap for additive manufacturing. Also, it broadens the research by encompassing several studies found in the literature that connected additive manufacturing and roadmaps by realizing an exploratory overview. Despite the potential shown by the roadmap tool, there are still many limitations. The lack of available practical support and the need to adapt it to the particular requirements of the corporation and its business context are some of the barriers limiting its application. The overload of initiatives, distraction due to short-term tasks, and inaccessibility of information and knowledge are all factors that can compromise the application of this tool. Furthermore, the application of this tool requires the commitment of the entire organization and the constant technological evolution and development are two major barriers that mean there is still a long road ahead full of challenges to be overcome.

Acknowledgments Radu Godina acknowledges Fundação para a Ciência e a Tecnologia (FCT-MCTES) for its financial support via the project UIDB/00667/2020 (UNIDEMI). Florinda Matos acknowledges Fundação para a Ciência e a Tecnologia (FCT-MCTES) for its financial support via KM3D (PTDC/EME-SIS/32232/2017).

References

1. Ghobakhloo, M. (2018). The future of manufacturing industry: A strategic roadmap toward Industry 4.0. *Journal of Manufacturing Technology Management, 29*(6), 910–936. https://doi.org/10.1108/JMTM-02-2018-0057
2. Santos, C., Mehrsai, A., Barros, A. C., Araújo, M., & Ares, E. (2017). Towards industry 4.0: An overview of European strategic roadmaps. *Procedia Manufacturing, 13*, 972–979. https://doi.org/10.1016/j.promfg.2017.09.093
3. Dilberoglu, U. M., Gharehpapagh, B., Yaman, U., & Dolen, M. (2017). The role of additive manufacturing in the era of Industry 4.0. *Procedia Manufacturing, 11*, 545–554. https://doi.org/10.1016/j.promfg.2017.07.148
4. Anand, P. B., Lokesh, N., Buradi, A., & Santhosh, N. (2021). A comprehensive review of emerging additive manufacturing (3D printing technology): Methods, materials, applications, challenges, trends and future potential. *Materials Today: Proceedings*. https://doi.org/10.1016/j.matpr.2021.11.059
5. Garza Ramos, A., Daim, T., Gaats, L., Hutmacher, D. W., & Hackenberger, D. (2022). Technology roadmap for the development of a 3D cell culture workstation for a biomedical industry startup. *Technological Forecasting and Social Change, 174*, 121213. https://doi.org/10.1016/j.techfore.2021.121213
6. Despeisse, M., Yang, M., Evans, S., Ford, S., & Minshall, T. (2017). Sustainable value roadmapping framework for additive manufacturing. *Procedia CIRP, 61*, 594–599. https://doi.org/10.1016/j.procir.2016.11.186
7. Hussain, I., Qurashi, A., Mujtaba, G., Waseem, M. A., & Iqbal, Z. (2019). Knowledge management: A roadmap for innovation in SMEs' sector of Azad Jammu & Kashmir. *Journal of Global Entrepreneurship Research, 9*(1), 9. https://doi.org/10.1186/s40497-018-0120-8
8. Pradel, P., Zhu, Z., Bibb, R., & Moultrie, J. (2018). A framework for mapping design for additive manufacturing knowledge for industrial and product design. *Journal of Engineering Design, 29*(6), 291–326. https://doi.org/10.1080/09544828.2018.1483011
9. Noorani, R. (2017). *3D printing: Technology, applications, and selection*. CRC Press. https://doi.org/10.1201/9781315155494
10. C. K. Chua, C. H. Wong, & W. Y. Yeong, «Chapter Two - Roadmap on additive manufacturing standards, Standards, quality control, and measurement sciences in 3D printing and additive manufacturing, C. K. Chua, C. H. Wong, W. Y. Yeong, Eds. Academic Press, 2017, pp. 31–55. doi: https://doi.org/10.1016/B978-0-12-813489-4.00002-7.
11. Chatterjee, K., & Ghosh, T. K. (2020). 3D printing of textiles: Potential roadmap to printing with fibers. *Advanced Materials, 32*(4), 1902086. https://doi.org/10.1002/adma.201902086
12. Guo, N., & Leu, M. C. (2013). Additive manufacturing: technology, applications and research needs. *Frontiers of Mechanical Engineering, 8*(3), 215–243. https://doi.org/10.1007/s11465-013-0248-8
13. Bhuvanesh Kumar, M., & Sathiya, P. (2021). Methods and materials for additive manufacturing: A critical review on advancements and challenges. *Thin-Walled Structures, 159*, 107228. https://doi.org/10.1016/j.tws.2020.107228

14. ISO/ASTM 52900. (2015). *Standard terminology for additive manufacturing – General principles – Terminology*, First edition 2015-12-15.
15. Phaal, R., Farrukh, C., & Probert, D. (2001). *Technology roadmapping: Linking Technology resources to business objectives* (Vol. 1–18). University of Cambridge.
16. Yang, M., & Despeisse, M. (2016). *Sustainable value roadmapping tool.*
17. Lahoti, G., Porter, A. L., Zhang, C., Youtie, J., & Wang, B. (2018). Tech mining to validate and refine a technology roadmap. *World Patent Information, 55*, 1–18. https://doi.org/10.1016/j.wpi.2018.07.003
18. Brown, R., & O'Hare, S. (2001). *The use of technology roadmapping as an enabler of knowledge management* (p. 7). https://doi.org/10.1049/ic:20010078
19. Kostoff, R. N., & Schaller, R. R. (2001). Science and technology roadmaps. *IEEE Transactions on Engineering Management, 48*(2), 132–143. https://doi.org/10.1109/17.922473
20. Arcos-Novillo, D. A., & Güemes-Castorena, D. (2017). Development of an additive manufacturing technology scenario for opportunity identification—The case of Mexico. *Futures, 90*, 1–15. https://doi.org/10.1016/j.futures.2017.05.001
21. Mohebalizadeh, S., & Ghazinoori, S. (2021). Developing a technology roadmap for regenerative medicine: A participatory action research. *Systemic Practice and Action Research, 34*(4), 377–397. https://doi.org/10.1007/s11213-020-09525-w
22. Kumar, S., & Czekanski, A. (2018). Roadmap to sustainable plastic additive manufacturing. *Materials Today Communications, 15*, 109–113. https://doi.org/10.1016/j.mtcomm.2018.02.006
23. Meng, L., et al. (2021). Correction to: From topology optimization design to additive manufacturing: Today's success and tomorrow's roadmap. *Archives of Computational Methods in Engineering, 28*(1), 269–269. https://doi.org/10.1007/s11831-020-09445-x
24. Otto, R., et al. (2021). Roadmap for additive manufacturing of HAYNES® 282® superalloy by laser beam powder bed fusion (PBF-LB) technology. *Materials and Design, 204*, 109656. https://doi.org/10.1016/j.matdes.2021.109656
25. Bourell, D. L., Rosen, D. W., & Leu, M. C. (2014). The roadmap for additive manufacturing and its impact. *3D Printing and Additive Manufacturing, 1*(1), 6–9. https://doi.org/10.1089/3dp.2013.0002
26. Butt, J. (2020). A strategic roadmap for the manufacturing industry to implement industry 4.0. *Designs, 4*(2), 1–31. https://doi.org/10.3390/designs4020011
27. Byrne, G., Ahearne, E., Cotterell, M., Mullany, B., O'Donnell, G. E., & Sammler, F. (2016). High performance cutting (HPC) in the new era of digital manufacturing – A roadmap. *Procedia CIRP, 46*, 1–6. https://doi.org/10.1016/j.procir.2016.05.038

Life Cycle Inventory of Additive Manufacturing Processes: A Review

Samruddha Kokare, Radu Godina, and João Pedro Oliveira

Abstract With a growing emphasis on environmentally friendly manufacturing, the evaluation of the environmental performance of additive manufacturing (AM) processes has gained attention. Life cycle assessment (LCA) is an important tool to compute the environmental impacts of AM processes. LCA studies of AM processes require life cycle inventory (LCI) data in the form of natural resources, energy consumed, and wastes and emissions associated with these processes. This chapter collects and summarizes the LCI data for different AM processes from the existing literature. It is observed that AM processes in general have higher energy demand than the conventional manufacturing processes. However, the better material utilization of the AM processes, particularly in manufacturing complex shapes, can make them more sustainable in comparison with the conventional manufacturing (CM) processes. The LCI data presented in this chapter can be useful to the manufacturers, researchers, and LCA practitioners in performing LCAs of AM processes and choosing the most sustainable process.

Keywords Life cycle inventory · Additive manufacturing · Life cycle assessment

---

S. Kokare (✉) · R. Godina
UNIDEMI, Department of Mechanical and Industrial Engineering, NOVA School of Science and Technology, Universidade NOVA de Lisboa, Caparica, Portugal
e-mail: s.kokare@campus.fct.unl.pt

J. P. Oliveira
CENIMAT/I3N, Department of Materials Science, NOVA School of Science and Technology, Universidade NOVA de Lisboa, Caparica, Portugal

J. C. de Oliveira Matias et al. (eds.), *Quality Innovation and Sustainability*, Springer Proceedings in Business and Economics,
https://doi.org/10.1007/978-3-031-12914-8_7

1 Introduction

The manufacturing sector is a major contributor to the economy but is also a prominent source of environmental emissions. In 2019, manufacturing accounted for about 14% of the European Union's gross domestic product (GDP) [1] but was responsible for roughly 21% of its total greenhouse gas emissions [2]. Tougher environmental regulations and increased consumer awareness about sustainability have led the companies to develop sustainable products and processes [3]. To reduce the environmental impacts of a product or process, it is essential to measure its environmental impacts first. Life cycle assessment (LCA) based on ISO 14044 standards is a well-known methodology used to calculate the environmental impacts of a product or a process over its life cycle [4]. Based on the results of LCA, recommendations to reduce the environmental impacts of a product or process can be made. LCA has also been applied to additive manufacturing (AM) processes to quantify their emissions [5]. In conducting LCA of a process, its life cycle inventory (LCI) data needs to be collected first. LCI involves input flows like raw material, energy, other resources consumed, and output flows like wastes and emissions. The aim of this chapter is to compile the LCI data for different AM processes. The LCI data for different AM processes is collected from the existing literature focusing on the environmental assessment of AM processes. The LCI data presented in this chapter can be useful to the manufacturers, researchers, and LCI practitioners in assessing the environmental impacts of AM processes and sustainable process selection. This chapter is structured as follows: Sect. 2 describes the systematic literature review methodology used in this chapter. Section 3 summarizes the LCI data collected for different AM processes. In Sect. 4, the LCI data of AM processes is compared to that of conventional manufacturing (CM) processes. The sustainability potential of AM processes is also discussed in this section. Finally, the main conclusions of this study are discussed in Sect. 5.

2 Methodology

A systematic literature review was carried out by entering the keyword string "life cycle assessment" AND "additive manufacturing" OR "3d printing" in the Scopus database. This search was limited to the research items written in English and published up to the year 2021. This initial search resulted in 267 matches. Then, title analysis of papers was performed and only papers whose title indicated LCA of an AM process were shortlisted. This resulted in 122 research items. Further, the abstracts of these items were analyzed, which led to 116 research items. Finally, the full paper analysis was performed, where 66 research items addressing the LCA of an AM process were selected. Out of these 66 research items, only 25 research items that presented the LCI data for AM processes such as raw material, electricity, other process consumables, and material wastes based on actual experiments and on-site measurements were considered for this study. The research items relying heavily on the existing literature and LCI databases were excluded from this review.

3 LCI of AM Processes

This section provides a summary of different types of LCI data available for selective laser melting (SLM), selective laser sintering (SLS), direct energy deposition (DED), electron beam melting (EBM), fused deposition modeling (FDM), material jetting (MJ), material jet fusion (MJF), binder jetting (BJ), and stereolithography (SL) processes. The LCI data compiled includes the raw material used, energy required in its production, specific energy consumption (SEC) of AM processes, portion of raw material wasted during the process, and other process consumables such as inert gas and compressed air.

3.1 Selective Laser Sintering

The studies on energy and material consumption in SLS are described in Table 1. Polyamide (PA) powders are used as raw materials in SLS. The energy required for the production of these powders are 85 MJ/kg and 129.1 MJ/kg as reported by Telenko et al. [6] and Kwon et al. [7], respectively. The SEC of the process varies between 29.2 MJ/kg and 145.1 MJ/kg. High powder wastage rates ranging between 20% and 97% of the input powder are seen in all the three studies.

3.2 Selective Laser Melting

Table 2 summarizes the LCI data for SLM process. In SLM, metal powders are used as the raw material. The energy required to produce these powders, generally using gas or water atomization process, ranges from 2.48 MJ/kg to 55.58 MJ/kg. The SEC values SLM process vary between 55.7 MJ/kg and 567.2 MJ/kg. The powder

Table 1 LCI data for SLS process

Machine tool	Raw material	Raw material production energy (MJ/kg)	SEC (MJ/kg)	Material wastage (% wt.)	Compressed air consumption	Reference
–	PA12	85	52.2	20–44	–	[6]
P770	PA12	129.1	29.2	97	–	[7]
EOSINT P760	PA2200	NA	145.1	44.1	20 m^3/h	[8]
	PA3200	NA	129.5	46		

Table 2 LCI data for SLM process

Machine tool	Raw material	Raw material production energy (MJ/kg)	SEC (MJ/kg)	Material wastage (% wt.)	Inert gas consumption	Reference
Reinshaw AM250	Al-Si	8.1	567.2	20	Ar: 208 l/build	[9]
SLM 280 HL	X2CrNiMo1712	2.48	383.1	6	Ar: 3.08 kg/build	[10]
Reinshaw AM250	SS 316 L	7.2	55.7	4	Ar: 10 l/build	[11]
EOSM 290	Ti6Al4V	NA	299.5	4	Ar: 1.45 kg/part	[12]
–	Inconel 718	55.58	291.3	NA	Compressed air: 1 Nm^3; Ar: 2 l/min	[13]

Table 3 LCI data for EBM process

Machine tool	Raw material	Raw material production energy (MJ/kg)	SEC (MJ/kg)	Material wastage (% wt.)	Inert gas consumption	Reference
ARCAM	Ti6Al4V	23.76	396.1	–	–	[15]
ARCAM A1	Ti6Al4V	23.76	41–135.8	–	–	[16]
–	Ti6Al4V	93.24	176.3	35 (machining allowance)	–	[14]
ARCAM A2X	Ti6Al4V	70	84.6–507.6	–	–	[17]
ARCAM AB	Ti6Al4V	30.1–33.3	59.9	–	He: 1 l/h	[18]

wastage relatively lower than the SLS process is seen between 4% and 20% of the input powder. Additionally, the amount of inert gas consumed during the process is also documented in Table 2.

3.3 Electron Beam Melting

In studies focusing on EBM process, Ti6Al4V alloy is commonly used as the raw material. The energy consumed in its powder production is reported in the range of 23.76–93.24 MJ/kg. The SEC for this process varies between 41 MJ/kg and 507.6 MJ/kg. Generally, the powder wastage is not described explicitly. Lyons et al. [14] estimated a machining allowance of 35% of the total powder consumed during the process (Table 3).

3.4 *Direct Energy Deposition Processes*

The LCI data for DED processes, namely, direct additive laser manufacturing (DALM), laser engineered net shaping (LENS), laser cladding forming (LCF), and wire arc additive manufacturing (WAAM) are used. The energy required for powder production in the case of laser-based processes varies between 0.97 MJ/kg and 14.4 MJ/kg. The SEC of these processes varies between 87.1 MJ/kg and 316.8 MJ/kg. WAAM demonstrated a relatively lower SEC of 9.77 MJ/kg and 10 MJ/kg. WAAM also reported lower material wastage than the laser-based DED processes but required postprocessing operations (Table 4).

3.5 *Fused Deposition Modeling*

The studies containing LCI data for FDM process are summarized in Table 5. The raw materials used in this case were polymeric filaments of acrylonitrile styrene acrylate (ASA), acrylonitrile butadiene styrene (ABS), and polylactic acid (PLA). The energy consumed in PLA filament production ranges between 50.4 MJ/kg and 61.2 MJ/kg [24]. Kwon et al. reported 93.04 MJ/kg energy requirement in the production of ABS [7].The SEC for FDM process ranges between 43.2 MJ/kg and 610.8 MJ/kg. The raw material wastage rates of 3 and 34% were reported by Kwon et al. [7] and Nagarajan & Haapala [25], respectively.

Table 4 LCI data for DED processes

Process	Machine tool	Raw material	Raw material production energy (MJ/kg)	SEC (MJ/kg)	Material wastage (% wt.)	Inert gas consumption	Reference
DALM	–	Steel powder	14.4	87.1	NA	Ar: 2.13 m^3/part	[19]
LENS	LENS 450	AISI 4140 powder	1.38	243.5	86	–	[20]
LCF	–	Ti alloy powder	0.97–2.34	316.8	20–30	Ar:10 l/min	[21]
WAAM	MX3D	SS3081 wire	–	9.77	1.10	98% Ar + 2% CO_2: 12 l/min	[22]
WAAM	–	EN S235JR steel wire	36.3	10.375	1.3	82% Ar + 18% CO_2; 271 l/part	[23]

Table 5 LCI data for FDM process

Machine tool	Raw material	Raw material production energy (MJ/kg)	SEC (MJ/kg)	Material wastage (% wt.)	Reference
–	PLA	50.4–61.2	43.2	–	[24]
uPrint SE plus	ABS	93.04	122.9	3%	[7]
Stratasys Dimension SST	Plastic	–	533.3–610.8	–	[26]
Stratasys Fortus 400 mc	ASA	–	457.2	34%	[25]

Table 6 LCI for MJ, MJF, BJ, and SLA processes

Process	Machine tool	Raw material	Raw material production energy (MJ/kg)	SEC (MJ/kg)	Material wastage (% wt.)	Reference
MJ	J750 PJ	ABS	92.2	115.8	–	[7]
MJF	HP MJF 4210	PA12 powder	–	98.7	5% (assumed)	[27]
BJ	–	17-4 PH SS powder	1.65	32.5	–	[29]
BJ	–	SS powder	2.28	67.1–99.66	–	[30]
SL	Form2	Clear 4 resin	–	86.54	–	[28]

3.6 *Other AM Processes*

AM processes like material jetting (MJ), multiple jet fusion (MJF), binder jetting (BJ), and stereolithography (SL) are not extensively studied in LCA. According to Kwon et al., the SEC of MJ process is 115.8 MJ/kg [7]. London et al. reported the SEC of 98.7 MJ/kg for MJF process with material wastage of 5% [27]. The SEC for BJ process lies in the range of 32.5–99.66 MJ/kg. Mele et al. calculated the SEC of 86.54 MJ/kg for desktop stereolithography process [28] (Table 6).

4 Comparison Between AM and CM Processes

Based on the above-compiled LCI data for AM processes, it can be concluded that powder bed fusion (PBF) AM processes, i.e., SLS, SLM, and EBM, have received more attention than other processes with regard to studies on energy and resource consumption. Although a high variation is seen in SEC data ranges for AM processes, they are significantly higher than SEC values of CM processes listed in

Table 7 Energy requirements for some conventional manufacturing processes

Process	Raw material	Raw material production energy (MJ/kg)	SEC (MJ/kg)	Reference
Green sand casting	Steel	–	10.65	[22]
Casting	Cast iron	–	3.45	[29]
Investment casting	Inconel 718	279.67	24.41	[13]
Machining	Stainless steel	81	3.9	[29]
Machining	Cast iron	17	3.7	[29]
Machining	Aluminum	127.1	Roughing: 1.4; finishing: 9.9	[33]
Machining	Titanium	556.2	Roughing: 2.7; finishing: 22.1	[33]
Injection molding	–	–	12.6–19	[34]

Table 7. Furthermore, AM processes need additional energy for the production of raw material in the form of powders, wires, or filaments. However, studies have shown more environmentally friendly performance of AM processes compared to conventional manufacturing processes [13–15, 22, 23, 26, 31, 32]. This is attributed to better material utilization by AM processes compared to CM processes, especially in manufacturing complex designs. For instance, in a case study demonstrated by Paris et al. [15], EBM and CNC milling consumed almost identical amount of electricity in the manufacturing of a turbine blade. The low material utilization of 14% by the CNC milling process caused more environmental impact than the EBM process. Similarly, Lyons et al. [14] observed material utilization of 15% for machining process and 65% for EBM process. Ingarao and Priarone [32] showed that EBM is more energy efficient than machining due to its higher material utilization. A similar conclusion was drawn by Campatelli et al. [23], where integrated WAAM and machining exhibited 34% energy saving enabled by raw material saving compared to pure machining approach. Bekker and Verlinden [22] showed that WAAM is environmentally friendly mainly due to its efficient material use than the CM processes. However, it must be noted that AM processes are not always more sustainable than CM processes as shown by some studies [10, 20, 21]. AM is sustainable for highly complex shapes where part weight reduction is enabled and used in transportation applications [31]. Therefore, it can be concluded that AM processes have higher sustainability potential than the CM processes in cases when the part geometry is highly complex, requiring high amounts of material removal by CM processes.

5 Conclusions

In this chapter, the LCI data for different AM processes were compiled by reviewing the existing literature focusing on the environmental performance of AM processes using LCA methodology. The LCI data collected includes mainly energy requirement, raw material wastage, and other process consumables for different AM processes. Based on the review, it is concluded that the LCI data for PBF-based AM processes like SLS, SLM, and EBM have been documented relatively better than the other AM processes. This is evident from the fact that 13 out of 25 articles considered in this study focused on PBF-based AM processes. However, high variability is observed in the collected LCI data for AM processes that can be attributed to the use of different machine tools, raw materials, process parameters, and lower number of studies. This emphasizes the need to collect and record more LCI data for AM processes based on experiments and on-site measurements. The energy consumption and material wastage of AM processes were also compared to conventional manufacturing processes. Although AM has higher energy requirements, it can be more sustainable than the conventional processes due to better material utilization, particularly in manufacturing complex geometries as indicated in the previous section. More studies on the sustainability of AM processes, including PBF processes, need to be conducted and its LCI data need to be documented to fully understand the environmental behavior of AM processes.

Acknowledgments The authors thank Fundação para a Ciência e a Tecnologia (FCT-MCTES) for its financial support via the project UIDB/00667/2020 (UNIDEMI). This study has also received funding from the European Institute of Innovation and Technology (EIT) – Project Smart WAAM: Microstructural Engineering and Integrated Non-Destructive Testing. This body of the European Union receives support from the European Union's Horizon 2020 research and innovation program. João P. Oliveira acknowledges the funding of CENIMAT/i3N by national funds through the FCT-Fundação para a Ciência e a Tecnologia, I.P., within the scope of Multiannual Financing of R&D Units, reference UIDB/50025/2020-2023.

References

1. Manufacturing, value added (% of GDP) - European Union. https://data.worldbank.org/indicator/NV.IND.MANF.ZS?locations=EU. Last accessed 9 Feb 2022.
2. EEA greenhouse gases - data viewer. https://www.eea.europa.eu/data-and-maps/data/data-viewers/greenhouse-gases-viewer. Last accessed 9 Feb 2022.
3. Nörmann, N., & Maier-Speredelozzi, V. (2016). Cost and environmental impacts in manufacturing: A case study approach. *Procedia Manufacturing, 5*, 58–74.
4. ISO: ISO 14044:2006-Environmental management—Life cycle assessment—Requirements and guidelines. (2006).
5. Saade, M. R. M., Yahia, A., & Amor, B. (2020). How has LCA been applied to 3D printing? A systematic literature review and recommendations for future studies. *Journal of Cleaner Production, 244*, 118803.

6. Telenko, C., & Seepersad, C. (2010). Assessing energy requirements and material flows of selective laser sintering of nylon parts. In *2010 Presented at the international solid freeform fabrication symposium*. University of Texas at Austin.
7. Kwon, J. Y., Kim, N., & Ma, J. (2020). Environmental sustainability evaluation of additive manufacturing using the NIST test artifact. *Journal of Mechanical Science and Technology, 34*, 1265–1274.
8. Kellens, K., Renaldi, R., Dewulf, W., Kruth, J., & Duflou, J. R. (2014). Environmental impact modeling of selective laser sintering processes. *Rapid Prototyping Journal, 20*, 459–470.
9. Faludi, J., Baumers, M., Maskery, I., & Hague, R. (2017). Environmental impacts of selective laser melting: Do printer, powder, or power dominate? *Journal of Industrial Ecology, 21*, S144–S156.
10. Guarino, S., Ponticelli, G. S., & Venettacci, S. (2020). Environmental assessment of selective laser melting compared with laser cutting of 316L stainless steel: A case study for flat washers' production. *CIRP Journal of Manufacturing Science and Technology, 31*, 525–538.
11. Peng, T., Wang, Y., Zhu, Y., Yang, Y., Yang, Y., & Tang, R. (2020). Life cycle assessment of selective-laser-melting-produced hydraulic valve body with integrated design and manufacturing optimization: A cradle-to-gate study. *Additive Manufacturing, 36*, 101530.
12. Cappucci, G. M., Pini, M., Neri, P., Marassi, M., Bassoli, E., & Ferrari, A. M. (2020). Environmental sustainability of orthopedic devices produced with powder bed fusion. *Journal of Industrial Ecology, 24*, 681–694.
13. Torres-Carrillo, S., Siller, H. R., Vila, C., López, C., & Rodríguez, C. A. (2020). Environmental analysis of selective laser melting in the manufacturing of aeronautical turbine blades. *Journal of Cleaner Production, 246*, 119068.
14. Lyons, R., Newell, A., Ghadimi, P., & Papakostas, N. (2021). Environmental impacts of conventional and additive manufacturing for the production of Ti-6Al-4V knee implant: A life cycle approach. *International Journal of Advanced Manufacturing Technology, 112*, 787–801.
15. Paris, H., Mokhtarian, H., Coatanéa, E., Museau, M., & Ituarte, I. F. (2016). Comparative environmental impacts of additive and subtractive manufacturing technologies. *CIRP Annals - Manufacturing Technology, 65*, 29–32.
16. Le, V. T., Paris, H., & Mandil, G. (2017). Environmental impact assessment of an innovative strategy based on an additive and subtractive manufacturing combination. *Journal of Cleaner Production, 164*, 508–523.
17. Lunetto, V., Priarone, P. C., Kara, S., & Settineri, L. (2021). A comparative LCA method for environmentally friendly manufacturing: Additive manufacturing versus Machining case. *Procedia CIRP, 98*, 406–411.
18. Baumers, M., Tuck, C., Wildman, R., Ashcroft, I., & Hague, R. (2017). Shape complexity and process energy consumption in electron beam melting: A case of something for nothing in additive manufacturing?: Shape complexity and energy usage in 3D printing. *Journal of Industrial Ecology, 21*, S157–S167.
19. Bourhis, F. L., Kerbrat, O., Hascoet, J.-Y., & Mognol, P. (2013). Sustainable manufacturing: Evaluation and modeling of environmental impacts in additive manufacturing. *International Journal of Advanced Manufacturing Technology, 69*, 1927–1939.
20. Liu, Z., Jiang, Q., Cong, W., Li, T., & Zhang, H.-C. (2018). Comparative study for environmental performances of traditional manufacturing and directed energy deposition processes. *International Journal of Environmental Science and Technology, 15*, 2273–2282.
21. Peng, S., Li, T., Wang, X., Dong, M., Liu, Z., Shi, J., & Zhang, H. (2017). Toward a sustainable impeller production: Environmental impact comparison of different impeller manufacturing methods. *Journal of Industrial Ecology, 21*, S216–S229.
22. Bekker, A. C. M., & Verlinden, J. C. (2018). Life cycle assessment of wire + arc additive manufacturing compared to green sand casting and CNC milling in stainless steel. *Journal of Cleaner Production, 177*, 438–447.
23. Campatelli, G., Montevecchi, F., Venturini, G., Ingarao, G., & Priarone, P. C. (2020). Integrated WAAM-subtractive versus pure subtractive manufacturing approaches: An energy

efficiency comparison. *International Journal of Precision Engineering and Manufacturing--Green Technology, 7*, 1–11.
24. Cerdas, F., Juraschek, M., Thiede, S., & Herrmann, C. (2017). Life cycle assessment of 3D printed products in a distributed manufacturing system. *Journal of Industrial Ecology, 21*, S80–S93.
25. Nagarajan, H. P. N., & Haapala, K. R. (2018). Characterizing the influence of resource-energy--exergy factors on the environmental performance of additive manufacturing systems. *Journal of Manufacturing Systems, 48*, 87–96.
26. Krishna, L. S. R., & Srikanth, P. J. (2021). Evaluation of environmental impact of additive and subtractive manufacturing processes for sustainable manufacturing. *Materials Today: Proceedings, 45*, 3054–3060.
27. London, M. B., Lewis, G. M., & Keoleian, G. A. (2020). Life cycle greenhouse gas implications of multi jet fusion additive manufacturing. *ACS Sustainable Chemistry and Engineering, 8*, 15595–15602.
28. Mele, M., Campana, G., & D'Avino, G. (2020). Life cycle impact assessment of desktop stereolithography. *Journal of Cleaner Production, 244*, 118743.
29. DeBoer, B., Nguyen, N., Diba, F., & Hosseini, A. (2021). Additive, subtractive, and formative manufacturing of metal components: A life cycle assessment comparison. *International Journal of Advanced Manufacturing Technology, 115*, 413–432.
30. Tang, Y., Mak, K., & Zhao, Y. F. (2016). A framework to reduce product environmental impact through design optimization for additive manufacturing. *Journal of Cleaner Production, 137*, 1560–1572.
31. Ingarao, G., Priarone, P. C., Deng, Y., & Paraskevas, D. (2018). Environmental modelling of aluminium based components manufacturing routes: Additive manufacturing versus machining versus forming. *Journal of Cleaner Production, 176*, 261–275.
32. Ingarao, G., & Priarone, P. C. (2020). A comparative assessment of energy demand and life cycle costs for additive- and subtractive-based manufacturing approaches. *Journal of Manufacturing Processes, 56*, 1219–1229.
33. Priarone, P. C., Pagone, E., Martina, F., Catalano, A. R., & Settineri, L. (2020). Multi-criteria environmental and economic impact assessment of wire arc additive manufacturing. *CIRP Annals, 69*, 37–40.
34. Thiriez, A., & Gutowski, T. (2006). An Environmental Analysis of Injection Molding. In *Proceedings of the 2006 IEEE International Symposium on Electronics and the Environment*, pp. 195–200. IEEE, Scottsdale.

Critical Success Factors of TQM for Sustainability in Higher Education Institutions: A Theoretical Contribution

Teresa Nogueiro, Margarida Saraiva, and António Ramos Pires

Abstract Total quality management (TQM) is a management approach that was initially used by the industry, but that over time has been adapted to other types of institutions, including higher education institutions (HEI). Currently, the topic of TQM associated with HEIs and higher education is certainly no longer a novelty. However, considering the competitiveness of these institutions for the best students, best researchers, and best means, the sustainability of these institutions has become almost a requirement. This chapter intends to analyze the critical success factors (CSF) in the implementation of total quality management and the critical success factors for sustainability in HEI as a way of contributing to a theoretical reflection on the sustainability of higher education. This work was based on the 11 CSFs for the implementation of TQM, following the studies of Bayraktar et al. (Total Quality Management 19:551–574, 2008), corroborated/validated by Nadim and Al-Hinai (International Journal of Applied Sciences and Management 1:147–156, 2016), in HEI: leadership, vision, measurement and evaluation, process control and improvement, program design and resources, quality system improvement, employee involvement, recognition and reward, education and training, student focus, and other stakeholders' focus. After identifying the keywords or phrases of the CSFs in TQM implementation, we identified many other aligned CSFs for a better understanding of the critical success factors for sustainability implementation in HEIs.

T. Nogueiro
School of Social Sciences, Universidade de Évora, Évora, Portugal

M. Saraiva (✉)
School of Social Sciences, Universidade de Évora, Évora, Portugal

BRU—Business Research Unit-Iscte-Instituto Universitário de Lisboa, Lisboa, Portugal
e-mail: msaraiva@uevora.pt

A. R. Pires
UNIDEMI – Universidade Nova de Lisboa, Portugal and Instituto Politécnico de Setúbal, Setúbal, Portugal

J. C. de Oliveira Matias et al. (eds.), *Quality Innovation and Sustainability*, Springer Proceedings in Business and Economics,
https://doi.org/10.1007/978-3-031-12914-8_8

Keywords Total quality management (TQM) · Critical success factors (CSF) · Sustainability · Higher education institutions (HEIs)

1 Introduction

The higher education sector has become very competitive and subject to market forces; as a result of these issues, higher education has been considered a worldwide business. To deal with these developments, higher education institutions (HEIs) have discovered that total quality management (TQM) is an unavoidable approach to achieving corporate goals [2] and increasing a company's overall effectiveness, efficiency, cohesiveness, adaptability, and competitiveness [3]. The overall structure of higher education institutions is designed to achieve research and teaching goals. The operational personnel of the institutions reinforce these goals [4–6].

While many organizations around the world have implemented TQM, its implementation in nonprofit organizations and higher education institutions offers more problems and difficulties than in corporate firms. The process of client identification is a crucial stage in TQM implementation [7]. "Who are your core customers?" is a crucial question for any business. Higher education makes the matter more challenging because its services are provided to a variety of groups, such as students, businesses/employers, societies/governments, and teachers. Higher education institutions have a large variety of customer groups that have been recognized by many authors who have examined the subject.

Students are the key customers, despite the fact that higher education has numerous customers and stakeholders (future employers, governments, and societies). Education is described as the process of learning knowledge, as well as the knowledge, skill, and understanding gained via attendance at a school, college, or university. The nature of education (as a service) is that it offers the foundation for learning and demonstrating that learning has occurred. Whether students see themselves as co-creators or not, it is important to remember that education can only happen if learning happens [8].

These customers have a variety of requirements that are either complementary or discordant [9]. According to these authors, the first key finding of their research is that "education" is divided into two distinct processes: teaching/education, which is mostly the responsibility of university teachers, and learning/knowledge, which is primarily the responsibility of students. For each process, the main internal and external customers were identified. For these authors, in the process of "teaching/education," the internal customers are the faculty members and the external customers are the students; in the process of "learning/knowledge," the internal customers are the students and the external customers are the employers/companies; and in the process of "research/investigation," the internal customers are the faculty members and the external customers are the society/government. Therefore, the work of these authors illustrates five opposing viewpoints among several client groups: (1) the student as a customer in classroom teaching activities; (2) the student's participation in his own learning; (3) future employers' expectations of the student as a product; (4) scientific productivity indicators in research; and (5) the double duty of teachers: teaching and research.

In a different approach, Reavill [10] created a framework for identifying higher education stakeholders, focusing on developing customer requirements as a key component of TQM (total quality management). Students, employers, the student's family and dependents, universities and their employees, suppliers, the secondary education sector, other universities, commerce and industry, the nation, the government, taxpayers, and finally professional bodies are among the 12 stakeholders who contribute to or benefit from higher education.

Akinyele [11] has classified HEIs' customers as internal customers (academic and administrative) and external customers (direct and indirect). More recently, Guilbault [8] referred that the student, employer, and other stakeholders might all be considered customers. Excluding the student from the role of customer can have an impact on student happiness/satisfaction and retention since who is perceived as the consumer determines policies and procedures. Students perceive themselves as customers, and there are many actions/activities/practices by HEIs that treat students as such.

Universities must be accountable to society, employers, students, and each other. Therefore, higher education must be of high quality [2, 12].

TQM has the ability to increase educational institutions' quality while also achieving continual improvement and refers to an organization's culture of continuous improvement aimed at increasing customer satisfaction [1]. For the objective of constructing a comprehensive TQM model, a synthesis of TQM philosophy and system aspects is possible. Aspects of a unit or organization that must work well if the unit or organization as a whole is to prosper are known as key issues or critical success factors (CSF). They must be addressed in order to achieve the level of quality that customers need in order to meet their demands and expectations. They have the following characteristics [13]: customer joy, external customer satisfaction, internal customer satisfaction, fact-based management, measurement, people-based management, continuous improvement, and continuous improvement cycle.

The CSFs for the implementation of TQM are critical in higher education institutions since they will assist the organization in improving its performance assessment [7].

TQM's CSF of strategic quality planning is crucial. It covers, among other things, the creation of vision/mission statements, quality policies, quality control, and other management tools [14].

Other challenges, including leadership, culture, and organizational issues, can make TQM implementation more difficult in higher education [15]. To properly adopt TQM, it is critical to ensure that everyone is completely engaged and dedicated to the process [7]. These authors identify the following CSF for the implementation of TQM: management commitment and leadership, total customer satisfaction, employee involvement, continuous improvement, training, communication, and teamwork.

Identifying critical success factors is a crucial step in incorporating them into the processes. It gives an organization a way to assess hazards and possibilities in its surroundings. CSFs also provide a set of criteria for evaluating organizations' strengths and weaknesses [13].

This chapter intends to analyze the critical success factors in the implementation of total quality management and the critical success factors for sustainability in HEI as a way of contributing to a theoretical reflection on the sustainability of higher education.

Aside from this introduction, this chapter is divided into four sections: "Framework of the themes"; "Methodology of the study"; "Results obtained"; and "Final considerations, limitations to the study, and future research."

2 Exploring the Critical Success Factors in the Context of Higher Education Institutions

TQM can be adopted and implemented in HEIs, according to Nadim and Al-Hinai [2], but there is still a knowledge gap about how to do it successfully. The knowledge of TQM's CSF may aid in the removal of implementation difficulties.

Over the last 15 years, sustainability has gained traction and is still a hot topic in management circles. Scholars and practitioners agree on the significance of eliminating all trade-offs between business and society, as well as their inextricable interplay [16]. Indeed, there are many parallels between sustainability and TQM, particularly in terms of beliefs and methods. Both sustainability and TQM are focused on proactive and preventive techniques with the purpose of achieving long-term objectives and maintaining performance levels. On a practical level, both of these ideologies have established technical and organizational methods to ensure the true governance of their processes, such as (1) continuous improvement; (2) zero defects; (3) life cycle evaluation; (4) waste reduction; and (5) employee involvement and training [17].

TQM as a management approach might be broadened to include elements of sustainability development, according to Isaksson [18], identifying "process management" as the most appropriate tool for enhancing not only the economic firm's performance, but also its environmental and social performance.

Zink [19] highlighted the main TQM principles (i.e., leadership, continuous improvement) that can easily be discounted in this theoretical framework, stating that excellence models are valuable tools to transfer the concept of corporate sustainability into practice, according to Edgeman's theory [20], where excellence for sustainable development can be built on the triangle "person–organization–society."

Apart from the TQM CSFs already described and "naturally" linked to sustainability, environmental sustainability is frequently emphasized in the literature as a fundamental antecedent of sustainability [17].

In HEI, Bayraktar et al. [1] identified 11 TQM CSFs. While Nadim and Al-Hinai [2] gathered data from 144 academics from 22 HEIs in Turkey, the 11 CSFs of Bayraktar et al. [1] were validated for reliability and validity. Leadership, vision, measurement and evaluation, process control and improvement, program design and resources, quality system improvement, employee involvement, recognition and reward, education and training, student focus, and other stakeholders' focus are all reliable CSFs, according to their study's findings. Employees' involvement was found to be the most important success element, followed by stakeholder focus and other stakeholder focus, according to Nandim's and Al-Hinai [2] findings.

The 11 CSFs in HEI are presented next, according to Bayraktar et al. [1] corroborated/validated by Nadim and Al-Hinai [2].

2.1 Leadership

The European Quality Award and the Malcolm Baldrige Quality Award both emphasize the need for leadership commitment to TQM efforts as a well-accepted TQM implementation premise. HEIs'top management should be aware of TQM's requirements, recognize the value of employee involvement, and focus on long-term stable performance measures while actively supporting TQM procedures through their activities [1]. If top management is not committed to TQM, it will not be implemented efficiently [21].

Vision, mission, and values are components of leadership dimension [13]. TQM believed leadership to be a primary CSF. In order to reach the given aims, leaders are responsible for defining and articulating the company's vision, as well as building organization-wide commitment in the workforce [2, 22, 23]. The dedication of the top management is critical to the success of a quality improvement program. Management commitment entails defining a clear and appealing future vision as well as offering strategic leadership [24].

It is vital for the top management to commit to developing an organizational atmosphere that empowers employees [25].

It is critical for HEIs'top management/leaders to pay more attention to employee participation in decision-making and ensure that key performance indicators (KPIs) are clearly communicated to all employees [2].

To foster a sustainable culture, HEIs, like other companies, rely on competent leadership [26].

According to Weiss et al. [27] (pp. 51), "There is a significant effect that no leadership is associated with a low-level ('bolt-on') of sustainability curricula implementation." Their findings imply that a curriculum change is only achievable if leadership support is at least reasonable.

2.2 Vision

A vision is a declaration of one's wish to evolve into a preferred state in the future. All personnel should be allowed to focus on how they can help realize the goal. Visions are linked to what is known as transformational or charismatic leadership, which refers to leaders who have a significant impact on their organizations [13].

An HEI's vision is a public expression of what kind of organization it wants to be in the future. The vision statement shared and practiced throughout the organization can be seen in the values, beliefs, and business practices of the organization. Obviously, different vision statements will result in varied TQM implementation policies and performance measurements [1] and that may prove misleading [2]. To these authors, leadership's innovative approach and goals in the growth of HEIs are determined by vision.

To develop a prosperous environment for sustainability, university leadership must promote a vision of sustainability on campus and continually support it in order for it to take root and grow. Building a team with a shared vision, acting on that vision, and leveraging the vision to empower the campus are all part of leadership's responsibilities [26].

2.3 Measurement and Evaluation

These are tools for determining the quality of HEI performance. As a result, areas of weakness will be identified, and performance adjustments will be pursued [2].

In any implementation, determining the degree of success is essential for identifying areas for improvement. Even though it is difficult to identify globally acceptable performance measures for all HEIs, measurement and assessment are almost impossible without them [1].

Before monitoring and evaluating staff performance in HEIs, Bayraktar [28] emphasized that it is critical to precisely establish the key performance indicators (KPIs).

2.4 Process Control and Improvement

This is a natural result of measuring and evaluating things. Because HEIs are viewed as service organizations with many processes, they may require a unique organizational structure. Measurement, evaluation, control, and improvement of administrative and academic procedures in HEIs should be done on a regular basis [1].

Any complete quality approach must include process management. Processes that have an impact on the quality of products and services should be given special attention. Process management is concerned with ensuring that processes run smoothly as expected [14].

Process control and later improvement are required at each step to relieve pressure on the quality improvement system and contribute to meeting stakeholder demands [2].

2.5 Program Design and Resources

Academic programs are HEIs'principal products for attracting and satisfying the needs/requirements of stakeholders like students, industry/firms, academia, and the general public. These initiatives should be examined on a regular basis, taking into account the requests of stakeholders and technology advancements, and revised as needed. Interdisciplinary study fields, as well as the facilities required to perform such research, should be taken into account while developing curriculum and programs [1].

The designed programs must be reassessed on a frequent basis in response to any internal or external pressure to modify them [2].

2.6 Quality System Improvement

To ensure the consistency of quality-related concerns in HEIs, a well-documented quality assurance system is required [1].

Any organization's quality processes must be established and reviewed on a regular basis. It was clear in HEIs that there are two recommended instruments for improving the consistency of quality systems: process flow charts and quality criteria checklists [2, 29].

2.7 Employee Involvement

A good TQM implementation will not be achieved without the unambiguous support and involvement of the workforce. TQM is an organization-wide endeavor to create a high-quality culture. Employees' negative attitudes about TQM implementation may be eliminated with active participation. Employee involvement in HEIs can be measured by cross-functional team formations, employee collaboration, voluntary employee participation in TQM research, and system acceptance of employee proposals [1].

Teamwork and cross-functional interactions are emphasized in TQM techniques, which give numerous chances for social engagement and reinforcement [30].

Employee participation is widely cited as a critical TQM CSF. Employees that are more involved have a greater knowledge of the value of product quality and are more devoted to improving it. Employees should feel a sense of belonging to the organization [14].

Employee empowerment is a motivational strategy that encourages administrators, academics, staff, and others to advocate for environmental sustainability [26].

2.8 *Recognition and Reward*

Any employee, department, or school that demonstrates success in TQM-related initiatives should be rewarded as a means of bolstering a specific performance level. Performance measurements for HEIs may need to be updated to take quality efforts into account in order to encourage employee commitment to TQM adoption [1].

According to Zhang [31] quoted by [2], a regular and open method for evaluating employee performance and selecting award criteria is required. As a result, it is advised that the recognition and reward system criteria be designed to reflect the HEIs' staff involvement in this process in order to create and strengthen the level of organizational commitment among the employees.

2.9 *Education and Training*

Even in higher education institutions, educating and training personnel on TQM adoption and its consequences is critical to the program's success. Academic and nonacademic staff training needs should be recognized individually and included as part of a quality awareness workshop. Missing skill sets should be identified, and additional training should be organized to bridge the gaps. Such activities should have access to the necessary financial resources [1].

Staff should be trained and informed of all relevant sectors, and a mandatory financial arrangement should be in place to support this effort [32].

One of the most crucial elements for a successful TQM deployment is training. Employees must be knowledgeable and skilled in order to provide high-quality services and products [14].

2.10 *Student Focus*

To Zhang et al. [33], each organization's main mission is to understand, serve, and exceed client demands and expectations. Effective TQM implementation necessitates a strong focus on the customer, resulting in excellent customer satisfaction.

Recognizing students' needs requires a close interaction with them within the boundaries of academic ethics. Some of the primary issues of a successful TQM program that is deemed to be student-focused are the collection and review of student complaints, careful examination of course evaluations, support of student club activities, and alumni follow-ups [1, 34].

2.11 Other Stakeholders' Focus

The demands and expectations of a certain business or industry, as well as society as a whole, should be systematically observed and used to inform HEI academic and continuing education programs. Employees of higher education institutions play a critical role in providing practical services to HEI clients. The ability of employees to grasp the entire process and goal of the organization is critical to the success of a TQM implementation program [1].

To Bayraktar [28], employees, students, society, governing bodies, and other stakeholders all play a role in any HEI. As a result, it is recommended that you be aware of the demands of these various stakeholder groups and work hard to meet their needs [2].

3 Methodology

This research is based on a qualitative analysis of the themes. Thus, regarding the critical success factors associated with total quality management in higher education institutions, a literature review was carried out.

The actions undertaken to process the data were as follows:

1. Literature review.
2. Collection of systematized information on the CSFs related to TQM, identifying the keywords or the key expressions per CSF, based on the studies of Bayraktar et al. [1], which were corroborated/validated by Nadim and Al-Hinai [2], who identified 11 CSF in HEI.
3. Analysis of the systematized information, from the identification of keywords or key expressions of the CSF in the implementation of TQM for a better understanding of the critical success factors for sustainability in HEIs.
4. All information about CSFs, obstacles/barriers, and challenges to the implementation of sustainability in higher education institutions in various countries was analyzed in each study offered by the following authors: Aleixo et al. [35], Disterheft et al. [36], Larrán Jorge et al. [37], Nadim and Al-Hinai [2], Salleh et al. [6], Velazquez et al. [38], Verhulst and Lambrechts [39], Blanco-Portela et al. [40], and E Akins et al. [26]. Following the collection of these elements, a preliminary screening was conducted to eliminate duplicates. The elements were

then classified as closely as possible according to their nature (processes, resources, structure, or others). Because the bulk of these studies make no direct reference to the CSFs connected with sustainability, their assessment was based on an examination of the obstacles/barriers and challenges to the implementation of this component in higher education institutions.

The data were analyzed in a qualitative way, done manually without using any other analysis tool, and the results obtained were organized in tables for a better understanding.

The purpose of this study is to study the critical success factors in the implementation of total quality management and the critical success factors for sustainability in higher education institutions, in order to contribute to a theoretical reflection on higher education sustainability.

4 Results

Section 2 clearly identifies critical success factors in the implementation of total quality management in HEIs. These elements can be summarized for a better understanding of the critical success factors of the implementation of total quality management identified in that section. In order to group, it was necessary to define per TQM's CSF implementation keywords or key expressions as presented in Table 1.

The 11 critical success factors for the implementation of total quality management were analyzed in Sect. 2, and Table 1 presents the keywords and key expressions per CSF, in order to enable a better identification and characterization of each CSF.

The perspective of sustainability in higher education institutions was added due to the importance of this dimension to their quality and excellence. Therefore, the critical success factors for sustainability were identified by taking into consideration the studies of Aleixo et al. [35], Disterheft et al. [36], Larrán Jorge et al. [37], Nadim and Al-Hinai [2], Salleh et al. [6], Velazquez et al. [38], Verhulst and Lambrechts [39], Blanco-Portela et al. [40], and E Akins et al. [26]. In each study presented by the authors, all information concerning CSFs, and obstacles/barriers and challenges to the implementation of sustainability in higher education institutions in various countries was analyzed. After collecting these elements, a first screening was carried out in order to eliminate repetitions. Subsequently, the elements were grouped as far as possible according to their nature (processes, resources, structure, or others). Since the majority of these studies make no direct reference to the CSFs associated with sustainability, their determination was based on the analysis of the obstacles/barriers and challenges to the implementation of this dimension in HEIs. The main results obtained are presented in Table 2.

As main findings, we have identified more CSF for the implementation of sustainability than for the implementation of TQM. For sure and analyzing Table 2, we were able to observe the perfect match or alignment between CSFs for the implementation of TQM and those for the implementation of sustainability.

Table 1 Keywords/key expressions per CSF

CSF for TQM in HEI	Keywords/key expressions
Leadership	Commitment; top management; vision; mission; values; empower employees; employee participation in decision-making; clear communication to all employees
Vision	Focus of employee in the future; leadership association to vision; can be seen in values, beliefs and business practices of HEIs; innovative approach determined by vision
Measurement and evaluation	Tools for determining quality of HEI performance; establish key performance indicators
Process control and improvement	Organizational structure; procedures; measurement; evaluation, control and improvement; done on a regular basis; process management; meet stakeholders' demands
Program design and resources	Academic programs; interdisciplinary study fields; facilities required to perform research; development of curricula and programs
Quality system improvement	Quality assurance system; quality processes; instruments for improving the consistency of quality systems: process flow charts and quality criteria checklists
Employee involvement	Support and involvement of workforce; attitudes; quality culture; participation; teamwork and cross-functional interactions; social engagement and reinforcement; sense of belonging; knowledge of the value of the product/service
Recognition and reward	Reward as a mean of bolstering performance level
Education and training	Education and training; awareness; skill sets; financial resources; knowledge
Student focus	Customer focus and satisfaction; recognition of students' needs; complaints; course evaluations; club activities; alumni follow-ups
Other stakeholders' focus	Demands and expectations; society; employees of HEIs; needs of stakeholders

Source: Own elaboration

5 Final Considerations

5.1 Discussion

It might seem at first sight that the themes of total quality management, higher education, and sustainability have nothing in common or related. However, synergies are required in these areas. With this study and based on the studies of Bayraktar et al. [1], corroborated/validated by Nadim and Al-Hinai [2], we identified the alignment between the CSFs for the implementation of TQM and those for the implementation of sustainability in HEIs.

Sustainability is still a fresh political agenda according to Aleixo et al. [41], but it is also critical for all institutions to fulfill their responsibilities to create proactive interactions among institutions, agents, and people. To attain sustainability in HEIs,

Table 2 CSF for TQM and sustainability

Higher education institutions	
Critical success factors for TQM	Critical success factors for sustainability
Leadership	Support of top management, policymakers and administrators and empowerment; support from university leaders and policymakers
Education and training	Education and training and knowledge
Measurement and evaluation	Measurement instruments, monitoring and qualitative and quantitative performance indicators
Process control and improvement	Communication, more dialogue, making sure that the right people are at the table and that they are heard and find out what people are caring about; inefficient communication; long-term planning, systematization and continuity
Program design and resources	Organizational structure and workplace, fund-raising, financial resources and factors, investment in R&D and socioeconomic factors, interdisciplinarity of courses, programs and research; available resources; internal organizational structure
Quality system improvement	Enough time and starting on time and autonomy, positive image of the university and optimism, excellence and quality
Employee involvement	Personal strength and persistence, empowerment and confidence, dedication, engagement, awareness, interest and involvement, non-judging attitude, change resistance, increase of acceptance; education staff involvement; assumption of responsibilities
Recognition and reward	Stimulate positive feelings and sense of belonging; recognition
Vision	Identification with goals, strategy with a goal and tangible objectives
Student focus	Raising champions and attract students; student's engagement in extracurricular activities
Other stakeholders' focus	Needs of enterprises and labor market
	Outcomes/benefits
	Capacity building
	Rigorous and consistent regulation and legislation; inconsistent institutional legislation and implementation; institutional framework for sustainability
	Collaboration and networking
	Social legitimacy; social, economic and political context of the country

Source: Own elaboration

the Sustainable Development Goals, according to Leal Filho et al. [42], may provide a chance to overcome challenges. Because of their function as centers of learning, innovation, and research, universities can now make a significant contribution to the sustainability challenge. Universities, on the other hand, can approach sustainability issues in a variety of ways, all of which should be clearly recognized in their strategy [43].

We consider that HEIs should take into consideration, in addition to their own ambitions and strategies in terms of quality and sustainability, also the alignment with the European university strategy published by the European Commission and the national strategy for higher education defined per country.

5.2 Conclusions

This study offers a list of CSFs based on studies and experiences in several countries with different realities and that could serve as guidelines for higher education institutions that are starting the process of integrating sustainability in their system, or for those that are looking for models and referential support from others to define the best approaches and practices.

We believe that the strong alignment that exists between the critical success factors for the implementation of TQM and sustainability will allow us to create a model that will help higher education institutions to more easily contribute to the SDGs that they consider strategic and to have a performance that will aim for excellence in these dimensions. We aim with this chapter to contribute to the discussion that has been going on for at least two decades around the literature on sustainability and, more recently, the SDGs and the alignment to TQM approach.

5.3 Limitations to the Study and Future Research

The limitation of this study is the fact that it is only a theoretical contribution, which needs to be validated. Therefore, it is suggested that future research should validate this study by applying surveys or other tools that are considered more appropriate to obtain results that may corroborate the theoretical foundations described herein.

It is suggested in future studies to analyze the sustainability critical success factors that apparently have no relation with the CSFs of TQM implementation in order to assess the need to consider them as CSFs in TQM implementation, too. It can be done through analysis of the HEIs'strategic documents, webpages, application of surveys, and other documentary information.

In order to confirm/corroborate the existence of common CSFs and their importance for the implementation of TQM and sustainability, and their contributions to the SDGs, identified through documentary analysis, it is suggested to apply surveys to the various stakeholders (internal and external) of higher education institutions.

Equally, as future research it is intended to understand how the TQM and sustainability CSFs can relate to the SDGs and how HEIs can contribute to them in order to continuously improve toward sustainability.

References

1. Bayraktar, E., Tatoglu, E., & Zaim, S. (2008). An instrument for measuring the critical factors of TQM in Turkish higher education. *Total Quality Management, 19*(6), 551–574.
2. Nadim, Z. S., & Al-Hinai, A. H. (2016). Critical success factors of TQM in higher education institutions context. *International Journal of Applied Sciences and Management, 1*(2), 147–156.
3. Brookes, M., & Becket, N. (2007). Quality management in higher education: A review of international issues and practice. *International Journal of Quality Standards, 1*(1), 85–121.
4. Davim, J. P., & Filho, W. L. (2016). *Challenges in higher education for sustainability*. Springer International Publishing.
5. Galdeano, D., Ahmed, U., Fati, M., Rehan, R., & Ahmed, A. (2019). Financial performance and corporate social responsibility in the banking sector of Bahrain: Can engagement moderate? *Management Science Letters, 9*(10), 1529–1542.
6. Salleh, M. I., Habidin, N. F., Noor, K. M., & Zakaria, S. Z. S. (2019). The development of higher education for sustainable development model (HESD): Critical success factors, benefits, and challenges. *International Journal of Academic Research in Progressive Education and Development, 8*(4), 47–54.
7. Salleh, N. M., Zakuan, N., Ariff, M. S. M., Bahari, A. Z., Chin, T. A., Sulaiman, Z., … & Saman, M. Z. M. (2018, November). Critical success factors of total quality management implementation in higher education institution: UTM case study. In *AIP conference proceedings*, Vol. 2044, No. 1, p. 020007. AIP Publishing LLC.
8. Guilbault, M. (2018). Students as customers in higher education: The (controversial) debate needs to end. *Journal of Retailing and Consumer Services, 40*, 295–298.
9. Carvalho Pereira, M. A., & Terra Da Silva, M. (2003, April). A key question for higher education: Who are the customers? In *Annual conference of the production and operations management society*.
10. Reavill, L. R. P. (1998). Quality assessment, total quality management and the stakeholders in the UK higher education system. *Managing Service Quality: An International Journal, 8*(1), 55–63.
11. Akinyele, S. T. (2010). Customers: Identifying the needs in higher education. *Educational Research, 1*(7), 210–218.
12. Frazer, M. (1994). Quality in higher education: An international perspective. In D. Green (Ed.), *What is quality in higher education*. SRHE and Open University Press.
13. Tambi, A. M. (2000). *Total quality management in higher education: Modelling critical success factors*. Sheffield Hallam University.
14. Jamali, G., Ebrahimi, M., & Abbaszadeh, M. A. (2010). TQM implementation: an investigation of critical success factors. In *2010 international conference on education and management technology*, pp. 112–116. IEEE.
15. Ariff, M. S. M., Zaidin, N., & Sulong, N. (2007) Total quality management implementation in higher education: Concerns and challenges faced by the faculty. In *12th international conference on ISO 9000 & TQM*. Republic of China.
16. Aquilani, B., Silvestri, C., Ruggieri, A., & Gatti, C. (2017). A systematic literature review on total quality management critical success factors and the identification of new avenues of research. *The TQM Journal, 29*(1), 184–213.
17. Aquilani, B., Silvestri, C., & Ruggieri, A. (2016). Sustainability, TQM and value co-creation processes: The role of critical success factors. *Sustainability, 8*(10), 995.
18. Isaksson, R. (2006). Total quality management for sustainable development: Process based system models. *Business Process Management Journal, 12*, 632–645.
19. Zink, K. J. (2007). From total quality management to corporate sustainability based on a stakeholder management. *Journal of Management History, 13*, 394–401.
20. Edgeman, R. L. (2000). BEST business excellence: An expanded view. *Measuring Business Excellence, 4*, 15–17.

21. Pheng, L. S., & Teo, J. A. (2004). Implementing total quality management in construction firms. *Journal of Management in Engineering, 20*(1), 8–15.
22. Kehoe, D. K. (1996). *The fundamentals of quality management*. Chapman & Hall.
23. Waldman, D. A. (1994). The contributions of total quality management to a theory of work performance. *Academy of Management Review, 19*(4), 510–536.
24. Tsang, J. H. Y., & Antony, J. (2001). Total quality management in UK service organisations: Some key findings from a survey. *Managing Service Quality, 11*(2), 132–141.
25. Ugboro, I. O., & Obeng, K. (2000). Top management leadership, employee empowerment, job satisfaction, and customer satisfaction in TQM organizations: An empirical study. *Journal of Quality Management, 5*, 247–272.
26. Akins, E. E., Giddens, E., Glassmeyer, D., Gruss, A., Kalamas Hedden, M., Slinger-Friedman, V., & Weand, M. (2019). Sustainability education and organizational change: A critical case study of barriers and change drivers at a higher education institution. *Sustainability, 11*(2), 501.
27. Weiss, M., Barth, M., Wiek, A., & von Wehrden, H. (2021). Drivers and barriers of implementing sustainability curricula in higher education-Assumptions and evidence. *Higher Education Studies, 11*(2), 42–64.
28. Bayraktar, E. (2006). Designing higher education institutions as service organizations: A process oriented approach. *International Journal of Business, Management and Economics, 2*(5), 15–27.
29. GoÈzacËan, N. B., & Ziarati, R. (2002). Developing quality criteria for application in the higher education sector in Turkey. *Total Quality Management, 3*(7), 913–926.
30. Behara, R. S., & Gunderson, D. E. (2001). Analysis of quality management practices in services. *International Journal of Quality and Reliability Management, 18*(6), 584–604.
31. Zhang, Z. (2000). *Implementation of total quality management, an empirical study of Chinese manufacturing firms* (PhD theses). Faculty of Management and Organization of the University of Groningen.
32. Dale, B. G. (2003). *Managing quality* (4th ed.). Blackwell Publishing, Ltd.
33. Zhang, Z., Waszink, A., & Wijngaard, J. (2000). An instrument for measuring TQM implementation for Chinese manufacturing companies. *International Journal of Quality & Reliability Management, 17*(7), 730–755.
34. Sirvanci, M. B. (2004). TQM implementation: Critical issues for TQM implementation in higher education. *The TQM Magazine, 16*(6), 382–386.
35. Aleixo, A. M., Leal, S., & Azeiteiro, U. M. (2018). Conceptualization of sustainable higher education institutions, roles, barriers, and challenges for sustainability: An exploratory study in Portugal. *Journal of Cleaner Production, 172*, 1664–1673.
36. Disterheft, A., Caeiro, S., Azeiteiro, U. M., & Leal Filho, W. (2015). Sustainable universities–A study of critical success factors for participatory approaches. *Journal of Cleaner Production, 106*, 11–21.
37. Larrán Jorge, M., Herrera Madueño, J., Calzado Cejas, M. Y., & Andrades Peña, F. J. (2015). An approach to the implementation of sustainability practices in Spanish universities. *Journal of Cleaner Production, 106*, 34–44.
38. Velazquez, L., Munguia, N., & Sanchez, M. (2005). Deterring sustainability in higher education institutions: An appraisal of the factors which influence sustainability in higher education institutions. *International Journal of Sustainability in Higher Education, 6*(4), 383–391.
39. Verhulst, E., & Lambrechts, W. (2015). Fostering the incorporation of sustainable development in higher education. Lessons learned from a change management perspective. *Journal of Cleaner Production, 106*, 189–204.
40. Blanco-Portela, N., R-Pertierra, L., Benayas, J., & Lozano, R. (2018). Sustainability leaders' perceptions on the drivers for and the barriers to the integration of sustainability in Latin American higher education institutions. *Sustainability, 10*(8), 2954.
41. Aleixo, A., Azeiteiro, U., & Leal, S. (2020). Are the sustainable development goals being implemented in the Portuguese higher education formative offer? *International Journal of Sustainability in Higher Education, 21*(2), 336–352.

42. Leal Filho, W., Azeiteiro, U., Alves, F., Pace, P., Mifsud, M., Brandli, L., Caeiro, S. S., & Disterheft, A. (2018). Reinvigorating the sustainable development research agenda: The role of the sustainable development goals (SDG). *International Journal of Sustainable Development & World Ecology, 25*(2), 131–142.
43. Sisto, R., Sica, E., & Cappelletti, G. M. (2020). Drafting the strategy for sustainability in universities: A backcasting approach. *Sustainability, 12*(10), 4288.

Customer Loyalty in Hospitality: Can the Quality Perception Be a Key Factor?

Mariana Inácio Marques , Ana Afonso Alcântara ,
Gabriel Guerreiro e Carreira, and João Caldeira Heitor

Abstract This chapter aims to analyze the features related to the perception of service quality and understand whether they are somehow related to the customers' loyalty. In fact, it is quite complex to analyze the quality of a service so broad such as hospitality, and sometimes the quality perception varies according to personal intrinsic aspects, which makes the evaluation a subjective process. Since quality perception is something complex and subjective, it is important to understand whether customer loyalty is related to the perception of the quality of the service they enjoy. For this study, we applied a questionnaire to the Estoril Coast hotel managers since this is a region of a major importance for Portuguese tourism, and it is even considered an international brand. Questionnaires were also applied to tourists to better understand the demand behavior. Models of quality in services are also presented, some of them applied to hospitality. The analysis of the guests' answers showed just a few claims to fulfill hotels' questionnaires, so it is important to consider other mechanisms for evaluating the service quality. In fact, this chapter helps to verify the existence of a relationship between the perception of quality in the hotel industry and guest loyalty and the finding can provide some inputs to hotel

M. I. Marques (✉)
ISG – Business & Economics School, GOVCOPP – Research Unit on Governance, Competitiveness and Public Policies, Lisbon, Portugal
e-mail: mail@marianamarques.com

A. A. Alcântara
Instituto de Geografia e Ordenamento do Território, Master's in Strategic Management and Tourism Development by Instituto Superior Técnico, Lisbon, Portugal

G. G. e. Carreira
Escola Superior Hotelaria e Turismo Estoril, National Interpreter Guide and Tourism Mail Certified by the National Union of Tourist Activity, Translators and Interpreters (SNATTI), Estoril, Portugal

J. C. Heitor
ISG – Business & Economics School, CiTUR – Centre for Tourism Research, Development and Innovation, Lisbon, Portugal

J. C. de Oliveira Matias et al. (eds.), *Quality Innovation and Sustainability*, Springer Proceedings in Business and Economics,
https://doi.org/10.1007/978-3-031-12914-8_9

managers and marketeers. The major results show that hotels recognize the importance of keeping customers loyal and that loyalty increases sales (90%), helps to get free advertising (80%), and importantly constantly improves service (70%).

Keywords Services · Hospitality · Quality · Loyalty

1 Introduction

This chapter aims to verify the existence of a relationship between the perception of quality in the hotel industry and guest loyalty.

The research question for the study was: "Does the quality provided during the hotel service influence customer loyalty?"

This is a current issue as companies increasingly must make an increased effort to please and retain their customers. This loyalty sometimes involves ensuring that the guest has a quality service that meets their needs and expectations. It should also be noted that new technologies and new media have made the tourist a more informed and, consequently, more demanding consumer. Hotels therefore must live up to the expectations of customers who expect differentiated treatment when using the service [8].

It is therefore important that the hospitality sector is aware of the aspects to which the customer gives importance when providing the service and that it studies ways to meet their expectations. However, it must be borne in mind that the customer does not base his assessment of the service solely on the physical characteristics of the hotels and their amenities. It remains an important aspect, but the customer gives more and more importance to the intangible aspects of the service, namely, the way they are treated and how they feel special. Thus, loyalty should be seen as a time horizon for building a long-term relationship, in which the consumer buys repeatedly and bases his decision on stimuli provided by the company [6]. Also, it should be noted that customers who are loyal to companies buy more products if there is an effective total quality management (TQM) approach [20].

Thus, the objective should be to maintain a long-term relationship while creating added value for the customer, which allows creating and preserving their loyalty [9].

The Estoril Coast was chosen within the scope of this work because it is a region whose hotel sector has an economic performance above the national average. In addition, it is a region that has seen great investment in the creation and improvement of essential infrastructure for tourism. One cannot ignore the entire historical heritage of the region as it has been chosen as a royal seaside resort for years on end.

The motivation of this study is to try to understand whether the quality perception can be a key factor in customer loyalty and provide some inputs to hotel managers and marketeers on that theme.

In terms of structure, the chapter presents, in a first phase, a theoretical contextualization, followed by the presentation and discussion of the results obtained in the applied surveys.

2 Theoretical Background

2.1 *Quality Perception in Hospitality*

Luís Saias [16] states that quality management has been a concern since the second half of the twentieth century and led to the creation of total quality management (TQM), which consists of a control system for products at the exit of the factory or a line of production, thus avoiding that defective products are offered for sale and that, because they are not in good condition, the customer is dissatisfied with them. However, in services this quality is more difficult to guarantee. It is not enough to have the intention of offering quality customer service in the missions and objectives of the company. To obtain results, these intentions must get off the ground and create a series of actions that ensure that the customer is satisfied. For this, they should start by ensuring good employees are in direct contact with the customer. In hospitality, it is necessary to ensure that employees adjust to what the customer is looking for at that moment. It is therefore necessary to pay special attention to the hiring process of employees, especially those who are in direct contact with the customer. The author warns that what the company defines as quality does not always match with the way the customer perceives it. So, there is a difference between having quality and appearing to have quality. In fact, the term quality can be consensually defined as what customers or consumers say is quality; in other words, it is not what the company thinks it is, but what the customer says it is, with an emphasis on customer opinion.

A quality management system is, then, the management practice within companies that translates into the involvement of all those who work in the organization who cooperate to provide products and services that meet the needs and expectations of customers [13]. In fact, according to Talapatra et al. [21], management practices must involve employees in providing products and services to satisfy customers' needs and expectations. Also, employees must be trained on TQM to prevent problems, and when they are listened, they feel empowered, perform better, and it reflects on the service they provide to the customer.

Being aware of the difficulty in defining quality in services, Parasuraman, Zeithaml, & Berry [12] carried out an exploratory investigation on this topic using interviews and focus groups and developed a model (SERVQUAL) in which they were able to identify gaps, that is, gaps between what marketers understand as quality and what consumers understand as quality. It is important to know these gaps to verify what can be done to deliver the service that the customer wants and that they perceive as having quality.

Five gaps were identified, as shown in Fig. 1, as follows:

- Gap 1: discrepancy between the customer's expectation and what the service provider understands as the customer's expectation.
- Gap 2: discrepancy between the provider's understanding and the specifications it draws up to serve the customer.

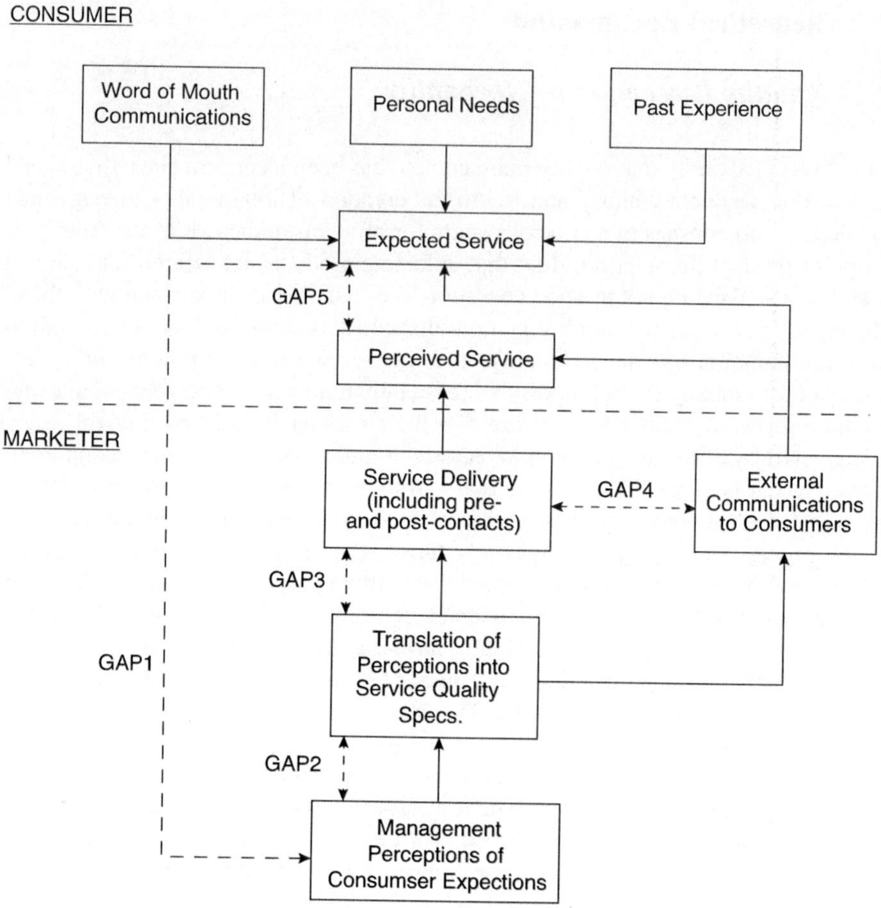

Fig. 1 SERVQUAL model gaps. ([12]:44)

- Gap 3: discrepancy between the specifications drawn up and the service provided.
- Gap 4: discrepancy between the service provided and communication with the customer.
- Gap 5: discrepancy between the expected service and the service provided to the customer.

This model provides important information to understand that the perception of quality is not a linear issue and that there are two important components to consider: the consumer and the service provider. By knowing the identified gaps, the provider will be able to outline more assertive strategies to minimize the damage of the identified discrepancies.

Babić-Hodović et al. [3] applied this SERVQUAL model to the perception of service quality in the hotel industry. For this purpose, they also refer to the SERVPERF model developed by Cronin and Taylor in 1992. For these authors, the perceived quality of services precedes customer satisfaction, which influences purchase intention. Therefore, the result (performance represented by satisfaction) is what really matters. Babić-Hodović et al. [3] present some models of quality previously studied by other authors and developed a study with the objective of investigating the perception of the quality of service provided in hotels in Bosnia and Herzegovina. The results validate the SERVQUAL model. Respondents have high expectations regarding the service provided, mainly through previous experiences, word of mouth, and marketing strategies that allow for creating a positive reputation. It should be noted that this study proved that customer expectations are then exceeded as there is a commitment to creating a pleasant customer experience.

In terms of business survival in the services area, it is important to note that, increasingly, the businesses that last the longest are those that work on the relationship with the customer, and this factor ends up becoming an important differentiating factor [10].

Regarding the hotel industry, it is a fact that the management and measurement of quality in hotels is quite complex as they provide more than one product to their guests, such as accommodation and food service, among others. It is therefore a constant challenge. One highlighted factor that facilitates the provision of a quality service concerns the hotel staff. In fact, there is a direct correlation between the performance of employees and the degree of customer satisfaction, and their provision is a factor that they evaluate when talking about perceived quality in relation to the service provided [23].

The complexity of quality management in the hotel sector leads to greater pressure on human resources managers to find ways to enhance this relationship between employees and customers to make the experience quite pleasant. It is therefore necessary to motivate employees so that they feel important and understand, in fact, their importance for the perception of the quality of the hotel they represent. One of the main sources of motivation for employees is the good atmosphere within the hotel, between employees and managers. One problem identified is the issue of employee turnover, and hotels must ensure that good employees do not want to leave. To this end, it is necessary to introduce motivational methods that enhance recognition, a sense of belonging and respect among all [2].

Is important for hotel management to improve internal service quality (ISQ) because this is a strategic tool to enhance operational efficiency in hospitality, so, a collaborative culture involving employees must be built [22].

2.2 Customers Loyalty

The concept of quality is, therefore, very complex as it varies a lot according to what each person believes and likes, whether a consumer or a service provider. Absolute quality does not exist; hence, it is very important to correctly segment the market to ensure that the characteristics of the product or service are as close as possible to what the segment understands as having quality. In this way, the service provider can guarantee that the service meets what the consumer wants, needs, and likes, and even so, satisfaction of all those who enjoy the service or consume the product is not guaranteed [14].

We must speak about the perceived quality concept that influences the perceived price, so managers must be aware of customers' quality perception and see it as a strategic targeting and positioning tool [18].

We cannot forget all the complexities of customer satisfaction these days, in which the use of the Internet predominates as a way of doing business. In fact, online businesses are more complex when it comes to understanding the satisfaction of their customers, given that it is more complicated to know the customer and there is no physical relationship with them, which can lead to a series of failures as sometimes the salesperson can physically adapt to the customer's behavior and make a purchase. In the online environment, there is no possibility of immediately adapting to the customer, which often translates into a withdrawal from the purchase. Another problem that companies should be aware of is the fact that, in the online environment, it is more difficult to be aware of existing competitors [8].

In this context, it is also important to present the concept of loyalty. In fact, although it may seem synonymous with faithful, they are different notions, although directly related. Thus, loyalty implies intention and predisposition for the customer to buy, implying a repetition of the purchase behavior on the part of the consumer. It is, therefore, a conscious decision to continually buy the same brand [11].

According to Agudo et al. [1], tourists prefer to stay in a quality-certified hotel, and they are willing to pay more for it. So, this is important to tourists' loyalty to hospitality.

In fact, in hospitality, regarding customer loyalty, priority should be given to reservations made by regular customers while at the same time facilitating their realization. In addition, a certain number and type of room must be reserved internally for this type of customer, meeting their preferences. Associated with customer loyalty will be their satisfaction. In this sense, managers should pay attention to the type of customers they want for their unit, avoiding, whenever possible, accepting all customers, bearing in mind that habits and tastes may not be compatible. Thus, managers should consider customer satisfaction and not just occupancy rates [4].

Saias [16] defines what can be considered financially as a loyal customer and the consequences that arise in terms of sales and profit. In fact, the following aspects can be mentioned:

(i) Loyal customer allows repeat sales, that is, the loyal customer buys again, which means that each time he buys, the fixed cost of obtaining the customer will be progressively covered by the margin of the units sold.
(ii) Cross-selling is nothing more than selling more products or services through the same distribution channel, adding value to the customer. Thus, a loyal customer likes the company and is more open to trying other products or services from the company, rather than trying products or services from competing companies.

(iii) Loyal customers tend to be cheaper to serve because, with repeated purchases, they end up having progressively more dexterity in using the product or service.
(iv) Loyal customers are easier to manage because the company knows them better and vice versa.
(v) A loyal customer is less sensitive to price, which makes it more profitable in the medium to long term. In fact, normally the first time the company does business with the customer, the price is discussed, but then, if the service meets their expectations, they do not discuss prices again.

It is worth emphasizing that the influence of consumer satisfaction has certain outputs (effects) for the company and that consumer satisfaction leads to consumer loyalty and, the two together, can represent the company's economic success. However, loyalty depends on the degree of satisfaction with the product/service [10].

In the online environment, online reviews are also important in the hotel evaluation and in a long term it is a way to enhance loyalty. In fact, online we saw loyal customers speaking well about the hotel and recommend it [7]. Also, by creating an online relationship and a connection with the customer, it is easier to accomplish their loyalty [19]. According to Sharma et al. [17], the information technology used in hospitality such as hardware, software, and business analytics has the potential to satisfy customer needs and reinforce loyalty. So, it is important to think about these and use those tools.

3 Methodology

The study objectives were the 37 hotels on the Estoril Coast, regardless of star rating, distributed as shown in Fig. 2.

In terms of sample, a total of 20 responding hotels were obtained. This sample represents 68.9% of the universe, which means that it can be considered representative, since it presents more than 51% of responses [15].

It should be noted that the questionnaire was applied before the Covid-19 pandemic, in February 2017, and it was necessary to go in person to each hotel to guarantee answers since sending it by email was not very effective. The completion date of the application of the surveys was in December 2017.

Fig. 2 Study object. (Personal elaboration)

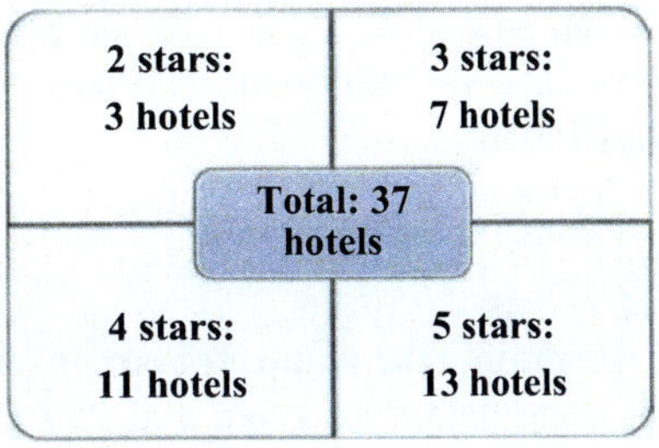

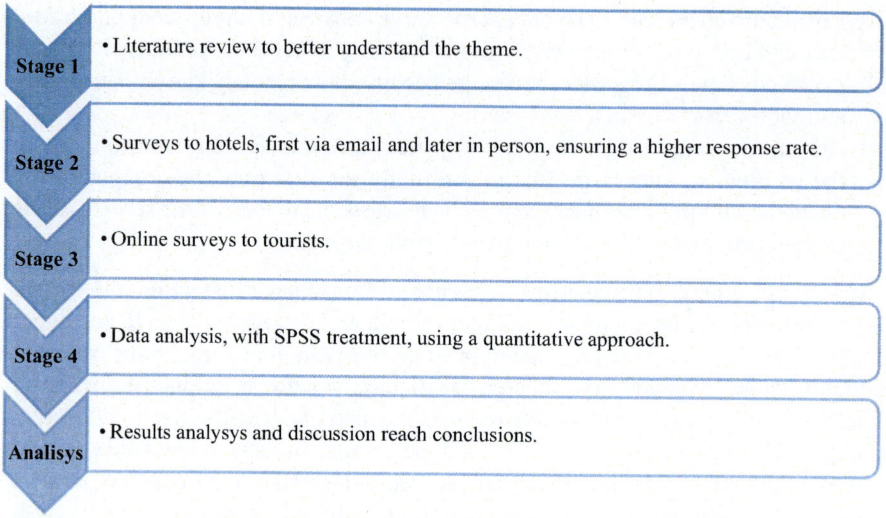

Fig. 3 Research methodology. (Personal elaboration)

The purpose of collecting these surveys was to understand the perception that hoteliers have of the importance of customer loyalty and verify how they work on this issue in their activity.

The methodology used was quantitative, in the form of statistics that allowed to draw conclusions from the responses to a survey applied to hotel directors (or someone on their behalf). For the statistical treatment, the SPSS program was used.

As a way of complementing the study, a brief survey was carried out among people who have the habit of traveling by asking about the collaboration of participants in travel groups on social networks. The survey carried out aimed to understand the main aspects that make the customer perceive the service as a quality service and test, in a way, what makes the customer feel loyal to a hotel. It should be noted that 296 valid questionnaires were obtained in November 2019. In Fig. 3, the authors present a flow diagram with the research methodology.

4 Results

As can be seen in Table 1, it was found that the highest percentage of respondents (40%) are five-star hotels, followed by three stars (35%), four stars (20%), and two stars (5%).

As for the number of rooms, it appears that 35% of respondents have between 40 and 80 rooms. However, there is also a significant percentage (25%) with more than 120 rooms.

Regarding the number of employees in the sample, it was found that the answers are quite balanced, given that 25% have less than 20 employees, also 25% have

Table 1 Sample distribution: number of stars (personal elaboration)

		Frequency	Percent	Valid percent	Cumulative percent
Valid	2 stars	1	5.0	5.0	5.0
	3 stars	7	35.0	35.0	40.0
	4 stars	4	20.0	20.0	60.0
	5 stars	8	40.0	40.0	100.0
	Total	20	100.0	100.0	

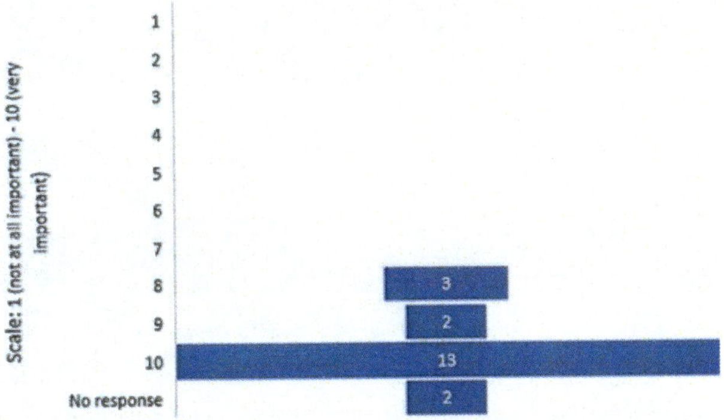

Fig. 4 Keeping customers loyal: scale. (Personal elaboration)

between 20 and 40, 25% more than 60, and only 20% have between 40 and 60 employees. It should be noted that one hotel did not respond.

According to Fig. 4, when asked about the importance of keeping customers loyal, on a scale of 1–10 (in which 1 is "not at all important" and 10 is "very important"), it was found that the minimum scale score was 8, with 65% of respondents saying to consider very important, thereby assigning the maximum on the scale.

Regarding the rate of customers who usually return to the hotel, it was found that in 55% of respondents this rate can be set between 30% and 60%, followed by a rate of less than 30%, corresponding to 35% of respondents, as we can see in Table 2.

Using a scale of 1–5 (in which 1 is "not at all important" and 5 is "very important"), the following results were verified regarding how to maintain a relationship with customers and, in addition, retain them: 85% consider it "very important" to keep the website updated; 55% rate the use of social networks as "very important"; as for newsletters, 20% consider their use as "very important"; 45% classify the updating of databases as "very important"; 50% of respondents did not answer the question about the use of loyalty cards, but 25% classified it as "very important"; 35% consider personalized offers "very important"; 65% of respondents rate the response to complaints as "very important"; 70% consider the analysis of satisfaction surveys "very important"; and 65% consider it "very important" to train employees in direct contact with customers. These results can be seen in Fig. 5.

Table 2 Rate of customers who usually return to the hotel (personal elaboration)

		Frequency	Percent	Valid percent	Cumulative percent
Valid	Less than 30%	7	35.0	35.0	35.0
	Between 30% and 60%	11	55.0	55.0	90.0
	More than 60%	1	5.0	5.0	95.0
	Don't know	1	5.0	5.0	100.0
	Total	20	100.0	100.0	

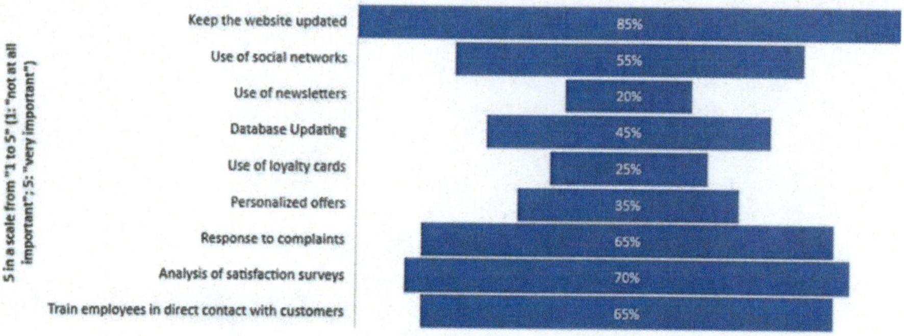

Fig. 5 How to maintain a relationship with customers and retain them – "very important." (Personal elaboration)

When asked about whether they consider that there is a direct relationship between customer satisfaction and employee performance, it was found that all respondents (100%) consider that this relationship exists.

When asked about the existence of any customer satisfaction study, 95% said yes and only 5% said no. Regarding the way they study customer satisfaction, 80% use satisfaction surveys, 80% state they have a conversation with customers at checkout, and only one hotel reports having conversations with the customer during their stay.

About the advantages of having loyal customers, as we can see in Fig. 6 the results were as follows: 90% believe that having loyal customers increases sales; 75% say that loyal customers do not allow them to reduce costs; 80% state that it allows them to get free advertising; 65% do not recognize that having loyal customers allows them to retain employees; 70% recognize the importance of loyal customers to constantly improve the service; and one hotel claims that having loyal customers also allows them to predict long-term revenue.

Listening to the demand side, a survey was carried out with tourists, assessing some aspects related to their choices and preferences when choosing a hotel. We have 296 valid questionnaires, 82.1% female and 17.9% male. The majority (72.3%) work for someone.

As we can see in Fig. 7, 57.1% travel between 3 to 5 times a year.

76.4% say that the establishment's classification is a major factor in decision-making (Fig. 8), with 76% of respondents favoring stays in three or four-star hotels (Fig. 9).

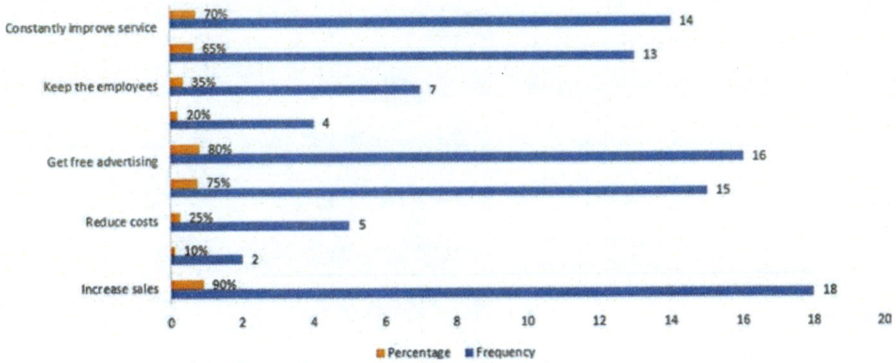

Fig. 6 Major advantages of having loyal customers. (Personal elaboration)

Fig. 7 How many times do they travel a year? (Personal elaboration)

Fig. 8 Is the establishment's classification a major factor in decision-making? (Personal elaboration)

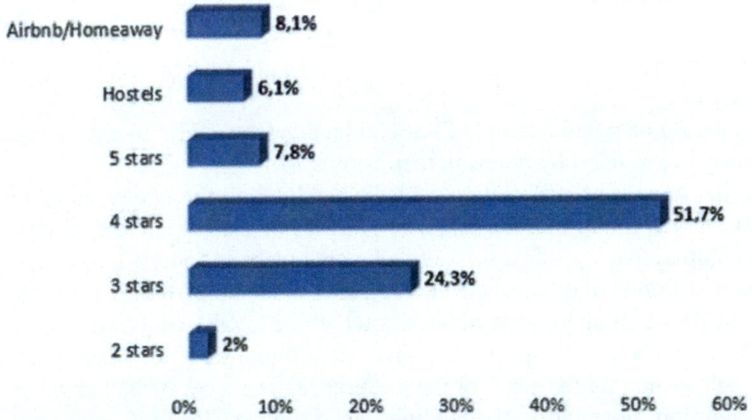

Fig. 9 Preferences for accommodation. (Personal elaboration)

Fig. 10 Sources of information that influence hotel choice. (Personal elaboration)

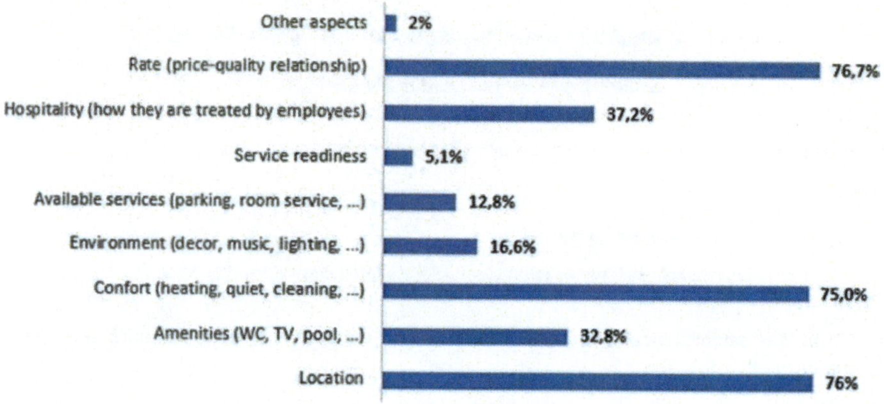

Fig. 11 Most valued aspects when choosing hotels. (Personal elaboration)

It should be noted that 23% of respondents have already stayed in hotels on the Estoril Coast.

It also should be noted that only 39.9% of respondents admit to answering satisfaction surveys provided by hotels in their rooms.

As for the sources of information valued when looking for a hotel, as we can see in Fig. 10, it was found that the majority (71.3%) of the respondents give importance to comparative websites, to the hotel's own website (54.1%), not neglecting the opinion of family and friends, which together make up 62.8%.

Only 32.1% say they have stayed in a hotel with a loyalty card or points system.

When asked about the aspects they give more importance to, when choosing a hotel, it was found that the top 5 of the preferences were, by attributed value, rate, location, comfort, hospitality, and facilities, according to Fig. 11.

Finally, the question regarding whether they consider that the quality of the service provided influences their loyalty as customers, on a scale from 1 to 5 (in which

1 is "I do not agree at all" and 5 is "I totally agree"), 92% of respondents assigned "5" and the remaining 8% assigned "4" on the same scale.

5 Conclusion

This chapter reflects the main aspects related to the perception of quality in services such as hotels, as well as customer loyalty.

Thus, it is important to mention that there is a direct relationship between the quality of the product/service and customer satisfaction. In this sense, there are studies that state that quality is responsible for about 60% of changes in consumer attitude and loyalty, and the loyal consumer translates into higher income and consequent growth margin for the company [5].

It is true that customer loyalty is closely linked to the quality they experience when using the service. Therefore, hotel managers should (i) dedicate more attention to getting to know their customers and (ii) provide a service of solid quality as this perception can influence the loyalty of their customers and boost a long-term relationship between the hotel and the guest.

Through the analysis of responses to guest surveys, it was found that few claim to respond to surveys, so it is particularly important to consider other mechanisms for measuring the quality of the service provided.

It should be noted, however, that hotels should not neglect the importance of the opinion of family and friends because, although importance is given to comparative websites and the hotel itself (mentioned by tourists as a channel considered and by hotel directors as a need to update constant), the truth is that the traditional channel continues to be a privileged source, hence the importance of an excellent service for recommendation.

Regarding loyalty cards or points system, it would be pertinent for hotels to communicate the existence of this tool, when it exists. However, in hotels that do not have this, it should be considered to create a points system to improve loyalty.

According to the results, the website is important for these issues. In fact, 54.1% of visitors assume that the website is an information source on the hotel choice and 85% of the hotels in the sample consider it very important to have an updated website to maintain a relationship with customers and even to retain them.

In what regards the last question related to the fact that the quality provided in the use of the service is considered to influence customer loyalty, the truth is that almost all respondents attributed the maximum rate on the defined scale, so hotels should keep in mind that, more important than attracting new customers, it is above all important to value each guest, providing them with the best possible service, since they can be loyal and also recommend the service to potential customers.

This study contributes to the existing literature because it allowed to respond to a research gap. In fact, there are some studies about loyalty and quality themes, but there is not one that clearly states a relationship between what visitors value in a hotel and what makes them return and what hotels' perspectives about quality should do to make tourists loyal.

As practical implications, this chapter provides some inputs to hotel managers and marketeers mostly about what visitors value in a hotel and what makes them return. So, they can improve on these aspects to try to improve customer loyalty and make better the hotel performance.

Regarding future line of research, it is suggested to deepen the study of aspects related to their quality and influence on the loyalty of customers who are staying in hotels of the sample at the time of the study.

As limitations, we can point out the representativeness of the data collected from the visitors' side.

References

1. Agudo, J., Leaniz, P., Crespo, A., & Gómez-López, R. (2021). Quality-certified hotels: The role of certification bodies on the formation of customer behavioral intentions. *Sustainabilty, 13*, 12733.
2. Azic, M. (2017). The impact of hotel employee satisfaction on hospitability performance. *Tourism and Hospitality Management, 23*(1), 105–117.
3. Babić-Hodović, V., Arslanagić-Kalajdžić, M., & Banda, A. (2015). Hotel service quality in developing countries measured by SERVQUAL model. *Acta geographica Bosniae et Herzegovinae, 4*, 5–24.
4. Costa, R. (2008). *Introdução à Gestão Hoteleira*. Lidel Edições Técnicas.
5. Dai, H., & Salam, A. F. (2014). Does service convenience matter? An empirical assessment of service quality, service convenience and exchange relationship in electronic mediated environment. *Electron Markets, 24*(4), 269–284.
6. Feng, N., Zhang, M., & Tang, S. (2010). *Impact of loyalty programs on relationship benefits and customer loyalty: A customer perspective*. Proceedings of international conference on management science & engineering, Vol. 17, pp. 533–538, Melbourne, VIC, Australia.
7. Gunasekar, S., Purani, D., Sudhakar, S., Dixit, S., & Menon, D. (2021). Understanding service quality attributes that drive user ratings: A text mining approach. *Journal of Vacation Marketing, 27*(4), 400–419.
8. Hashemi, J., & Abassi, F. (2017). The impact of e-service quality on customer satisfaction. *Arabian Journal of Business and Management Review, 6*(9), 10–16.
9. Kim, H., Lee, J., Choi, D., Wu, J., & Johnson, K. (2013). Perceived benefits of retail loyalty programs: Their effects on program loyalty and customer loyalty. *Journal of Relationship Marketing, 12*(2), 95–113.
10. Marquardt, K., Olaru, M., & Ceausu, I. (2017). Study on the development of quality measurements models for steering business services in relation to customer satisfaction. *Quality Management and Business Excellence, 19*(44), 95–109.
11. Ornelas, S (2013). Determinantes da Fidelização: qualidade percebida e satisfação dos utentes das farmácias na Região Autónoma da Madeira. Tese de doutoramento.
12. Parasuraman, A., Zeithaml, V., & Berry, L. (1985). A conceptual model of service quality and its implications for future research. *Journal of Marketing, 49*, 41–50.
13. Pinto, A., & Soares, I. (2011). *Sistemas de gestão da qualidade – Guia para a sua implementação*. Edições Silabo.
14. Pires, A. (2012). *Gestão da qualidade: ambiente, segurança, responsabilidade social, indústria, serviços, administração pública e educação*. Edições Silabo.

15. Ramos, S., & Naranjo, E. (2014). *Metodologia da Investigação Científica*. Escolar Editora.
16. Saias, L. (2007). *Marketing de serviços – qualidade e fidelização de clientes*. Universidade Católica Editora.
17. Sharma, M., Antony, R., Sehrawat, R., Cruz, A., & Daim, T. (2022). Exploring post- adoption behaviors of e-service users: Evidence from the hospitality sector/online travel services. *Technology in Society, 68*, 101781.
18. Solakis, K., Peña-Vinces, J., & Lopez-Bonilla, J. (2022). Value co-creation and perceived value: A customer perspective in the hospitality context. *European Research on Management and Business Economics, 28*, 100175.
19. Song, B., Liew, C., Sia, J., & Gopal, K. (2021). Electronic word-of-mouth in travel social networking sites and young consumers' purchase intentions: An extended information adoption model. *Young Consumers, 22*(4), 521–538.
20. Talapatra, S., & Uddin, M. (2019). Prioritizing the barriers of TQM implementation from the perspective of garment sector in developing countries. *Benchmarketing: An International Journal, 26*(7), 2205–2224.
21. Talapatra, S., Uddin, M., Antony, J., Gupta, S., & Cudney, E. (2019). An empirical study to investigate the effects of critical factors on TQM implementation in the garment industry in Bangladesh. *International Journal of Quality & Reliability Management, 37*(9), 1209–1232.
22. Wu, X., Wang, J., & Ling, Q. (2021). Managing internal service quality in hotels: Determinants and implications. *Tourism Management, 86*, 104329.
23. Zylfijaj, R., & Pina, B. (2017). Assessment of critical success factors of TQM culture in hospitality sector in Kosovo. *International Journal of Research in Business and Social Science, 6*(1), 65–77.

The Role of Enabling Technologies from Industry 4.0 in the Formulation of Public Policies for Smart Cities

Pedro Henrique Souza da Silva, Marcelo Albuquerque de Oliveira, Gabriela de Mattos Verenoze, and Raimundo Kennedy Vieira

Abstract Smart cities emerge as a viable solution for the design and management of urban centers increasingly concerned with issues related to improving sustainability and resilience. The chapter seeks to enrich discussions on the subject of smart cities by carrying out a systematic review of the literature on the importance of enabling technologies of the Fourth Industrial Revolution for the formulation of public policies for smart cities. On the occasion, issues related to the advancement of enabling technologies for Industry 4.0 in Brazil and how these technologies are understood, from the point of view of their importance, by public agents responsible for formulating public policies for smart cities in the country, were addressed. As a result of the investigation, it became clear that, in the case of emerging countries, such as Brazil, the definition of objectives and approaches for the solution of urban problems that can be addressed through public policies based on technologies for smart cities must, above all, take into account the various human and environmental factors inherent to society, whose demands are intended to be met. In addition, the conscious action of public agents, mainly in the political sphere, can eventually contribute to the failure to conduct projects categorized in the literature as smart-washed cities.

Keywords Smart cities · Public policies · Enabling technologies

1 Introduction

To increase the competitiveness of the German manufacturing industry, representatives of commerce, politics, and academia created the Industry 4.0 initiative in 2011. Supported by the German federal government, Industry 4.0 became part of the

P. H. S. da Silva · M. A. de Oliveira (✉) · G. de Mattos Verenoze · R. K. Vieira
Federal University of Amazonas, Manaus, Brazil
e-mail: marcelooliveira@ufam.edu.br; gveroneze@ufam.edu.br

J. C. de Oliveira Matias et al. (eds.), *Quality Innovation and Sustainability*, Springer Proceedings in Business and Economics,
https://doi.org/10.1007/978-3-031-12914-8_10

High-Tech Strategy for Germany 2020 in the year 2012. In summary, the main objective of the program was to make Germany the main provider of scientific and technological solutions in the areas of climate, energy, health, nutrition, mobility, security, and communication. In 2013, the Industry 4.0 Working Group published its final report, where it discussed the first application in the field of Industry 4.0, which consisted of adopting a totally new approach to production and manufacturing processes, through the implementation of smart factories [1, 2].

Although it has its origins in the German productive systems, Industry 4.0 soon caught the attention of other developed nations around the world due to its strong potential for accelerating economic growth. As the Fourth Industrial Revolution, Industry 4.0 is based on three main aspects, namely, (1) digitization and greater integration of horizontal and vertical value chains; (2) digitization of the offer of products and services; and (3) introduction of innovative digital business models [3]. In the Fourth Industrial Revolution, the ability to spread digitalization and information technology are common features among technological innovations that integrate the physical, digital, and biological worlds [4].

According to the report released by the Global Agenda Council of the World Economic Forum in October 2015, the Fourth Industrial Revolution is a phenomenon of unprecedented scales in human history. International organizations, such as the World Economic Forum, believe that the Fourth Industrial Revolution is the result of the joint action of a network of smart cities, countries, and regional clusters that guide their efforts to guarantee the wide access and use of communication and information technologies [4].

Based on what has been exposed so far, this chapter aims to enrich discussions on the topic of smart cities through the analysis of cases of implementation of public policies based on the application of technologies that facilitate the Fourth Industrial Revolution as fundamental elements for the advancement of projects of smart cities. In addition, the chapter also seeks to contribute to the advancement of projects underway in Brazil by evaluating the importance of enabling technologies of the Fourth Industrial Revolution from the expectations and perceptions of the Brazilian public sector. To fulfill these objectives, the chapter will carry out a systematic literature review to provide answers to the following research questions:

1. What are the enabling technologies of Industry 4.0?
2. How are these technologies perceived by the Brazilian public sector?
3. What is the contribution of these technologies to the elaboration of public policies aimed at the development of smart cities?

The chapter is organized as follows: Sect. 2 presents the theoretical framework; Sect. 3 presents the methodology used in the systematic literature review; Sect. 4 presents the results of the research, as well as the discussions derived from the analyses carried out; and Sect. 5 concludes the work, including proposals for future work on the topic of smart cities.

2 Background

2.1 Enabling Technologies of Industry 4.0

According to [4, 5], Industry 4.0 is based on intelligent production achieved by reducing sensor prices, the ubiquity of the Internet, and advances in the field of artificial intelligence and machine learning [1, 6]. agree that Industry 4.0 is a goal to be achieved by consolidating different digital-based technologies. The set of distinct technologies that make up Industry 4.0 can vary according to the perception of each analyst. However, based on [6], the most relevant technologies are sensors and actuators, the Internet of Things, Big Data, and computing [1] states that the technologies that constitute Industry 4.0 comprise cyber-physical systems, cloud systems, machine-to-machine communication, smart factories, augmented reality and simulation, data mining, the Internet of Things, resource planning systems, enterprise and business intelligence, and virtual manufacturing.

For [7], the technological factors that lead to Industry 4.0 are Big Data, cloud computing, Internet of Services, Internet of Things, cyber-physical systems, and three-dimensional printing technologies [8] concluded that the essential elements of Industry 4.0 are cyber-physical systems, additive manufacturing, virtual and augmented reality, cloud computing, Big Data analytics, and data science [9] assumes that the nine technological advances that stimulate Industry 4.0 are autonomous robots, augmented reality, Big Data, data analysis, additive manufacturing, industrial Internet of Things, horizontal and vertical systems integration, cloud computing, simulation, and cybersecurity.

Da Costa et al. [10] conclude that the Internet of Things and cyber-physical systems are the true technological pillars behind Industry 4.0. For them, simulation, virtualization, and cloud computing are directly linked to the development of cyber-physical systems, while the Internet of Things would be the basic architecture for digitization, Big Data, real-time systems, and other information and communication technologies [11] proposed that the constituent technologies of Industry 4.0 are divided into technology base layers and application interface layers. According to them, the technological base layer would have the Internet of Things, Big Data, cloud computing, and data analysis. The application interface layer includes smart production, smart products, smart work, and smart supply chain.

Guimarães and Sant'Anna [5] states that Industry 4.0 is a productive logic that contributes to increased productivity through the incorporation of information and communication technologies in production processes [1] understands Industry 4.0 as a philosophical transformation of society. When evaluated from a broader perspective, Industry 4.0 takes on the contours of the Fourth Industrial Revolution, that is, a movement that aims at the philosophical transformation of both the industrial environment and the social environment [1–5].

In this sense, [1] emphasizes that discussions related to Industry 4.0 should not be limited to the industrial scope since there is also ongoing research outside the factory domain. As suggested by [12], Industry 4.0 is a topic of global relevance that also needs to be monitored by Brazil as the subject gains more and more prominence due to productivity gains and the opening of new fronts of economic and social development through the integration of technologies and knowledge previously restricted to factories.

2.2 *Industry 4.0 from the Brazilian Public Sector Perspective*

Comparing the industry revitalization policies of Germany, China, and the United States through Industry 4.0, [13] highlighted the importance of the State in industrial renewal since it is the government that provides the economic agents necessary for the creation of favorable environments for the implementation of Industry 4.0. The authors state that this environment is provided by open and democratic societies that have advanced national innovation systems, mechanisms for technological diffusion, satisfactory transport and communication infrastructure, free and competitive markets, and regulatory frameworks.

Industry 4.0 is perceived by [6] as a set of digitally based technologies aimed at the development of industrial production led by the United States, Germany, and China that vie for world leadership, with regard to the modernization of industry and production enhancement through support programs such as the Advanced Manufacturing Partnership (AMP), Industrie 4.0, and Made in China [5, 6]. For [12], Brazil must follow the development of Industry 4.0 so as not to be left behind in the race for innovation.

However, [5] highlights the lack of a sense of urgency and the late and timid nature of government actions undertaken by Brazil, with regard to the creation of a favorable environment for the dissemination of Industry 4.0. In this sense, they also state that there are no concrete initiatives aimed at building a regulatory framework that encourages the development of Industry 4.0. In addition, the authors emphasize the Brazilian deficiency with regard to planning, resource allocation, and execution of strategic programs and public policies for regional development with a focus on Industry 4.0.

Even so, [5] states that the number of national initiatives aimed at developing Industry 4.0 is not very distant from those carried out by other countries. According to them, these actions have their origin in initiatives conducted by universities, research institutes, S System, and industry federations, which aim at the development of laboratories, research projects, and the construction of intellectual capital. Nevertheless, [6] adds the efforts undertaken by public institutions in general, business entities, and development agencies that have implemented actions and public policy proposals for Industry 4.0.

Lima and Russo [12] highlights the work of the former Ministry of Science, Technology, Innovation and Communication (current Ministry of Science, Technology, and Innovation), the Brazilian Agency for Industrial Development, the National Confederation of Industry, and the National Service for Industrial Learning in the development of public policies and strategic actions for Industry 4.0.

According to [5], incentive actions coordinated by the Brazilian government and made possible through modern regulatory frameworks, public policies, and purchases would have the potential to create the necessary environment for innovation. In the view of [12, 13], the social and economic effects of an innovative environment are undeniable. According to them, economic growth in the most developed countries is favored by the prioritization of innovation, a fact that can be proven by the existence of several public policies that stimulate innovation in the productive sectors, education institutions, science and technology, and in society as a whole.

Oztemel and Gursev [1] cites as advantages of the transformation derived from the implementation of Industry 4.0 the development of increasingly customized products, faster production, improvement in innovation capacity, productivity gains, greater flexibility, cost reduction, decision-making real-time and knowledge-based impartiality, ease of diagnosis of multifunctional systems, greater capacity for self-awareness and maintenance of systems, contributions to the national economy, increase in electronic commerce, ease of access to personal information, easier access to public services, and the development of smart and remotely controlled cities, buildings, and factories.

3 Methodology

The systematic review described by [14] is an efficient and high-quality method available for the evaluation of extensive literature, being comprehensive and unbiased. The method initially consists of defining the keywords and terms that will guide the construction of the search strings. Next, the research strategy must be carefully detailed so that it can be replicated when necessary.

In these terms, the search should only return works that meet all the inclusion criteria and that do not present any of the exclusion criteria. The method presupposes ensuring that relevant and high-quality sources are obtained, which must be retrieved later when reading the full articles and choosing those that will be part of the systematic review. However, the number of works included and excluded in each step and the reasons that led to the exclusions must be recorded in the research protocol.

As explained by [14, 15], the systematic review of the literature is a scientific, replicable, and transparent process that aims to minimize biases by carrying out exhaustive searches in published and unpublished literature and offering traceability to reviewers.

To carry out this research, three sets of keywords were initially defined, as described in Fig. 1, to guide the searches for scientific articles in the Scopus and Web of Science databases.

The search strings were assembled by combining the keywords described in the first set, which corresponds to a list of terms that refer to the importance of enabling technologies in Industry 4.0, together with the keywords of the second and third

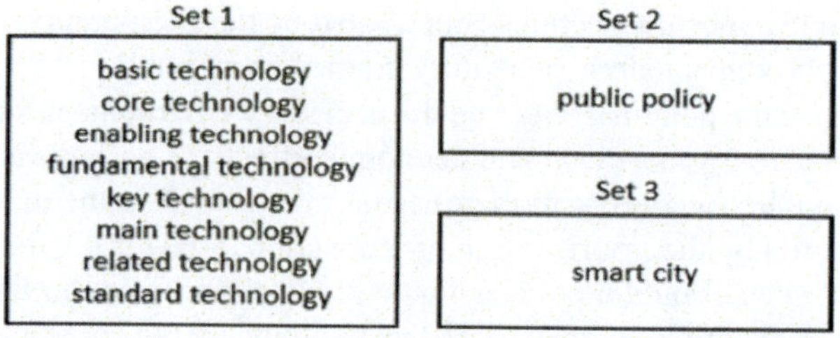

Fig. 1 Keywords sets

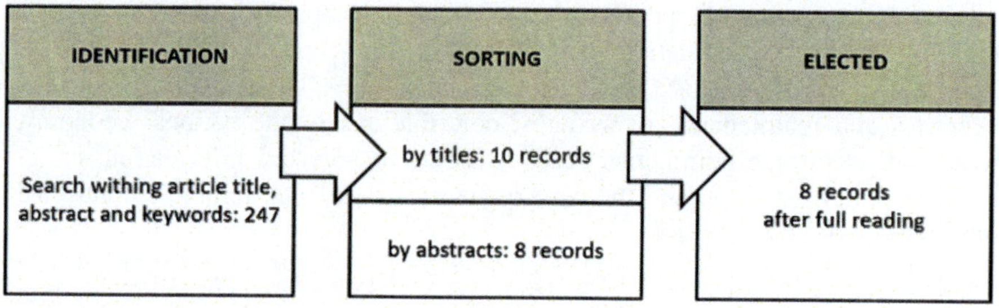

Fig. 2 Selection and election of articles in the Scopus database

sets, which correspond to the terms that refer to the application of these technologies in the context of the formulation of public policies for smart cities.

The use of Scopus and Web of Science as sources of bibliographic references of good quality, relevance, and consistency for formatting the systematic literature review was based on the work developed by [16], so that the selection of these databases to carry out the research in question considered characteristics such as focus, scope, coverage period, and database update, in addition to factors related to the usability of each system and the offering of citation analysis.

At first, we carried out searches in the Scopus database in order to select scientific articles published in journals, in the last 5 years, written in English, duly reviewed by peers and that showed themselves to be adherent to the research topic. As shown in Fig. 2, the search carried out in Scopus returned 247 records in the first stage of the review. Then, in the step, 10 articles were selected, after reading the titles. Finally, eight articles were approved for systematic review, after reading the abstracts.

In a second moment, the research focused on the search for bibliographic references in the Web of Science database using the same combinations of keywords used in the research carried out in Scopus. Thus, Fig. 3 illustrates that the search in the Web of Science database returned 475 records. Subsequently, 10 articles were selected based on the reading of their titles, which resulted in the selection of 2 articles, considering that the remaining 8 articles were previously found in the search carried out in the Scopus database.

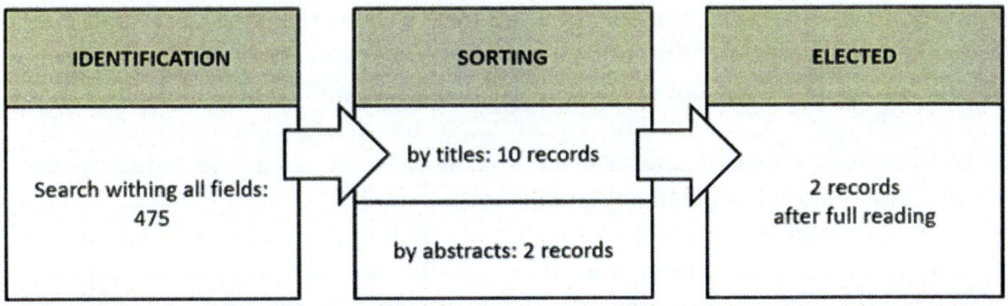

Fig. 3 Selection and election of articles in the Web of Science database

The complete reading of the 10 articles selected from the reading of their titles and abstracts confirmed the relevance of all records for the conduct of this research so that all were considered for the systematic review of the literature.

4 Analysis and Discussion

Through document analysis [17], investigated age-friendly smart city initiatives at local, regional, and national levels, delving deeper into the study of the characteristics of smart city initiatives in Romania and concluding that the current smart city concept in the country is restricted to prioritizing improvements in infrastructure and sustainability, to the detriment of long-term public policies for the exploitation of technologies such as the Internet of Things and sensors in the collection of public and private data with potential use in the development of more efficient and intelligent services.

As a product of their research [18], proposed an adaptable system for the management of public policy measures for the development of smart cities, mainly considering the human point of view with regard to the challenges of the political sphere and the principles of smart governance such as the active participation of citizens and stakeholder involvement.

When analyzing technological trends and discussing the challenges related to cybersecurity, cyber privacy, and investments in infrastructure, then, public and private policy targets concerning smart city projects [19], highlights that smart city engineers and developers should also pay attention to the benefits and risks associated with the adoption of intelligent systems. For the authors, the smart city is not just a favorable environment for technological development and economic growth. Before that, the smart city must have humanistic, environmentally friendly, and livable characteristics.

Research conducted by [20] investigated the relationship between public policies for the development of smart cities and the total productivity of urban green factors. In the case in question, data corresponding to the years 2007–2016 were evaluated, obtained from statistical panels of 200 Chinese cities. The results suggest that the expansion and integration of communication technologies and the Internet of Things contribute to intelligence gains in the transport, energy, resource allocation, and waste disposal sectors. For the authors, the smart city presents itself as a new standard for urban development by establishing operational command centers and employing Internet of Things technologies in information collection and data analysis, so that green concepts and advanced technology are always in harmony and duly integrated into urban planning guidelines.

The investigation conducted by [21] focused on discussing how Big Data strategies help society and government to deliver public services and produce increasingly responsive, accountable, transparent, and citizen-centered public policies. The effectiveness of evidence-driven public policymaking is demonstrated through the analysis of cases of application of Big Data technologies by government agencies in the United States, China, the United Kingdom, and India working on fraud detection, crime-fighting, market analysis finance, government oversight, public health, public education, environmental protection, ecosystem management, energy exploration, agriculture, and weather forecasting. In addition, the research also identified Big Data applications focused on government and public policy and others focused on citizens and smart cities located in North America (Canada), Asia (Singapore, China, Korea, India, and Japan), and Europe (the United Kingdom).

The case study conducted by [22] assessed the state of smart city policies based on innovative technologies conducted by developing countries and concluded, through the analysis of the new Administrative Capital Policy underway in Egypt since 2015, that smart cities are characterized, above all, by the dichotomy between objectives and development approaches, when evaluated from the perspective of spatial strategies.

Assuming the perspective of public policies [23], examined how smart city initiatives overlap with priority issues consolidated, for example, in the social commitments assumed by the United Nations, when proposing the 2030 Agenda and the New Urban Agenda. In addition, the research considered primary data from Ghana, India, Indonesia, and the United Arab Emirates to formulate a typology of categorization of smart city projects in the political context in order to contribute to the formulation of multilayered public policies based on solidarity models of social and economic responsibility between citizens, companies, and government.

By employing data mining techniques in the evaluation of terms with the highest incidence in the literature concerning smart cities and sustainable cities, Gonçalves et al. [24] concluded that the Fourth Industrial Revolution will have a significant impact on determinants of the success of cities in achieving their ideals of sustainability. Among these factors, the generation of employment, industry, innovation, preservation of the environment, community involvement, and accessibility stands out. In addition, Big Data, artificial intelligence, augmented reality, and simulation emerge as technologies capable of helping public policymakers, even if they bring with them positive and negative externalities.

With the objective of identifying the main technologies that are necessary for the construction of smart cities [25], analyzed an electronic platform that optimizes the allocation of resources and the provision of services to citizens, through the use of communication technologies and information, such as Big Data, cloud computing, artificial intelligence, the Internet of Things, virtual reality, augmented reality, and 5G mobile communication. The study showed that solving urbanization problems is directly related to the ability to process intra-municipal information in real time.

In addition, the deployment of the smart city platform significantly contributes to improving the security of society. It was also concluded that the design and construction of smart cities must consider regional characteristics in the design of the specialized system. Furthermore, it was found that smart cities constitute two-way cooperation systems, in which the government and citizens participate in the operationalization of the city, making it clear that the active use of communication and information technologies can solve urban problems by predicting the demands of the city and society, effectively reducing the costs of public policies.

Mills et al. [26] explored the dynamics of the evidence-based decision-making process in the context of public policies for smart cities in order to verify whether the theoretical concepts described in the literature and the practice experienced in cities actually contribute to the expansion of relative knowledge of evidence-based decision-making in the public sector. Conclusively, the study found benefits to the decision-making process when supported by Big Data-based approaches.

However, problems and challenges related to inadequate collaborative decision-making, disregard of evidence in favor of the application of Big Data, controversial decision-making by artificial intelligence agents, and confusion and poor understanding of the concept of decision-making based on evidence when applied to the context of smart cities.

5 Conclusions

The Fourth Industrial Revolution, in principle, presents itself as a response of global companies to the problems and challenges related to the aging of the population in developed countries, reduction of the workforce, increase in operating costs, decrease in the life cycle of products and services, and rapid changes in market demands [13].

However, this phenomenon is not restricted to the industrial environment since changes in social demands and technological developments are factors that trigger significant changes in other dimensions of social dynamics as they bring with them new problems and challenges for increasingly overburdened urban centers and busy with the achievement of their ideals of sustainability that, not infrequently, are opposed to the choice of the correct approaches to the solution of their problems [23, 24].

In a scenario of sudden changes and great pressures, the key to meeting social demands seems to depend directly on the ability of municipalities to process public and private information in real time [25]. In this scenario, smart cities present themselves as a paradigm shift when compared to traditional city models by employing sensors, the Internet of Things, Big Data, cloud computing, and artificial intelligence as technological solutions in the design and offering of a more efficient smart public services, being flexible, accountable, transparent, and citizen-oriented [17–21].

Despite this, the experience recorded in the literature shows that it is foolhardy to guide the formulation or reformulation of public policies for smart cities solely on the implementation of technological solutions, without taking into account the particular characteristics of each of the stakeholders, with special emphasis on human factors and environmental issues involved [19, 21, 25]. In this sense, the proposition of long-term public policies for smart cities should emphasize a two-way approach to the governance of urban centers, with the participation of government representatives and citizens in the establishment of resource allocation priorities [5, 20, 25].

In emerging countries, such as Brazil, the reflection on the identification of objectives and selection of the correct approach to face the challenges must be prioritized, without prejudice to the increase of investments in the field of enabling technologies of Industry 4.0, whose dimensions, in scale and scope, must be duly clear, with regard to their importance, for the stakeholders responsible for the formulation of public policies for smart cities, above all, for those who participate in the discussions in the political sphere that, by definition, should protect public resources from improper application in smart-washed cities projects that only offer superficial solutions to urban problems [26].

Acknowledgments Special thanks to the Master's Program in Production Engineering at the Federal University of Amazonas for the support of its professors for the research carried out. This work was conducted in partnership with the Federal University of Amazonas (UFAM) within the scope of the Master's Program in Production Engineering (PPGEP).

The authors thank the project partner Amazon State Foundation for Research Support (FAPEAM – Fundação de Apoio à Pesquisa do Estado do Amazonas) for supporting this work under POSGRAD Project (Resolução no. 008/2021-POSGRAD).

References

1. Oztemel, E., & Gursev, S. (2020). Literature review of Industry 4.0 and related technologies. *Journal of Intelligent Manufacturing, 31*(1), 127–182.
2. Kagermann, H., Wahlster, W., & Helbig, J. (2013). *Recommendations for implementing the strategic initiative Industrie 4.0: Securing the future of German manufacturing industry; final report of the Industrie 4.0 Working Group*. Forschungsunion.
3. Petrillo, A., De Felice, F., Cioffi, R., & Zomparelli, F. (2018). *Digital transformation in smart manufacturing*. Intech.
4. Schwab, K. (2016). *A Quarta Revolução Industrial*. Edipro.
5. Guimarães, D. S., & Sant'Anna, C. H. (2019). Iniciativas de desenvolvimento da indústria 4.0 no Brasil: políticas públicas e perspectivas. *Boletim do Tempo Presente, 08*(02), 13–24.
6. Vermulm, R. (2018). *Políticas para o desenvolvimento da indústria 4.0 no Brasil*. Instituto de Estudos para o Desenvolvimento Industrial.
7. Chukwuekwe, D. O., Schjolberg, P., Rodseth, H., & Stuber, A. (2016). Reliable, robust and resilient systems: Towards development of a predictive maintenance concept within the industry 4.0 environment. In: *Athens: EFNMS Euro Maintenance Conference*, Athens.

8. Muhudi, P. K., Shukla, A. K., & Abraham, A. (2019). Industry 4.0: A bibliometric analysis and detailed overview. *Engineering Applications of Artificial Intelligence, 78*, 218–235.
9. Gilchrist, A. (2016). *Industry 4.0: The industrial internet of things*. Apress.
10. Da Costa, M. B., Santos, L. M., Schaefer, J. L., Baierle, I. C., & Nara, E. O. (2019). Industry 4.0 technologies basic network identification. *Scientometrics, 121*(2), 977–994.
11. Frank, A. G., Dalenogare, L. S., & Ayala, N. F. (2019). Industry 4.0 technologies: Implementation patterns in manufacturing companies. *International Journal of Production Economics, 210*, 15–26.
12. Lima, R. R., & Russo, S. L. (2019). *Políticas públicas e ações estratégicas para inovação no Brasil na era da indústria 4.0*. 10th International Symposium on Technological Innovation, pp. 264–273, Aapi, Aracajú.
13. Kuo, C. C., Shyu, J. Z., & Ding, K. (2019). Industrial revitalization via industry 4.0: A comparative policy analysis among China, Germany and the USA. *Global Transitions, 1*(1), 3–14.
14. Tranfield, D., Denyer, D., & Smart, P. (2003). Towards a methodology for developing evidence-informed management knowledge by means of systematic review. *British Journal of Management, 14*, 207–222.
15. Cook, D. J., Mulrow, C. D., & Haynes, R. B. (1997). Systematic reviews: Synthesis of best evidence for clinical decisions. *Annals of Internal Medicine, 126*(5), 376–380.
16. Falagas, M. E., Pitsouni, E. i., Malietzis, G. A., & Pappas, G. (2008). Comparison of PubMed, Scopus, Web of Science, and Google Scholar: Strengths and weakness. *The FASEB Journal, 22*(2), 338–342.
17. Ivan, L., Beu, D., & Van Hoof, J. (2020). Smart and age-friendly cities in Romania: An overview of public policy and practice. *International Journal of Environmental Research and Public Health, 17*(14), 5202.
18. Javed, B., Khan, Z., & McClatchey, R. (2018). An adaptable system to support provenance management for the public policy-making process in smart cities. *Informatics, 5*(1), 3.
19. Law, K. H., & Lynch, J. P. (2019). Smart city: Technologies and challenges. *IT Professional, 21*(6), 46–51.
20. Xin, B., & Qu, Y. (2019). Effects of smart city policies on green total factor productivity: Evidence from a quasi-natural experiment in China. *International Journal of Environmental Research and Public Health, 16*(13), 2396.
21. Lee, J. W. (2020). Big data strategies for government, society and policy-making. *Journal of Asian Finance, Economics and Business, 7*(7), 475–487.
22. Ali, M. A. (2021). Smart city policy in developing countries: Case study of the new administrative capital in Egypt. *Journal of Public Affairs, e2774*, 1–16.
23. Anand, P. B. (2021). Assessing smart city projects and their implications for public policy in the global south. *Journal of the Academy of Social Sciences, 16*(2), 199–212.
24. Gonçalves, G. L., Leal Filho, W., Neiva, S. S., Deggau, A. B., Veras, M. O., Ceci, F., Lima, M. A., & Guerra, J. B. S. O. A. (2021). The Impacts of the fourth industrial revolution on smart and sustainable cities. *Sustainability, 13*(13), 7165.
25. Kim, J. S., & Choi, H. Y. (2021). Analysis of core technologies and implementation services for smart city. *Turkish Journal of Computer and Mathematics Education, 12*(5), 278–285.
26. Mills, D., Pudney, S., Pevcin, P., & Dvorak, J. (2022). Evidence-based public policy decision-making in smart cities: Does extant theory support achievement of city sustainability objectives? *Sustainability, 14*(1), 3.

Impact of TQM and Industry 4.0 on Sustainable Performance: An Empirical Study on the Bangladeshi Garment Industry

Palash Saha, Subrata Talapatra, José Carlos Sá, and Gilberto Santos

Abstract The purpose of this research is to analyze the structural relationship between total quality management (TQM), Industry 4.0 (I4.0), and sustainability performance in the ready-made garments (RMG) sector of Bangladesh. An online survey was used to collect data from 240 employees working in ready-made garments located in Bangladesh. The data was analyzed employing structural equation modeling (SEM) to determine whether TQM practices and I4.0 technologies increase organizational sustainability performance. The findings indicate that both TQM and I4.0 positively contribute to increasing sustainability performance. Additionally, I4.0 technologies also positively contribute to a successful TQM implementation. This research presents a research framework for other manufacturing sectors where TQM approaches and I4.0 technologies can be implemented. This work makes a significant addition by addressing inadequacies in the existing literature. Very few empirical studies were conducted to examine the probable relationship between TQM, I4.0, and sustainability performance in the RMG sector, particularly in developing nations such as Bangladesh, where TQM and I4.0 adaptation are still in their early phases.

Keywords Total quality management · Sustainability performance · Industry 4.0 · Ready-made garment sector · Structural equation modeling

P. Saha · S. Talapatra
Khulna University of Engineering & Technology, Khulna, Bangladesh
e-mail: palash@iem.kuet.ac.bd; subrata@iem.kuet.ac.bd

J. C. Sá
ISEP – School of Engineering, Polytechnic of Porto, Porto, Portugal
e-mail: cvs@isep.ipp.pt

G. Santos (✉)
Polytechnic Institute Cavado Ave, Barcelos, Portugal
e-mail: gsantos@ipca.pt

J. C. de Oliveira Matias et al. (eds.), *Quality Innovation and Sustainability*, Springer Proceedings in Business and Economics,
https://doi.org/10.1007/978-3-031-12914-8_11

1 Introduction

In today's age, our world is experiencing sustainability difficulties in all three dimensions: economic, environmental, and social [1]. Unfortunately, over 1 billion people are living in extreme poverty, and economic disparity is increasing. The escalation of the problem is mostly due to unsustainable consumption and production, which is causing major socioeconomic crises and posing a threat to life on earth [2]. Greater customer demand leads to higher items' manufacture, which leads to increased production and output. This system is likely to create severe ecological damage, primarily through the use of nonrenewable inputs, as well as faster, non-sustainable consumption; and as a result of higher manufacturing emissions and improper disposal of life products [3]. The only way to solve these issues is to follow the road of sustainable improvement [4]. The goal of achieving sustainable performance in businesses is to decrease the negative impact that production, services, and logistics have on the ecosystem [5]. Bangladesh has indeed risen to the top of global apparel manufacturing. To gain business competitive advantages over competitors and decrease the negative impact of production and logistics on the ecosystem by obtaining sustainability, Bangladeshi garment authorities should take into consideration the impact of total quality management (TQM) and Industry 4.0 (I4.0) implementation. Depending on the market scenario, the research on the impact of TQM and I4.0 technologies on the sustainability performance in the Bangladeshi ready-made garments (RMG) industry has become critical as it would make it easier for the garment authorities to take decisions regarding TQM and I4.0 implementation. This is the key motivation for this research.

Previously, researchers had provided the theoretical framework of the impact of TQM and I4.0 technologies on sustainability performance separately [6, 7]. But no single study has integrated these two approaches empirically yet with respect to sustainability performance. Therefore, no single study has investigated the relationship of TQM, I4.0, and sustainability performance in developing countries like Bangladesh. As a result, in order to fill these research gaps, this study aims to answer the following research questions:

RQ1: What are the distinct impacts of TQM and I4.0 on sustainable performance?
RQ2: What is the impact of I4.0 technologies on TQM implementation?

This study developed a theoretical model to answer these research questions. Then to answer these research questions, this study formulates research hypotheses that are then tested with survey data collected from the garment industries located in Bangladesh using the partial least–square structural equation modeling (PLS-SEM) technique.

The remainder of the chapter is organized as follows. Section 2 presents a brief discussion of the theoretical background. Section 3 presents the theoretical model and hypothesis development to answer the research questions. Sections 4 and 5 present research methodology and data analysis using structural equation modeling (SEM). Then Sect. 6 presents the findings of this empirical research. Finally, theoretical and practical implications of this study, limitations, and future scopes are also presented.

2 Literature Review

2.1 TQM and Sustainability Performance

There are very few studies that empirically investigated the relationship between TQM and sustainability performance; here, the number of studies investigating the relationship between TQM and organizational performance or operational performance is widespread [8]. However, while many pieces of research have found a substantial impact of TQM on organizational performance [9], other studies have not found any such impact [10, 11]. In addition, Soltani et al. [12] suggested any noticeable results have not been gained through TQM in most organizations. Furthermore, some studies have investigated the relationship between TQM and sustainable performance. Most of these studies found that TQM has a substantial and positive impact on the sustainable performance of an organization [13]. According to Androwis et al. [14], TQM procedures are the most effective approach to enhance organizational performance across competitive advantage dimensions, including development, delivery and lead time, cost, and pricing. As a result of these improved competitive advantages, optimal sustainability performance may be achieved. On the other hand, Tasleem et al. [6] suggested that TQM does not only positively impact sustainable performance but also helps to improve each dimension of sustainable performance (economic, social, and environmental). Despite some varying results on the impact of TQM's key enabling factors on organizational performance and a paucity of research on TQM and sustainability performance, available literature suggests that TQM application adoption can positively affect sustainability performance [13]. Now, on the basis of the above arguments of TQM and sustainable performance, the next hypothesis is proposed:

H1: TQM positively impacts sustainable performance.

2.2 Industry 4.0 and Sustainability Performance

A few studies focused on the relationship between Industry 4.0 technologies and sustainability performance, namely, with respect to social, economic, and environmental perspectives [15]. Most of these studies provide insight into the effect of Industry 4.0 technologies on the sustainable development of any organization [16, 17]. Some of these studies have conceptualized specific industrial sustainability challenges, such as the circular economy [18]. In addition, some of these studies emphasize the relevance of sociotechnical factors for technology deployment in order to improve organizational sustainability [19]. Overall, most of the researchers are trying to explore the potential positive impact of Industry 4.0 technologies on the sustainable performance of an organization [20]; environmental sustainability can be improved by integrating the Industry 4.0 technologies and Sustainable Development Goals by providing ecological assistance that ensures excellent environmental performance with a greater positive effect than previously achieved. Haseeb et al. [21] suggested that Industry 4.0 plays a key role in the growth of sustainability. In addition, Kamble et al. [22] established that Industry 4.0 has a direct influence on organizations' sustainability, affecting all the economic, environmental, and social pillars. However, several studies bring significant concerns about the possible negative effects of Industry 4.0 on each of the sustainability pillars [17]. Some experts, for example, have raised concerns about the effect of Industry 4.0 adoption on social factors such as employment availability and work, health, and safety of employees [23, 24]. In summary, despite having some anomalies, Industry 4.0 technologies are likely to have a substantial impact on the sustainability performance of an organization [22, 25]. Now, based on the above about Industry 4.0 and sustainability performance, the following hypothesis is stated:

H2: Industry 4.0 positively impacts sustainability performance.

2.3 *Industry 4.0 in TQM Implementation*

Industry 4.0 has the ability to impact the best practices for applying total quality management concepts [26]. Following the ISO 9001:2015 model, the TQM practices are customer focus, leadership, engagement of people, process approach, improvement, evidence-based decision-making, and relationship management. Asif [27] suggested that in order to have a successful TQM principles implementation a business organization should focus on developing the integration of working personnel with I4.0 and using real-time control techniques. Illés et al. [28] emphasized how thanks to I4.0 technologies it is now feasible to obtain data for quality that was unattainable earlier. According to the authors, the difficulty is determining where, how, and what to gather, as well as how to evaluate the resulting massive data. Durana et al. [29] looked at how traditional TQM principles must adapt to I4.0 developments and challenges. The adoption of I4.0 and quality management, according to the authors, is strongly linked to the creation of a quality culture instead of a purely technological aspect. Along with highlighting the opportunities provided by I4.0, Zhou et al. [30] highlighted the contexts where the purpose of the TQM principles cannot be completely served by I4.0 adaptation for various challenges and complications of Industry 4.0. In summary, despite a paucity of studies on the impact of Industry 4.0 on TQM implementation and having some difficulties, Industry 4.0 plays a significant role in a successful TQM implementation [31, 32]. Now, based on the above discussions of Industry 4.0 and successful TQM implementation, the following hypothesis can be stated:

H3: Industry 4.0 has a positive impact on TQM implementation.

The three hypotheses in respect of TQM, sustainability performance, and Industry 4.0 are shown in Fig. 1.

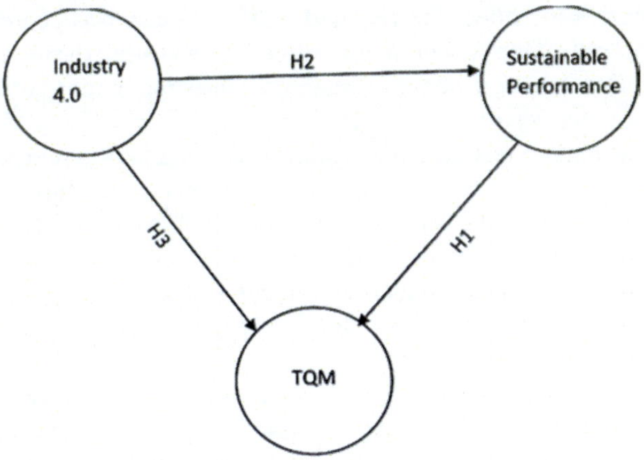

Fig. 1 The proposed research framework

3 Research Design

This empirical study was conducted using a questionnaire survey as a tool. The questionnaire was divided into four main parts. The first part contained general demographic pieces of information regarding respondents and their related industries. Four items (age, gender, designation, and the number of employees in the industry) were included in the demographic information section. The second part included information on Industry 4.0 technologies in use in the industry. The third and last parts were designed on required information about the TQM practices in use in the firm and the current status of the industry's sustainable performance, respectively. For all of the questions, respondents were requested to express their perceivable perspective on a five-point Likert scale, ranging from "1" (strongly disagree) to "5" (strongly agree).

Measurement of items of different variables was adapted from previous pieces of literature. TQM measurement items were adapted from Abbas [13] and Talapatra et al., [33]. Industry 4.0 measurement items were adapted from Machado et al. [15] and Bibby e Dehe [34]. Six items were used as sustainability performance indicators regarding social, economic, and environmental performance. These were adapted from Tasleem et al. [6] and Abbas [13]. A list of a total of 19 measured items was set for this study.

Bangladeshi ready-made garments (RMG) industries are the total population for this empirical study. Bangladesh has around 5000 export-oriented garments industries at the moment [35]. Taro Yamane [36] formula was used to calculate the sample size for this study. According to the calculations, a sample size of 355 is required to obtain a 95% confidence level. In recent literature, however, a sample size of 150 was shown to be sufficient in different fields of operations management [37]. The data was collected using random sampling techniques from various organizations across the country. Due to the Covid-19 pandemic, only electronic-based survey modes were applied to collect the required data. The data was acquired through email and phone calls. Don A. Dillman [38] method was used during the data collection process. Data was collected by sending out questionnaires to 600 employees who work in different departments of RMG industries, including production planning and control, human resource management, and marketing department. After 2 months, 220 complete and legible responses were selected for further study, with a 36.67% reply rate; hence, the response rate is acceptable and akin to other studies regarding the operations management field [39]. Therefore, the gathered responses can readily be used to test the hypothesis of this research [3]. The demographic information of the respondents, along with the response rate of each category, is shown in Table 1.

Every respondent's reply was sorted by receipt date and separated into early and late response sample groups. The *t*-test was performed to examine the statistical difference between early and late replies to see any nonresponse bias in the data [3]. The difference between the two response groups was measured using the wave analysis approach. On two sets of samples, a *t*-test was used to compare them. The test findings revealed no significant variations in the identification of TQM measures, Industry 4.0 measures, and sustainable performance measures between early and late replies. So, the data that were used for the model testing was comprised by merging the early and late responses.

4 Data Analysis

Numerous statistical techniques are available to confirm the relationship between Industry 4.0, TQM, and sustainability performance in the RMG sector in Bangladesh. In the previous research, structural equation modeling (SEM) and factor analysis techniques were broadly used [33]. Most of the scholars selected the first technique because of it being more sophisticated than the other [40]. Because of the capacity of combining factor analysis and multiple regression analysis, the SEM technique is regarded as a sophisticated technique [41]. Furthermore, there are two types of SEM techniques available: variance-based SEM and covariance-based SEM.

Table 1 Breakdown of respondents and their organization profile

Variables	Categories/class	Response (%)
Gender	Male	67.92
	Female	32.08
Age group	28–35 years	38.33
	36–45 years	61.67
Cadre	Top management	13.64
	Middle management	59.09
	Bottom management	27.27
Size of the organization	Small (<50 employees)	38.59
	Medium (50–500 employees)	61.41

Haseeb et al. [21] have effectively used this PLS-SEM technique to examine how Industry 4.0 technologies contribute to information technology implementation. Dubey et al. [42] have also used this technique to investigate the contribution of big data. The partial least squares (PLS) method is used in variance-based SEM. This study uses the PLS-SEM technique for analyzing collected data because of the following reasons:

(a) It is a successful technique to investigate and validate the link between the components in a complicated model.
(b) It can produce superior results independent of a sample or population size [41, 42].

Both scholars have preferred WrapPLS software for their research. This study also uses WrapPLS 7.0 software for analyzing data. Before analyzing, the data was standardized. There were no missing values and no data with a variance of zero. Likewise, the data also excluded rank-related issues. Moreover, for evaluating the feasibility of the collected data for SEM analysis, this research checked multicollinearity in the data set [13]. This study used the variance inflation factors (VIFs) to analyze the multicollinearity element. The result, illustrated in Table 4, shows that all the VIF values fully meet [41] the maximum requirement of 10. This study utilized three indices, as per [33], to determine whether the current measurement model had fitted with collected data: average block VIF (AVIF), average path coefficient (APC), and average R-squared (ARS). These variables were also used to assess the model's ability to explain the most significant variance. Table 2 shows that every one of these parameters' values is substantially within the allowable range. From the result, it can be said that the measurement model shows a great fit for collected data. It also reflects the capacity of the model better to retrieve variance.

Causality assessment was used as another way to verify the model's correctness. Three parameters were used in this research to check the model's correctness, which is consistent with [33]: Simpson's paradox ratio (SPR), R-squared contribution ratio (RSCR), and statistical suppression ratio (SSR). Table 3 shows that all the values of these parameters are within the allowable range.

Table 2 Model fix index and quality index

Indexes	Estimated value	Tolerable range
Average path coefficient (APC)	$p < 0.001$	$p < 0.001$
Average R^2 (ARS)	$p < 0.001$	$p < 0.001$
Average block VIF (AVIF)	1.583	Tolerable if $\leqslant 5$, best $\leqslant 3.3$

Table 3 Indexes of causality assessment

Indexes	Estimated value	Tolerable range
Simpson's paradox ratio (SPR)	1.000	Tolerable if $\geqslant 0.7$, best = 1
R^2 contribution ratio (RSCR)	1.000	Tolerable if $\geqslant 0.9$, best = 1
Statistical suppression ratio (SSR)	0.897	Tolerable if $\geqslant 0.7$

Table 4 Latent construct coefficients

	TQM	I4.0	SP
R^2 coefficients	0.240		0.353
Adjusted R^2 coefficients	0.210		0.331
Scale composite reliability (SCR)	0.889	0.765	0.831
Cronbach's α coefficients	0.873	0.760	0.944
Average variances extracted (AVE)	0.603	0.542	0.642
Variance inflation factors (VIFs)	3.363	6.472	4.581

Table 5 Outcomes of hypotheses assessment

Hypothesis for assessment	Estimate	Outcomes of assessment
H1: TQM positively impacts sustainable performance	$\beta = 0.54$ at $p < 0.01$	Supported
H2: Industry 4.0 positively impacts sustainability performance	$\beta = 0.30$ at $p < 0.01$	Supported
H3: Industry 4.0 has a positive impact on TQM implementation	$\beta = 0.37$ at $p < 0.01$	Supported

Moreover, to verify the reliability and consistency of the measurement scale, Cronbach's alpha (α) values were used. The Cronbach's alpha values for each latent variable are shown in Table 4, and it can be seen that all alpha (α) values considerably exceed the critical limit of 0.6, suggested by Molina et al. [43]. This result shows that the items have high internal consistency for the specified dimension. Combined and cross-loading values of each variable are also determined.

Then, to ensure the validity of the measurement model, convergent and discriminant validity tests were performed [13]. The convergent validity test was performed to check whether the observed perimeters were heavily loaded in their latent parent constructs. In order to obtain convergent validity, the coefficient values of the average variance extracted (AVE) and scale composite reliability (SCR) for each of the constructs must be greater than or equal to 0.5 and 0.7, according to Talapatra et al. [33]. All the values of AVE and SCR for each construct are shown in Table 4, and it shows that all the values are in an allowable range.

Again, to verify whether the latent constructs are distinctive and unrelated to each other, the discriminant validity test was performed. The discriminant validity test matrix is illustrated in Table 5. This table shows that no values in any column of the matrix beneath the diagonal element are higher than the diagonal value. It clearly demonstrates the discriminant validity of the model [44]. For testing the proposed hypotheses, the variance-based SEM approach was applied because of its ability to describe the maximum variance of latent variables [41]. WrapPLS 7.0 was used for analyzing the data.

Table 6 Discriminant validity test

	TQM	I4.0	SP
TQM	*0.776*		
I4.0	0.528	*0.736*	
SP	0.653	0.582	*0.801*

5 Discussion

This research was conducted to examine the impact of TQM and Industry 4.0 on sustainable performance along with the impact of I4.0 in TQM implementation in the RMG sector in Bangladesh. According to the result, the very first hypothesis was found to be statistically significant at the level of significance 0.01 ($p < 0.01$), having a β coefficient of 0.54. This hypothesis implies that the implementation of TQM principles makes a significant contribution to enhance the sustainability performance of an organization. TQM principles help an organization to increase customer satisfaction, reduce errors, and improve operational performance that directly contributes to increasing organizational sustainability. This finding confirms earlier research [13] (Table 6).

In terms of the second hypothesis, it is found that the hypothesis is statistically significant, having a β coefficient of 0.30 at the level of significance 0.01 ($p < 0.01$), suggesting that the adaptation of Industry 4.0 technologies has a positive impact on organizational sustainability performance. This result also implies that the use of I4.0 technologies encourages the organization to advance its operational processes and reduce energy consumption, waste, and pollution. Besides, the application of I4.0 technologies promotes reducing safety incidents and increasing employee morale, which all are directly related to the sustainability performance. This result also is confirmed by earlier studies [15, 45–48].

According to the result, the third hypothesis is also found to be statistically significant with a β coefficient of 0.37 ($p < 0.01$). This hypothesis recommends that the adoption of I4.0 technologies plays a substantial role in TQM implementation. Under an interconnected I4.0-TQM regimen, customer requirements and marketplace analysis would be directly transmitted to the manufacturing systems, and the quality of the product would be monitored and ensured utilizing advanced sensors and failure investigations analysis. Besides, Industry 4.0 could serve as a foundation for an organization's continuous development at both the product and process levels, which all are directly related to a successful TQM implementation [49–51]. This finding is also consistent with the previous research [26, 29, 52, 53].

6 Conclusion

The findings indicate that TQM practices and adoption of I4.0 technologies have a great impact on organizational sustainability performance in the RMG sector in Bangladesh.

Practically, a better knowledge of the possible association of TQM and I4.0 technologies on sustainability performance can assist practitioners, administrators, and policymakers, particularly in developing countries, like Bangladesh, in establishing appropriate expectations during the implementation process. Simply said, TQM practices and I4.0 technologies can assist the business organization in obtaining a competitive advantage that distinguishes them from their competitors and improves their market presence. Furthermore, the recognition of the synergistic linkage with combining effects assists executives in anticipating organizational sustainability improvement difficulties, avoiding wasted decisions and ineffective expenditures. This research also reveals that the positive impacts of TQM and I4.0 are not restricted to developed countries-based organizations; if organizations in developing countries use their methods in an effective manner, similar outcomes may be obtained.

Without a doubt, there are certain limitations to this study. For instance, the sample considered for this research was confined to 240 participants from the RMG sector in Bangladesh. Despite the fact that the sample size was enough for the analysis, a higher sample size would be beneficial to confirm the generalizability of this study. In addition, all the respondents regarding the data sample were from Bangladeshi RMG sectors that confine the generalization of this research. The addition of respondents from different sectors or countries to the data collection would enrich the sample both quantitatively and qualitatively. Finally, the number of dimensions of the latent variable used in this study could be added to the current measurement model for future studies.

References

1. Bulkeley, H., & Betsill, M. (2005). Rethinking sustainable cities: Multilevel governance and the 'Urban' politics of climate change. *Environmental Politics, 14*(1), 42–63.
2. UNIDO Homepage. https://www.unido.org/resources-publications-flagship-publications-industrial-development-report-series/industrial-development-report-2018. Last accessed 13 Feb 2022.
3. Dubey, R., Bag, S., & Ali, S. (2014). Green supply chain practices and its impact on organisational performance: An insight from Indian rubber industry. *International Journal of Logistics Systems and Management, 19*(1), 20–42.
4. Dubey, R., & Bag, S. (2013). Exploring the dimensions of sustainable practices: An empirical study on Indian manufacturing firms. *International Journal of Operations and Quantitative Management, 19*(2), 123–146.
5. Jabbour, L., Jabbour, C., Foropon, C., & Filho, M. (2018). When titans meet – Can industry 4.0 revolutionise the environmentally-sustainable manufacturing wave? The role of critical success factors. *Technological Forecasting and Social Change, 132*, 18–25.
6. Tasleem, M., & Khan, N. (2019). Impact of technology management on corporate sustainability performance. *International Journal of Quality & Reliability Management, 36*(9), 1574–1599.
7. Belhadi, A., Kamble, S., Gunasekaran, A., & Mani, V. (2021). Analyzing the mediating role of organizational ambidexterity and digital business transformation on industry 4.0 capabilities and sustainable supply chain performance. *Supply Chain Management: An International Journal.*, (ahead-of-print).

8. Rebelo, M., Santos, G., & Silva, R. (2015). Integration of standardized management systems: A dilemma? *Systems, 3*, 45–59.
9. Khan, R., Mirza, A., & Khushnood, M. (2020). The role of total quality management practices on operational performance of the service industry. *International Journal for Quality Research, 14*(2), 418–425.
10. Wynen, J., & Verhoest, K. (2016). Quality management in public-sector organizations: Evidence from six EU countries. *International Journal of Public Administration, 39*(2), 122–134.
11. Khalfallah, M., Salem, B., Zorgati, H., & Lakhal, L. (2021). Innovation mediating relationship between TQM and performance: Cases of industrial certified companies. *The TQM Journal*, (ahead-of-print).
12. Soltani, E., Meer, R., Williams, M., & Lai, C. (2006). The compatibility of performance appraisal systems with TQM principles – Evidence from current practice. *International Journal of Operations & Production Management, 26*(1), 92–112.
13. Abbas, J. (2020). Impact of total quality management on corporate sustainability through the mediating effect of knowledge management. *Journal of Cleaner Production, 244*, 118806.
14. Androwis, N., Sweis, J., Tarhini, A., Moarefi, A., & Amiri, H. (2018). Total quality management practices and organizational performance in the construction chemicals companies in Jordan. *Benchmarking: An International Journal, 25*(8), 3180–3205.
15. Machado, G., Winroth, P., & Silva, D. (2020). Sustainable manufacturing in Industry 4.0: An emerging research agenda. *International Journal of Production Research, 58*(5), 1462–1484.
16. Bag, S., & Pretorius, C. (2020). Relationships between industry 4.0, sustainable manufacturing and circular economy: Proposal of a research framework. *International Journal of Organizational Analysis*, (ahead-of-print).
17. Nara, E. O. B., da Costa, M. B., et al. (2021). Expected impact of industry 4.0 technologies on sustainable development: A study in the context of Brazil's plastic industry. *Sustainable Production and Consumption, 25*, 102–122.
18. Ramos, D., Fonseca, L., Gonçalves, G., Carvalho, R., Carvalho, S., & Santos, G. (2022). Cost-benefit analysis of implementing circular economy in a Portuguese company: From a case study to a model. *Quality Innovation Prosperity, 26*(1), 52–69.
19. Beier, G., Ullrich, A., Niehoff, S., Reißig, M., & Habich, M. (2020). Industry 4.0: How it is defined from a sociotechnical perspective and how much sustainability it includes–A literature review. *Journal of Cleaner Production, 259*, 841–856.
20. Bai, C., Dallasega, P., Orzes, G., & Sarkis, J. (2020). Industry 4.0 technologies assessment: A sustainability perspective. *International Journal of Production Economics, 229*, 107–118.
21. Haseeb, M., Hussain, I., Ślusarczyk, B., & Jermsittiparsert, K. (2019). Industry 4.0: A solution towards technology challenges of sustainable business performance. *Social Sciences, 8*(5), 154.
22. Kamble, S., Gunasekaran, A., & Dhone, C. (2020). Industry 4.0 and lean manufacturing practices for sustainable organisational performance in Indian manufacturing companies. *International Journal of Production Research, 58*(5), 1319–1337.
23. Ghobakhloo, M. (2020). Industry 4.0, digitization, and opportunities for sustainability. *Journal of Cleaner Production, 252*, 119869.
24. Yadav, G., Luthra, S., Jakhar, K., Mangla, S., & Rai, P. (2020). A framework to overcome sustainable supply chain challenges through solution measures of industry 4.0 and circular economy: An automotive case. *Journal of Cleaner Production, 254*, 120112.
25. Murmura, F., Bravi, L., & Santos, G. (2021). Sustainable process and product innovation in the eyewear sector: The role of industry 4.0 enabling technologies. *Sustainability, 13*, 1–16.
26. Souza, F., Corsi, A., Pagani, N., Balbinotti, G., & Kovaleski, L. (2021). Total quality management 4.0: Adapting quality management to Industry 4.0. *The TQM Journal*, ahead-of-print.
27. Asif, M. (2020). Are QM models aligned with Industry 4.0? A perspective on current practices. *Journal of Cleaner Production, 258*, 120–132.

28. Illés, B., Tamás, P., Dobos, P., & Skapinyecz, R. (2017). New challenges for quality assurance of manufacturing processes in industry 4.0. *Solid State Phenomena, 261*, 481–486, Trans Tech Publ.
29. Durana, P., Kral, P., Stehel, V., Lazaroiu, G., & Sroka, W. (2019). Quality culture of manufacturing enterprises: A possible way to adaptation to Industry 4.0. *Social Sciences, 8*(4), 124.
30. Zhou, K., Liu, T., & Zhou, L. (2015). Industry 4.0: Towards future industrial opportunities and challenges. In *12th International conference on fuzzy systems and knowledge discovery (FSKD)*. IEEE.
31. Talapatra, S., Santos, G., Udin, K., & Carvalho, F. (2019). Main benefits of integrated management systems through literature review. *International Journal for Quality Research, 13*(4), 1037–1054.
32. Carvalho, F., Santos, G., & Gonçalves, J. (2020). Critical analysis of information about integrated management systems and environmental policy on the Portuguese firms' website, towards sustainable development. *Corporate Social Responsibility and Environmental Management, 27*, 1069–1088.
33. Talapatra, S., Uddin, K., Antony, J., Gupta, S., & Cudney, A. (2020). An empirical study to investigate the effects of critical factors on TQM implementation in the garment industry in Bangladesh. *International Journal of Quality & Reliability Management, 37*(9/10), 1209–1232.
34. Bibby, L., & Dehe, B. (2018). Defining and assessing industry 4.0 maturity levels – case of the defence sector. *Production Planning & Control, 29*(12), 1030–1043.
35. BGMEA Homepage. https://www.bgmea.com.bd/page/Export_Performance. Last accessed 11 Feb 2022.
36. Taro Homepage. https://www.worldcat.org/title/statistics-an-introductory-analysis-2nd-ed/oclc/299725866. Last accessed 13 Feb 2022.
37. Aamer, A., Awlaqi, & Alkibsi, M. (2017). TQM implementation in a least developed country: An exploratory study of Yemen. *The TQM Journal, 29*(3), 467–487.
38. Dillman Homepage. https://sesrc.wsu.edu/about/total-design-method/. Last accessed 11 Feb 2022.
39. Baidoun, D., Salem, Z., & Omran, A. (2018). Assessment of TQM implementation level in Palestinian healthcare organizations. *The TQM Journal, 30*(2), 98–115.
40. Alharbi, K., Matari, E., & Yusoff, Z. (2016). The impact of Total Quality Management (TQM) on organisational sustainability: The case of the hotel industry in Saudi Arabia: Empirical study. *The Social Sciences, 11*(4), 3468–3473.
41. Gupta, S., Drave, A., Bag, S., & Luo, Z. (2019). Leveraging smart supply chain and information system agility for supply chain flexibility. *Information Systems Frontiers, 21*(3), 547–564.
42. Dubey, R., Gunasekaran, A., & Ali, S. (2015). Exploring the relationship between leadership, operational practices, institutional pressures and environmental performance: A framework for green supply chain. *International Journal of Production Economics, 160*, 120–132.
43. Molina, L. M., Lloréns-Montes, J., et al. (2007). Relationship between quality management practices and knowledge transfer. *Journal of Operations Management, 25*(3), 682–701.
44. Fornell, C., & Larcker, D. F. (1981). Evaluating structural equation models with unobservable variables and measurement error. *Journal of Marketing Research, 18*(1), 39–50.
45. Oláh, J., & Aburumman, N. (2020). Impact of industry 4.0 on environmental sustainability. *Sustainability, 12*(11), 46–74.
46. Santos, G., Rebelo, M., Barros, S., Silva, R., Pereira, M., Ramos, G., & Nuno Lopes, N. (2014). Chapter 6: Developments regarding the integration of the occupational safety and health with quality and environment management systems. In I. Kavouras & M.-C. G. Chalbot (Eds.), *Occupational safety and health – Public health in the 21st century* (pp. 113–146). The Nova Science Publishers. isbn:978-1-63117-698-2. (eBook).
47. Santos, D., Rebelo, M., & Santos, G. (2017). The integration of certified Management Systems. Case Study Organizations located at the district of Braga, Portugal. *Procedia Manufacturing, 13*, 964–971.

48. Bravi, L., Santos, G., Pagano, A., & Murmura, F. (2020). Environmental management system according to ISO 14001:2015 as a driver to sustainable development. *Corporate Social Responsibility and Environmental Management*, 1–16.
49. Costa, A. R., Barbosa, C., Santos, G., & Alves, M. R. (2019). Six sigma: Main metrics and R based software for training purposes and practical industrial quality control. *Quality Innovation Prosperity Journal, 23*(2), 83–100.
50. Sá, J., Vaz, S., Carvalho, O., Lima, V., Fonseca, L., Morgado, L., & Santos, G. (2020). A model of integration ISO 9001 with Lean Six Sigma and main benefits achieved. *Total Quality Management and Business Excellence*., 1–25, *33*(1–2), 218–242.
51. Zgodavova, K., Bober, P., Majstorovic, V., Monkova, K., Santos, G., & Juhaszova, D. (2020). Innovative methods for small mixed batches production system improvement: The case of a bakery machine manufacturer. *Sustainability (Switzerland), 12*(15), 1–20.
52. Santos, G., Sá, J. C., Félix, M. J., Barreto, L., Carvalho, F., Doiro, M., Zgodavová, K., & Stefanović, M. (2021). New needed quality management skills for quality managers 4.0. *Sustainability, 13*, 6149.
53. Santos, G., Doiro, M., Félix, M. J., Mandado, E., Sá, J. C., Gonçalves, J., & Teixeira, P. (2020). On the concept of an integrated and lean model of product development proposed for intellectual property creation and competitive economies. *International Journal of Intellectual Property Management, 10*(4), 409–435.

Academic Community Perceptions of Open Innovation: An Exploratory Study

Tiago Rodrigues-Sa and Manuel Au-Yong-Oliveira

Abstract This study seeks to assess the academic community's knowledge and perception of open innovation. While the so-called "closed innovation" is rooted in the discourse of future managers, open innovation, being a relatively recent paradigm, may be an unknown form of innovation and therefore its potential is underused. What is the perception of open innovation among students and faculty at a private Portuguese university? The technical procedure was based on a survey, and 64 responses were obtained (the population corresponds to 3666 people). It was shown that the theme of open innovation, despite being recognized in the academic world, is still little known in the community. Although Portugal is a country of meager funds for innovation, which should lead to a greater focus on open innovation, this may not be happening due to the lack of trust in strangers that exists culturally in Portugal, and that may be currently hindering open innovation partnerships. Finally, in order to identify possible relationships between gender and perceptions regarding open innovation, we applied the chi-square test of independence (X^2) in relation to gender. This exploratory study verified the existence of gender equality regarding the variables analyzed on open innovation.

Keywords Academic community · Open innovation · Perceptions · Higher education

T. Rodrigues-Sa (✉)
REMIT – Research on Economics, Management and Information Technologies, Universidade Portucalense, Porto, Portugal

M. Au-Yong-Oliveira
INESC TEC, GOVCOPP, Department of Economics, Management, Industrial Engineering and Tourism, University of Aveiro, Aveiro, Portugal
e-mail: mao@ua.pt

J. C. de Oliveira Matias et al. (eds.), *Quality Innovation and Sustainability*, Springer Proceedings in Business and Economics,
https://doi.org/10.1007/978-3-031-12914-8_12

1 Introduction

We live in an increasingly global and dynamic world, where ever-shorter innovation cycles and lower R&D costs are required [1]. Given the characteristics of technologies, the existence of complexity, and the need to respond to market needs [2], as well as due to scarce resources [3], the open innovation strategy is emerging, recognizing that not all good ideas derive from within the organization and that not all can be commercialized internally [4]. In this context, open innovation has progressively asserted itself as an approach to master innovation, allowing to save and solve the time and money problems of the innovation process, contributing to the maintaining of competitive advantage [5]. Companies that do not focus on an open innovation strategy will ultimately fail as rising development costs, as well as shorter product life cycles, make it increasingly difficult to justify investments in innovation [1, 6].

In 1977, von Hippel [7] presented for the first time a form of innovation stating that ideas could come from outside the organization, but concretely from the so-called lead users.

Albeit the concept was later popularized with the work of Chesbrough, Vanhaverbeke [8], which defines open innovation (OI) as "open innovation is the use of purposive inflows and outflows of knowledge to accelerate internal innovation, and expand the markets for external use of innovation, respectively (…) is a paradigm that assumes that firms can and should use external ideas as well as internal ideas (…) combine internal and external ideas into architectures and systems. Assumes that internal ideas can also be taken to market through external channels, outside the current businesses of the firm, to generate additional value. Open innovation suggests that valuable ideas can come from inside or outside the company and can go to market from inside or outside the company as well."

Open innovation proposes a democratic innovation process, given its influence on the acquisition of transferring capacities, resources, and technologies between the various organizations, allowing organizations of different sizes to compete against each other [9]. The knowledge of the inventive process becomes endless, not distinguishing a small organization from a bigger one, a young one from a mature one, a technologically advanced one from a more primitive one. This democratization of knowledge, i.e., the possibility of access for all, is the most remarkable feature of this contemporary innovation strategy. The motivation to understand how this innovation paradigm is rooted in the culture of academia encouraged the investigation of this study. The main objective of this study is to investigate the knowledge of an academic community about the type of contemporary innovation, open innovation, and more specifically to understand the degree of perception and importance that students linked to the areas of management (future managers and agents of change) attribute to open innovation.

A literature review is presented in the following section. Section 3 identifies the methodology of the work carried out; Sect. 4 presents the main results and main conclusions; Sect. 5 discusses the results, and, finally, Sect. 6 presents the main conclusions of the work and suggests future research tracks.

2 Literature Review

When performing the search in Scopus for publications, whose title contained the word "Open Innovation," 3161 documents were obtained. Restricting the search to the type of document "article," published in the English language and thematic "Management," "Business," and "Accounting," 1121 articles were obtained. Continuing the search, with the aim of obtaining a more specific sample to the theme under analysis, the search was refined based on the same filters, but now with the specificity for articles that encompassed "Open Innovation" and "perception" in the title. Based on these criteria, seven articles were gathered (Fig. 1).

In this sample, the oldest article is from 2010 and the most recent from 2022. Table 1 summarizes the research results, presenting the main contributions and focus of the articles analyzed.

By analyzing Table 1, we can see that the analysis of perceptions regarding the business world continues to dominate current scientific studies [12, 13, 15, 16], and there is a pathway to understand and explore the academic community's knowledge of open innovation. Previous studies can be divided on the basis of perceptions about open innovation: focus on employees' perceptions [12, 13], corporate perception [15], managers' perceptions [16], academic perception of knowledge

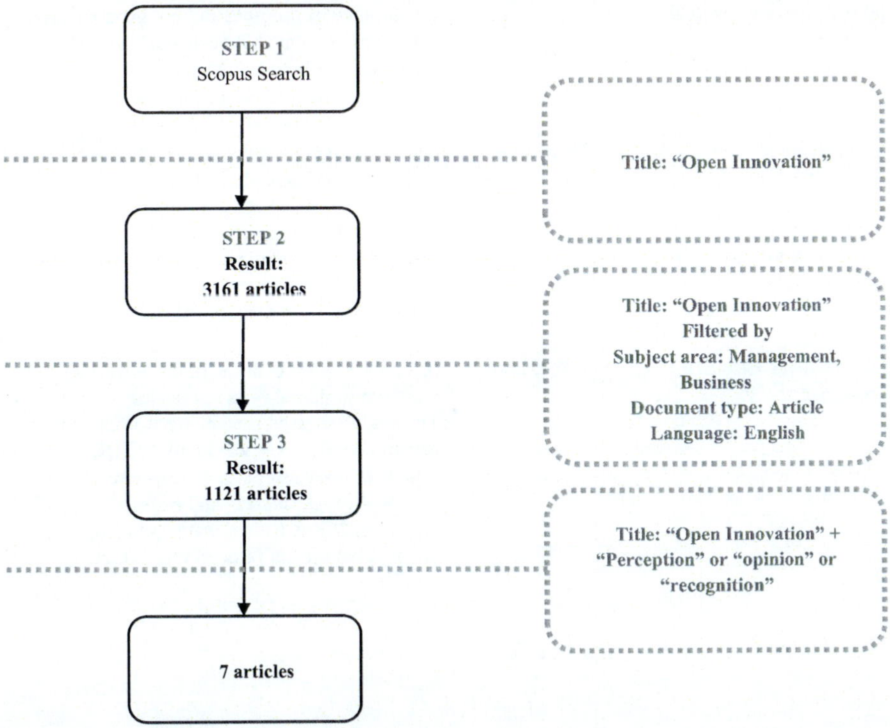

Fig. 1 Research protocol

Table 1 Analysis of Scopus articles based on the search for the title "Open Innovation" and "perception"

Title	Author	Focus of perception	Key findings
Strategic management of the Malaga brand through open innovation: tourists and residents' perception	[10]	Tourists and residents' perception	The results allow us to identify that the Malaga brand is being built and managed based on the cultural projection of the city, in which the ratings of tourists and residents serve as a basis for improving the management of Malaga as a tourist destination. Public entities that wish to transform their value creation and service delivery in a sustainable way should maintain a productive relationship with other public authorities and other external parties; exchange knowledge, skills, and experiences easily and securely with others to improve internal processes and deliver services to citizens and engage with citizens and other stakeholders to co-create new services
Higher education response in the time of coronavirus: perceptions of teachers and students, and open innovation	[11]	Teachers' and students' perceptions	The main objective of this work was to analyze how universities have managed the flow of knowledge during the pandemic situation. The results obtained showed that the absence of presence did not generate an increase in contact between teachers and students. Teachers and students showed a preference for the face-to-face method but recognize the potential of digital media
The intensity of organizational change and the perception of organizational innovativeness; with discussion on open innovation	[12]	Employee perceptions	This study assessed the relationship between investment in human resources (HR) and employees' perceptions of innovation. It sought to determine how the intensity of organizational change affects the relationship between HR investment and employees' perceptions of organizational innovation
Perceptions of open innovation at CERN: an explorative study	[13]	Employee perceptions	This study was designed to assess the perception of innovation, of employees of intergovernmental research institutes The study assessed how intergovernmental research institutes, specifically CERN, establish innovation as a driving force, mainly through internal and external openness. It was shown that most recognize the positive impact of open innovation

(continued)

Table 1 (continued)

Title	Author	Focus of perception	Key findings
What does open innovation mean? Business versus academic perceptions	[14]	Perception of the academic and business world	This paper compared the perceptions of innovation that exist academically and through business. The study showed that there are differences in the interpretation of open innovation between companies of different sizes. The results provide evidence of the issue of contextual ambiguity and dualism surrounding the OI concept
Mapping the perception and reality of open innovation	[15]	Companies' perceptions	This study assessed companies' perceptions of their degree of openness. As a result of the study, it was validated that companies' perceptions of their own openness differ from their actual situation and furthermore, each company has a different view on open innovation
Open innovation in secondary software firms: an exploration of managers' perceptions of open source software	[16]	Perceptions of managers	This paper examines how managers' perceptions of the benefits and drawbacks of open source software (OSS) affected the decision to adopt an open source policy in their companies The study reveals how perceptions about the business and technical benefits and disadvantages of OSS influenced the technological, organizational, environmental, and individual factors considered in the adoption process

management [11] and mixed academic and business perception studies [14], and focus on the perceptions of tourists and residents [10].

While also focusing on the perceptions of teachers and students, the article by Tejedor, Cervi [11] focuses on the perception of knowledge management in the context of the Covid-19 pandemic and not on knowledge about open innovation specifically. Only the article from Teplov, Albats [14] reflects an analysis on the perception of academics while continuing to focus mostly on the business world. This study has shown that there is a difference in the interpretation of open innovation, corroborating one of the results of the study of Dabrowska, Fiegenbaum [15].

3 Methodology

The main objective of this work is to assess the academic community's perception of open innovation and how it is rooted/familiarized in future professionals in society. This study is based on the adoption of a quantitative technical methodology through the elaboration of a questionnaire carried out on the GoogleForms platform.

The questionnaire is divided into closed response questions, open response (short), as well as encompassing questions structured in a Likert scale classification. Data processing was carried out using the Microsoft Excel tool.

This research is exploratory, with a convenience sampling focused on the population of a private Portuguese University, given the ease of access to the population and for covering a diversity of respondents that justifies the relevance of this study. The population of this university is represented by 3666 people. Students' limited knowledge (or not) about one of the most promising forms of innovation is important to know the potential that open innovation may have in the near future, as well as to understand what can be improved to share its potential.

The questionnaire was conducted on the GoogleForms platform and disseminated by the course coordinators of this private university via the e-learning platform Moodle. This form was shared by the academic community on June 2, 2021, and was open for responses until June 13, 2021. Sociodemographic information was gathered on the knowledge of open innovation, what is the biggest benefit, biggest drawback, and key partner they perceive in a process of open innovation, ending with the evaluation of the importance of innovation and more specifically open innovation. In total, 64 responses were obtained.

4 Results

The survey was addressed to the population of a private Portuguese university, and 64 answers were obtained. The questionnaire was answered by 62 students and 2 lecturers, with the majority of respondents belonging to the 18–25 age group (70.3%) and attending a degree course in management (53.1%). As regards representation by district, there was a predominance of responses from residents in the districts of Porto (78.1%) and Aveiro (17.2%).

With regard to knowledge about open innovation, 60.9% of the respondents are familiar with the topic, but 46.9% of those polled do not know any of the open innovation practices (inbound, outbound, or coupled), which shows that this type of innovation is still an unknown topic in the academic world.

Questions regarding the perception of the open innovation theme were also addressed, namely, asking about the perceived benefits and drawbacks, as well as the partners that they recognize as being most important in the adoption of this innovation paradigm. As far as benefits are concerned, 25% of respondents perceive as the main advantage the fostering of creativity and 15.6% the reduction of the time to find an innovation.

On the other hand, the majority of the respondents perceive as main drawbacks the leakage of information (31.3%) and the possibility of stealing the idea (23.4%). When asked about the key partner to promote open innovation (Fig. 2), respondents point to R&D centers (42.2%), followed by competitors (32.8%) and customers/ consumers (18.8%).

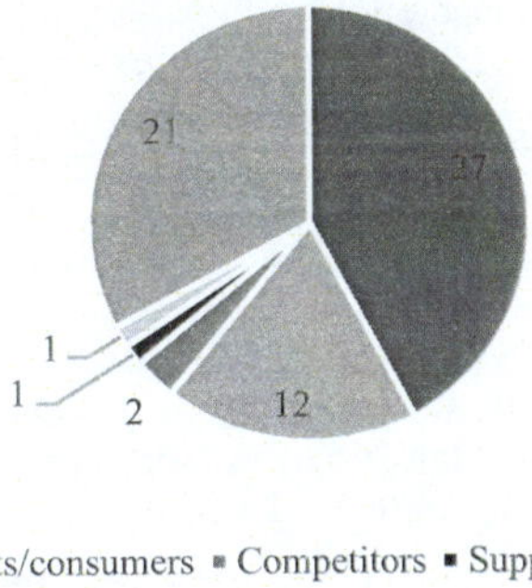

Fig. 2 Which partner do you consider most important in an open innovation process?

Fig. 3 Word cloud on open innovation

Regarding the word that respondents most identify with open innovation, the highlight goes to openness and sharing, followed by collaboration, and ideas and creativity. Figure 3 summarizes the words listed by respondents.

In order to identify possible relationships between gender and perceptions regarding open innovation, the chi-square test of independence was applied (X^2), in relation to gender.

With this purpose, four chi-square tests were performed, more specifically the association between the gender of the respondents and (1) Have you ever heard about open innovation (or open innovation)? ($X^2 = 0.18166073$); (2) Which partner do you consider most important in an open innovation process? ($X^2 = 3.837002828$); (3) What degree of importance do you attach to closed innovation (innovation carried out only within the organization)? ($X^2 = 0.000069827$); (4) How important do you consider open innovation to be? ($X^2 = 0.048092177$). All tests demonstrated independence in the relationship between gender and perception regarding the questions asked (variables studied). The contingency tables and the respective statistical tests for the remaining tests revealed that in terms of gender no pattern emerges in relation to the questions asked to the respondents. Tables 2, 3, 4, and 5 show clear examples where the percentages of men and women show similar patterns, indicating independence between the variables under study.

Table 2 Have you ever heard of open innovation?

	No	%	Yes	%	Total
Female	15	42.86	20	57.14	35
Male	10	34.48	19	65.52	29
Total	25		39		64

With regard to the questions in Tables 4 and 5, it should be noted that the data were grouped from a quantitative classification (scale of 1–7) to a qualitative classification, where low importance was attributed to classifications below 4 (including 4) and high importance was attributed to classifications above 5 (including 5).

It should be noted that for the chi-square calculation, in cases where the tables were formed in a two-by-two matrix (cases of Tables 2, 4, and 5), Yates's correction was applied due to the contingency tables being small, with only one degree of freedom. Table 6 shows the chi-square calculations in the case of relating gender with knowledge on the topic of open innovation.

Cronbach's alpha was calculated to validate the internal consistency and reliability of the questionnaire. The literature is not unanimous in defining a Cronbach's alpha that validates the reliability of the data, varying as to the minimum limit beyond which the questionnaire may be considered to have consistency. Davis [17] and Nunnally [18] report that a value above 0.5 may be considered acceptable under certain conditions (Table 7).

$$\alpha = \frac{k}{k-1} \times \left(1 - \frac{\Sigma\sigma^2}{\sigma T^2}\right) \tag{1}$$

where k represents the number of questions, σ^2 the variance, and σT^2 the variance of totals.

In this study, a Cronbach's α of 0.58 was obtained, which validates the consistency and reliability of the questionnaire.

The margin of error obtained taking into account the size of the population and the sample is ±10%.

5 Discussion

This section is dedicated to the reflection on the results obtained, and it is the result of the analysis of this study, as well as of the existing literature on the subject under study. We seek to indicate suggestions so that the community may rethink new contents that may be the object of study by the academic community.

Table 3 Which partner do you consider most important in an open innovation process?

	R&D centers	%	Customers	%	Competitors	%	Suppliers	%	Startup	%	Universities	Total
Female	13	37.14	8	22.86	1	2.86	0	0.00	0	0.00	13	35
Male	14	48.28	4	13.79	1	3.45	1	3.45	1	3.45	8	29
Total	27		12		2		1		1		21	64

Table 4 On a scale of 1–7, how important is closed innovation (innovation carried out only within the organization)?

	1–4 – low	%	5–7 – high	%	Total
Female	12	34.29	23	65.71	35
Male	9	31.03	20	68.97	29
Total	21		43		64

Table 5 On a scale of 1–7, how important is open innovation?

	1–4 – low	%	5–7 – high	%	Total
Female	3	8.57	32	91.43	35
Male	2	6.90	27	93.10	29
Total	5		59		64

Table 6 Chi-square calculation – Have you ever heard of open innovation?

O	E	O-E	IO-EI-0.5	(IO-EI-0.5)^2	((IO-EI-0.5)^2)/E
15	13.671875	1.328125	0.828125	0.685791016	0.050160714
20	21.328125	−1.328125	0.828125	0.685791016	0.032154304
10	11.328125	−1.328125	0.828125	0.685791016	0.060538793
19	17.671875	1.328125	0.828125	0.685791016	0.038806919
					0.18166073

Table 7 Recommended reliability levels for Cronbach's α

Author	Condition	Recommended Cronbach's α
Davis [17]	Forecast for groups of 25–50 individuals	Above 0.5
	Forecast for groups of more than 50 individuals	Below 0.5
Nunnally [18]	Preliminary research	0.5–0.6

From Peterson [19]

5.1 *Open Innovation Knowledge*

Starting by highlighting one of the strong points and that reinforces the robustness of this study, 65% of the respondents have a degree in management or economics, being students familiar with management and innovation.

This section is dedicated to the reflection on the results obtained and is the result of the analysis of this study, as well as of the existing literature on the subject under analysis, seeking to indicate suggestions so that the community may rethink new contents that may be the object of study by the academic community. However, one of the interesting and surprising results of this study is that 50% of these students do not know the practices of open innovation, which will allow us to conclude that open innovation is not a subject addressed in the academic environment, nor possibly in society in general, so there is a potential for this subject to be introduced in future courses of economics or management. Future decision-makers/managers need to know about this type of innovation because the success of companies, and in turn of nations, will depend on the degree of competitiveness that managers stimulate in the business environment. Studies show that the adoption of an open innovation practice has a positive correlation with the performance of an organization Future decision-makers/managers need to know about this type of innovation because the success of companies, and in turn of nations, will depend on the degree of competitiveness that managers stimulate in the business environment [20–22].

5.2 *Protection of Innovation*

One of the results of this study indicates that more than 50% of the respondents' fear/perceive that open innovation may lead to the theft of ideas and information leakage from the organization. These results are associated with studies that relate the adoption of open innovation practices (coupled, inbound, or outbound) with the degree of innovation protection. According to Freel and Robson [23], the extent of cooperation for innovation and networking, i.e., the choice of OI strategy, is strongly related to the type of ownership strategy chosen, with firms that emphasize informal and strategic methods of protection recording higher rates of coupled and inbound open innovation. As a result of the fear of imitation when exploiting technology externally, a defensive strategy is associated with superior OI outbound performance, although a collaborative strategy is preferable in terms of overall innovation performance [24].

5.3 *Creativity Phenomenon*

The fostering of creativity presents itself as a great potential of the adoption of OI. About 25% of the respondents mentioned that this is the great advantage of OI and that it allows for the promotion of the creative process. This perception reinforces the results of studies that relate the typology of innovation with the OI. For Hecker and Ganter [25], companies that want process innovations should choose to obtain knowledge through the market by hiring specialized consultants or new workers. If the objective is to innovate by launching new products, they should use a strategy based on collaboration [26].

5.4 *Key Partners*

One of the important points in the creation of innovation networks is that in order to maintain high levels of performance and a sustainable competitive advantage, companies must maintain a balance in the search for innovation between the exploration of external knowledge and the exploitation of internal knowledge [27]. The study showed that approximately 43% of the respondents favor R&D centers as key partners, followed by competitors and consumers, and allied to this, respondents relate OI to sharing and collaboration. This co-creation of knowledge is validated by previous works, emerging as the great idiosyncrasy and potential of open innovation. In their paper, Su, Lin [28] state that co-created technological knowledge is more exploratory and pioneering and has shorter technology cycle times than exclusively non-co-created knowledge.

5.5 *The Importance of Innovation*

The results of this study show that over 60% of the respondents attribute high importance to closed innovation and over 90% to OI, which contributes to validate the relevance of studying the perception that academics and future policymakers have about OI. Already in the last century, Solow [29] emphasized the importance of the physical accumulation of capital and of technological progress as forces for economic growth. This driving force is commonly called innovation and is the force that explains the progress of nations and the competitiveness of companies.

In today's constantly changing world, the winner is whoever is best prepared and whoever can absorb the most information. We are in the "Age of Knowledge," and only those who possess the dynamic capabilities that enable them to permanently leverage competitive advantage will survive [30, 31].

These results allow us to validate that the academic community is familiar with the importance of innovating, attributing high importance to this competitive factor in organizations.

5.6 *Economic and Cultural Factors*

Portugal, being a low-wage country [32], a lover of low-cost products/services (see the success, in Portugal, of companies such as IKEA and McDonald's), and with scarce funds available for innovation – in a country little oriented toward performance and high standards [33] –should embrace more open innovation as it is more economical than traditional closed innovation [34]. However, this study found that open innovation is not getting the prominence it deserves and that would be expected in Portugal. This may be happening due to the lack of trust in strangers that exists, culturally, in Portugal, and that may be hindering open innovation partnerships today. Culture appears to be a fundamental element in the development of nations and economies [35]; its influence shows and explains why development happens – or not [35]. "Cultural heritage provides the artifactual structure – beliefs, institutions, tools, instruments, technology – which not only plays an essential role in shaping the immediate choices of players in a society but also provides us with clues to the dynamic success or failure of societies through time" [35].

6 Conclusions

The world is increasingly volatile, and companies are increasingly global and more interdependent. Borders no longer transform companies into hard cores; they are now more a source of knowledge for organizations. Today's society is a global network where knowledge flows, is absorbed, and transformed at an immeasurable speed, namely, to and from the company.

6.1 Contributions to Theory

The literature on the perception of the academic community on the subject of open innovation is scarce, focusing mainly on the business world. This chapter seeks to initiate a debate on the understanding of the academic community's knowledge about this contemporary paradigm of innovation, contributing to the exploration of knowledge on this topic.

6.2 Managerial Contributions

This study sought to analyze the literature on the perception of open innovation, subsequently analyzing the knowledge of the subject in academia. We verified that despite the high importance attributed to OI and the high benefits perceived, the theme is still an unknown subject in the university environment and may prove to be harmful in the country's entrepreneurial future with the neglect of the application of this practice in the business world. Thus, the introduction of curricular units or seminars in management or economics courses is suggested, which could expose the potentialities of this new paradigm.

Finally, through the chi-square test, no statistically significant results were found in the relationship between gender and knowledge of open innovation, indicating gender equality in the perception of this new paradigm. This study could serve as a basis for further in-depth studies with larger and more representative samples of a given population (this study analyzed approximately 2% of the total population of this university).

6.3 Limitations

One of the limitations of the study stems from the size of the sample obtained since only 2% of the population of this private university was analyzed.

Another limitation of this study stems from the analysis being restricted to students' perception, being scarce in terms of teachers' perception.

The results are only representative of one Portuguese private university and do not represent the entire academic population in Portugal.

6.4 Suggestions for Future Research

As lines for future research, it is suggested to extend the sample to other universities, enlarging not only the number of respondents but also the universe of analysis, allowing the results of this study to be compared with those obtained in other universities.

In future studies, the results obtained in private and public universities could be studied, ascertaining the degree of perception between different universities, and if different, the reasons for this divergence of perception.

A further clue for future research stems from the very limitation of the sample, where student results predominate. Hence, in future studies, the perception of the teaching community on this issue should be sought.

References

1. Chesbrough, H. W. (2007). Why companies should have open business models. *MIT Sloan Management Review, 48*(2), 22.
2. Toma, A., Secundo, G., & Passiante, G. (2018). Open innovation and intellectual property strategies: Empirical evidence from a bio-pharmaceutical case study. *Business Process Management Journal, 24*(2), 501–516.
3. Gassmann, O., & Enkel, E. (2004). *Towards a theory of open innovation: Three core process archetypes*.
4. Chesbrough, H., & Crowther, A. K. (2006). Beyond high tech: Early adopters of open innovation in other industries. *R&D Management, 36*(3), 229–236.
5. Matulova, P., et al. (2018). Open innovation session as a tool supporting innovativeness in strategies for high-tech companies in the Czech Republic. *Economies, 6*(4), 1–13.
6. Kolk, A., & Puumann, K. (2008). *Co-development of open innovation strategy and dynamic capabilities as a source of corporate growth*. Tallinn School of Economics and Business Administration, Tallinn University of Technology.
7. von Hippel, E. (1986). Lead users: A source of novel product concepts. *Management Science, 32*(7), 791–805.
8. Chesbrough, H., Vanhaverbeke, W., & West, J. (2006). *Open innovation: Researching a new paradigm*. Oxford University Press.

9. Franca, R. D., et al. (2019). Open innovation: Propelling strategy for value in technology based companies. *Navus-Revista De Gestao E Tecnologia, 9*(4), 94–110.
10. Cruz-Ruiz, E., et al. (2022). Strategic management of the Malaga Brand through open innovation: Tourists and residents' perception. *Journal of Open Innovation: Technology, Market, and Complexity, 8*(1), 28.
11. Tejedor, S., et al. (2021). Higher education response in the time of coronavirus: Perceptions of teachers and students, and open innovation. *Journal of Open Innovation: Technology, Market, and Complexity, 7*(1), 1–15.
12. Kim, J., & Choi, S. O. (2020). The intensity of organizational change and the perception of organizational innovativeness; with discussion on open innovation. *Journal of Open Innovation: Technology, Market, and Complexity, 6*(3), 1–14.
13. El-Kebir, H., Mazza, M. C., & Liu, J. (2019). Perceptions of open innovation at CERN: An explorative study. *CERN IdeaSquare Journal of Experimental Innovation, 3*(2), 57–61.
14. Teplov, R., Albats, E., & Podmetina, D. (2019). What does open innovation mean? Business versus academic perceptions. *International Journal of Innovation Management, 23*(1), 1950002.
15. Dabrowska, J., Fiegenbaum, I., & Kutvonen, A. (2013). Mapping the perception and reality of open innovation. *International Journal of Innovation Management, 17*(6), 1–25.
16. Morgan, L., & Finnegan, P. (2010). Open innovation in secondary software firms: An exploration of managers' perceptions of open source software. *Data Base for Advances in Information Systems, 41*(1), 76–95.
17. Davis, F. B. (1964). *Educational measurements and their interpretation*. Wadsworth Publishing Company.
18. Nunnally, J. C. (1967). *Psychometric theory* (1st ed.). McGraw-Hill.
19. Peterson, R. A. (1994). A meta-analysis of Cronbach's coefficient alpha. *Journal of Consumer Research, 21*(2), 381–391.
20. Liao, S., Fu, L., & Liu, Z. (2020). Investigating open innovation strategies and firm performance: The moderating role of technological capability and market information management capability. *Journal of Business and Industrial Marketing, 35*(1), 23–39.
21. Bagherzadeh, M., et al. (2020). How does outside-in open innovation influence innovation performance? Analyzing the mediating roles of knowledge sharing and innovation strategy. *IEEE Transactions on Engineering Management, 67*(3), 740–753.
22. Martinez, M. G., et al. (2014). Open innovation strategies in the food and drink industry: Determinants and impact on innovation performance. *International Journal of Technology Management, 66*(2–3), 212–242.
23. Freel, M., & Robson, P. J. (2017). Appropriation strategies and open innovation in SMEs. *International Small Business Journal: Researching Entrepreneurship, 35*(5), 578–596.
24. Grimaldi, M., Greco, M., & Cricelli, L. (2021). A framework of intellectual property protection strategies and open innovation. *Journal of Business Research, 123*, 156–164.
25. Hecker, A., & Ganter, A. (2016). Organisational and technological innovation and the moderating effect of open innovation strategies. *International Journal of Innovation Management, 20*(2), 1–31.
26. Santamaría, L., Nieto, M. J., & Barge-Gil, A. (2010). The relevance of different open innovation strategies for R&D performers. *Cuadernos de Economia y Direccion de la Empresa, 13*(45), 93–114.
27. Dittrich, K., & Duysters, G. (2007). Networking as a means to strategy change: The case of open innovation in mobile telephony. *Journal of Product Innovation Management, 24*(6), 510–521.
28. Su, C. Y., Lin, B. W., & Chen, C. J. (2015). Technological knowledge co-creation strategies in the world of open innovation. *Innovation: Management, Policy and Practice, 17*(4), 485–507.
29. Solow, R. M. (1956). A contribution to the theory of economic growth. *The Quarterly Journal of Economics, 70*(1), 65–94.

30. Prahalad, C. K., & Hamel, G. (1997). The core competence of the corporation. In D. Hahn & B. Taylor (Eds.), *Strategische Unternehmungsplanung / Strategische Unternehmungsführung: Stand und Entwicklungstendenzen* (pp. 969–987). Physica-Verlag HD.
31. O'Reilly, C. A., & Tushman, M. L. (2008). Ambidexterity as a dynamic capability: Resolving the innovator's dilemma. *Research in Organizational Behavior, 28*, 185–206.
32. Eurostat. (2021). *Minimum wage statistics*. [cited 2021 14-09-2021]. Available from: https://ec.europa.eu/eurostat/statistics-explained/index.php?title=Minimum_wage_statistics
33. House, R. J., et al. (2004). *Culture, leadership, and organizations: The GLOBE study of 62 societies*. Sage.
34. Chesbrough, H. W. (2003). The era of open innovation. *MIT Sloan Management Review, 44*(3), 35–41.
35. North, D. C. (2005). *Understanding the process of economic change/Douglass C. North*. Princeton economic history of the Western world. Princeton University Press.

Green Human Resource Management: The Performance of Women Researchers Based on Bibliometric Indicators

Ana Carolina Cosenza, Gilberto Santos, and Luis Cesar Ferreira Motta Barbosa

Abstract Green human resources management (GHRM) is an innovative approach for promoting sustainable thinking and practices within the environment of organizations. GHRM means rethinking basic concepts, strategies, policies, objectives, functions, processes, and activities of human resource management providing benefits for individuals, businesses, and society. It is an emerging field of study with a scientific community in formation. In recent years, much has been discussed about the role of women in science. The objective of the chapter is to contribute to this debate by presenting bibliometric indicators on the role of women in the scientific production of GHRM. The study was based on a survey of publications and other indicators from the Web of Science database. A comparison was made of productivity indicators, impact of scientific production and research time between men and women. Researchers with at least three articles on GHRM were evaluated. Among other results, this study showed that men are the majority, but that women also play an important role in terms of the quantity and quality of publications.

Keywords GHRM · Women in science · Researchers

1 Introduction

Green human resource management (GHRM) is an innovative and emerging approach that has emerged recently [1]. Studies on the subject have grown exponentially, reflecting the increased interest and diffusion of knowledge in academia [2].

A. C. Cosenza · L. C. F. M. Barbosa (✉)
CEFET/RJ, DIPPG-PPPRO, Av. Maracanã, Brazil

G. Santos
IPCA, campus IPCA, Barcelos, Portugal
e-mail: gsantos@ipca.pt

J. C. de Oliveira Matias et al. (eds.), *Quality Innovation and Sustainability*, Springer Proceedings in Business and Economics,
https://doi.org/10.1007/978-3-031-12914-8_13

GHRM means rethinking basic concepts, strategies, policies, objectives, functions, processes, and activities of human resource management to incorporate the needs of ecological sustainability [3]. The adoption of GHRM practices must be present in all activities of the human resources area, which cover five traditional processes: recruitment and selection; training and development; empowerment or involvement; performance evaluation; and payment and rewards [4–8].

As pointed out in the literature, GHRM can contribute to the adoption of best practices and increase employees' commitment in relation to environmental sustainability [9–13], providing benefits for individuals, businesses, and society [14].

As GHRM is a field of research still in formation, there are several gaps to be explored [15, 16]. Several studies have been carried out on the subject, but little has been addressed about the profile of researchers. In particular, no studies were found on gender differences in relation to this topic in the survey carried out by the authors. This is an important aspect since researchers are the protagonists of knowledge production. The increase in production and diffusion of knowledge in a new field of research is related to the formation of a scientific community that investigates the theme [17].

In recent years, much has been discussed about the women's underrepresentation in science [18]. This underrepresentation is more evident in disciplines such as STEM – science, technology, engineering, and mathematics [19]. As the participation of women in science varies according to the area of knowledge, the following question arises: What is the role of women in the development of the GHRM field?

The chapter aims to contribute to this debate by presenting indicators on the role of women in the scientific production of GHRM. Specifically, the performance of men and women researchers was compared considering the following aspects: scientific productivity; impact and quality of production based on citations; and time of experience in research considering the date of the first publication. To complement the analysis, this study compared not only the production of GHRM but also the entire scientific production of researchers indexed in the Web of Science (WoS) database.

2 Method

The study presents a bibliometric approach. The identification of researchers was based on a survey of publications on GHRM indexed in the WoS, which is a recognized and widely used database in this type of study [20]. The query "Green Human Resource* Management" was used in the field named "Topic," which covers the search by title, abstract, and keywords. This survey resulted in a total of 266 documents. The following filters were applied: (i) only publications classified as articles and reviews (document type); and (ii) documents published until 2021. The application of these filters resulted in a total of 227 documents.

Considering this universe of publications, the researchers who published the most articles on GHRM were identified. The criterion of at least three publications on GHRM was established to identify researchers with more consolidated research on this theme. Thirty-six (36) researchers were selected based on this criterion.

Subsequently, a search was carried out on several websites to identify the sex of these researchers: woman or man. The sex of four researchers was not identified on the websites consulted (WoS, Scopus, Google Scholars, Research-Gate, LinkedIn, ORCID, universities, and others). Therefore, the sample was reduced to 32 researchers (11 women and 21 men).

Complementary data on these researchers were collected, such as H-Index, citations, total production (number of documents published in WoS regardless of topic), and others. These data were collected to define the profile and allow a better comparison between the two groups of researchers.

3 Results

The universe of 227 articles on GHRM presented a total of 6530 citations, which corresponds to an average of 28.76 citations per article. Such publications covered 51 countries, 90 journals, and 595 authors.

In turn, the selected sample of 32 researchers published 89 articles, with 4234 citations (average of 47.57 citations per article), covering 29 countries, 46 journals, and 199 authors (including other coauthors).

Therefore, this sample of researchers was responsible for approximately 40% of the articles on GHRM, presenting a much higher average of citations. Such data reflect the impact and influence of these researchers.

Considering the total of 89 articles, 22 were coauthored by researchers from both groups (men and women), which demonstrates scientific collaboration between them. In turn, 55 articles were published exclusively by researchers from the men's group and 12 from the women's group, as shown in Fig. 1.

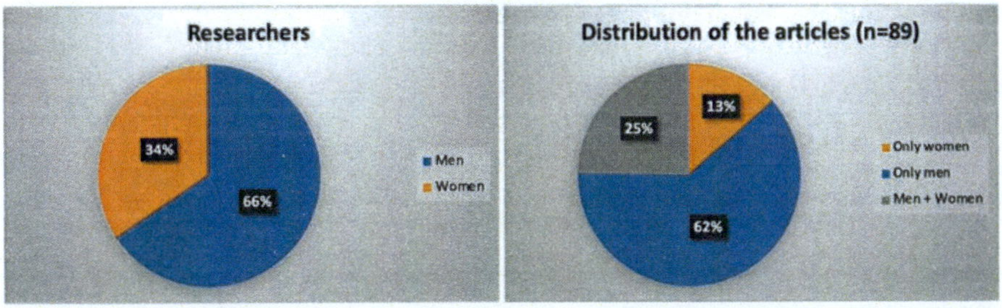

Fig. 1 Distribution of researchers and articles on GHRM considered in this study

3.1 *Scientific Productivity*

The number of articles is an indicator of academic productivity, which is important for obtaining funding, and for the prestige of researchers and institutions [21]. Table 1 presents the number of articles on GHRM written by the researchers of the two groups analyzed in this study.

The average number of articles per researcher is 4.36 among women researchers and 4.57 among men researchers. If the two most productive researchers (a man with 21 articles and a woman with 10 articles) are excluded from the sample, the averages are changed to 3.8 and 3.75, respectively.

Researchers who have articles on GHRM also publish on other topics. Aiming at a more complete analysis of the productivity of these two groups of researchers, a comparison was made considering all the publications of each researcher in WoS.

Among the men researchers, two stand out with 367 and 179 publications. Three researchers have more than 50 publications; 4 have between 20 and 50; and 12 below 20. In the case of women, two have more than 100 publications in WoS (107 and 105 documents). Only one is over 50; five have between 20 and 50; and three below 20. Figure 2 shows the comparison of the average of articles per researcher considering the production of articles on GHRM and the total of documents published in WoS.

As shown in Fig. 2, the average publication of men researchers was 4.57 and 46.29, respectively. In turn, the average of women was 4.36 and 44.82. As can be observed, the two groups of researchers have an average of more than 4 articles on GHRM and more than 44 documents in WoS.

3.2 *Quality and Impact of the Scientific Production*

The quality and impact of scientific production can be measured through the number of citations. This is an indicator that has been increasingly valued and that is related to the H-Index, which is an indicator of academic prestige [22].

The average citations per researcher and the average citations per publication were calculated to compare the quality of production of men and women researchers. The results are shown in Fig. 3.

Table 1 Number of articles on GHRM published by the researchers

Number of articles (GHRM)	Women	Men
10 or more	1	1
5 articles	3	4
4 articles	2	7
3 articles	5	9
Total researchers	11	21
Average (articles/researcher)	4.36	4.57
Average (without researcher with 10 or more articles)	3.8	3.75

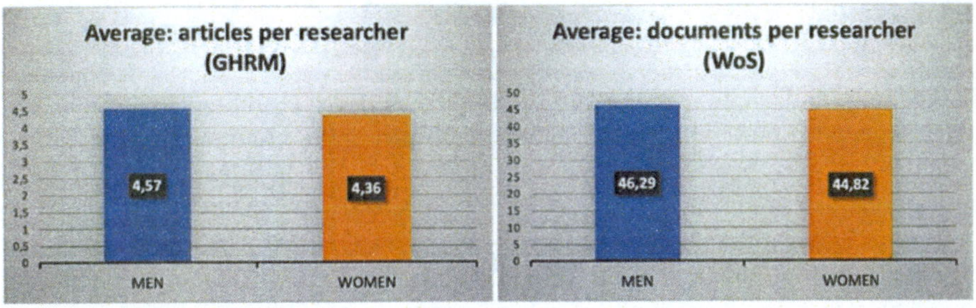

Fig. 2 Publications per researcher

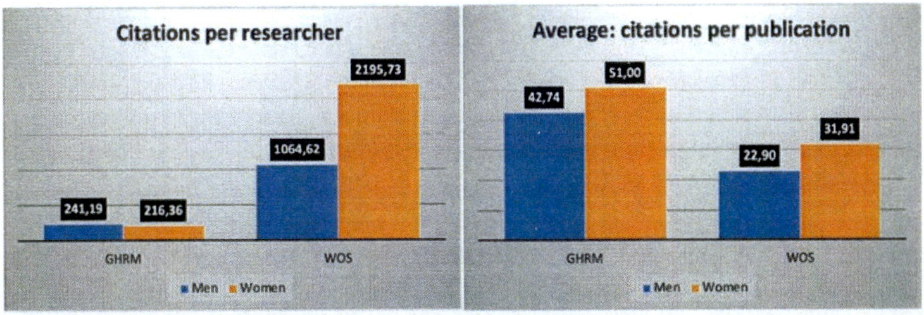

Fig. 3 Citations per researcher and per publication

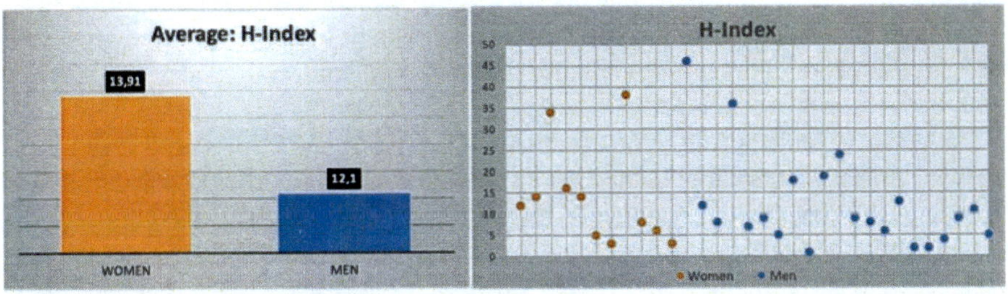

Fig. 4 H-Index

According to results, men and women had an average of citations per researcher greater than 200 considering the articles on GHRM, and greater than 1.000 (men) and 2.000 (women) in the total of documents published in WoS.

In relation to the indicator of citations per publication, the researchers had more than 40 (men) and 50 (women) citations per article on GHRM and more than 20 (men) and 30 (women) citations per document in WoS.

These results show that the scientific production of women also has been recognized and has influenced other researchers.

In turn, the average H-Index of the women was 13.91 and the men was 12.10. This is another indicator that also shows the prestige and quality of the publications of women researchers working in the GHRM area. Figure 4 shows the H-Index of the researchers analyzed in this study.

3.3 Time of Research Activities

Productivity and citation indicators are influenced by research activity time. An older researcher may be more productive and have more citations than an early-career researcher. Therefore, it is important to know how long researchers have been developing their studies.

Figure 5 shows the evolution of publications on GHRM over time. As can be observed, the first publication on GHRM dates from 2011. It was an article coauthored by researchers from both groups (men and women). After this article, women researchers only published again in 2016. In turn, men researchers published in 2013. In both groups, there was a significant increase in the number of publications from 2018 onward.

Figure 6 presents the year of the first article on GHRM as well as the year of the first publication in WoS of each researcher. Most men and women researchers published their first articles on GHRM as of 2016 and in WoS as of 2007.

With the exception of three men researchers, all others published on other topics in WoS before publishing on GHRM.

The researcher with the longest experience in research is a woman who published in WoS in 1978. This woman was also the first to publish an article on GHRM, which happened in 2011. In the case of men researchers, the first publication in WoS was in 1988, and the first article on GHRM was published by two other researchers in 2011 in coauthorship with the aforementioned woman researcher.

Six men researchers and only one woman had their first publication in WoS from 2017 onward. These data show that approximately 22% of the 32 researchers (especially men) started publishing in the last 5 years, which indicates that they may be young researchers. The average time of the first publication considering the two groups of researchers analyzed in this study is shown in Fig. 7.

The average years since the first publication are 2.81 years (men) and 3.18 years (women) in relation to GHRM, and 8.57 (men) and 14.91 (women) considering all documents in WoS. Even if the woman researcher with the longest period of publication in WoS (43 years) was excluded from the sample, the average of women researchers would remain high. These results point out that female researchers also have good experience in research activities considering the time of first publication.

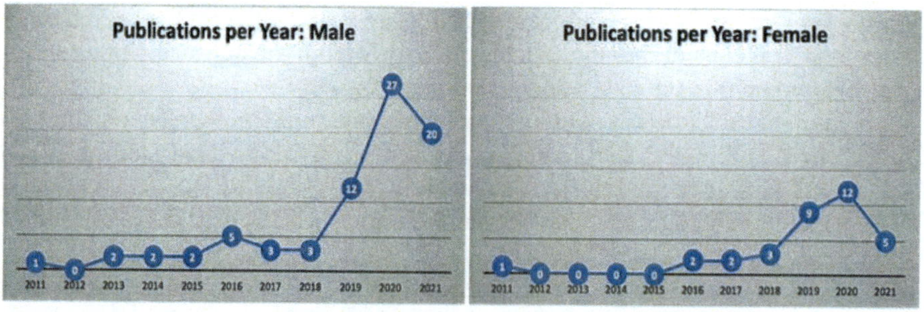

Fig. 5 Evolution of the publications on GHRM

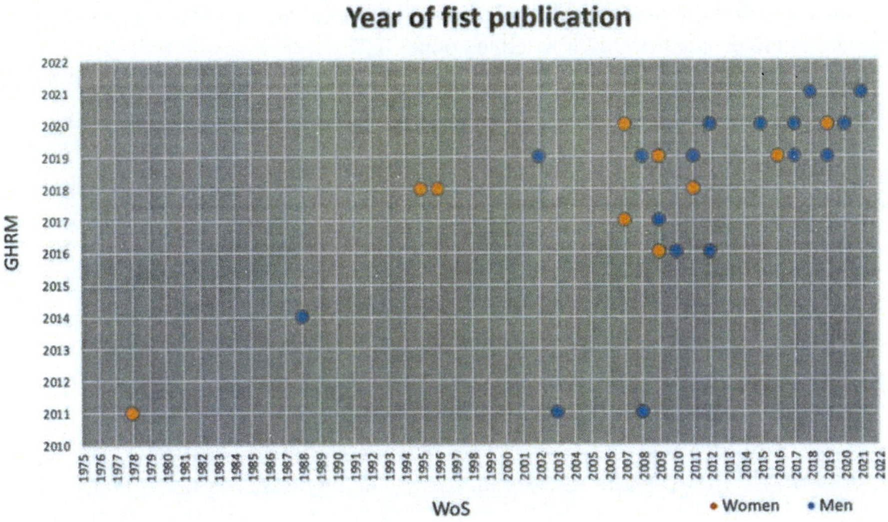

Fig. 6 Year of the first publication

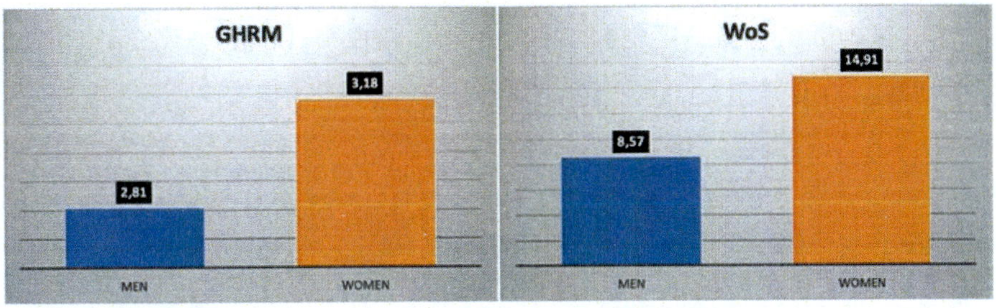

Fig. 7 Average years of the first publication

4 Final Considerations

This study sought to present the role of women in scientific production of GHRM through a bibliometric approach. Three aspects were analyzed: academic productivity, quality and impact of publications, and research experience. The performance of women researchers was compared with that of men researchers, considering the articles on GHRM, as well as the complete production of the researchers indexed in WoS.

Although there are a greater number of men researchers, the study showed that women also play an important role in terms of the quantity and quality of publications.

It should be noted that these results were based only on the average of publications, citations, H-Index, and time since the first publication. Other statistical metrics and other aspects such as coauthorship pattern and research focus can be considered in future studies to complement and deepen the analysis of the role of women in the production of knowledge about GHRM.

This study presents two main contributions. The first is a better understanding of the profile of researchers who are developing studies on GHRM, which is an emerging field with a scientific community in formation. It is these researchers who share concepts, values, and methodologies, consolidating a new field of research. The second is related to the role of women in science, which is a topic that has been widely discussed around the world. The authors believe that this is the first study with this approach related to GHRM.

References

1. Aust, I., Matthews, B., & Muller-Camen, M. (2020). Common Good HRM: A paradigm shift in Sustainable HRM? *Human Resource Management Review, 30*(3), 100705.
2. Khan, M. H., & Muktar, S. N. (2020). A bibliometric analysis of green human resource management based on scopus platform. *Cogent Business & Management, 7*(1), 1831165.
3. Arulrajah, A. A., Opatha, H. H. D. N. P., & Nawaratne, N. N. J. (2016). Green human resource management practices: A review. *Sri Lankan Journal of Human Resource Management, 5*(1), 1–16.
4. Gupta, H. (2018). Assessing organizations performance on the basis of GHRM practices using BWM and Fuzzy TOPSIS. *Journal of Environmental Management, 226*, 201–216.
5. Saed, B. B., Afsar, B., Hafeez, S., Khan, I., Tahir, M., & Afridi, M. A. (2019). Promoting employee's proenvironmental behavior through green human resource management practices. *Corporate Social Responsibility and Environmental Management, 26*(2), 424–438.
6. Tang, G., Chen, Y., Jiang, Y., Paillé, P., & Jia, J. (2018). Green human resource management practices: Scale development and validity. *Asia Pacific Journal of Human Resources, 56*, 31–55.
7. Jabbour, C. J. C., Santos, F. C. A., & Nagano, M. S. (2010). Contributions of HRM throughout the stages of environmental management: Methodological triangulation applied to companies in Brazil. *International Journal of Human Resource Management, 21*, 1049–1089.
8. Mishra, P. (2017). Green human resource management: A framework for sustainable organizational development in an emerging economy. *International Journal of Organizational Analysis, 25*(5), 762–788.
9. Jackson, S. E., Renwick, D. W., Jabbour, C. J., & Muller-Camen, M. (2011). State-of-the-art and future directions for green human resource management: Introduction to the special issue. *German Journal of Human Resource Management, 25*(2), 99–116.
10. Renwick, D. W. S., Redman, T., & Maguire, S. (2013). Green human resource management: A review and research agenda. *International Journal of Management Reviews, 15*(1), 1–14.
11. Jabbour, C. J. C. (2013). Environmental training in organisations: From a literature review to a framework for future research. *Resources, Conservation and Recycling, 74*, 144–155.
12. Bombiak, E., & Marciniuk-Kluska, A. (2018). Green human resource management as a tool for the sustainable development of enterprises: Polish young company experience. *Sustainability, 10*(6), 1739.
13. Masri, H. A., & Jaaron, A. A. M. (2017). Assessing green human resources management practices in Palestinian manufacturing context: An empirical study. *Journal of Cleaner Production, 143*, 474–489.
14. Opatha, H., & Arulrajaha, A. (2014). Green human resource management: Simplified general reflections. *International Business Research, 7*(9), 101–112.

15. Benevene, P., & Buonomo, I. (2020). Green human resource management: An evidence-based systematic literature review. *Sustainability, 12*, 1–25.
16. Pham, N. T., Hoang, H. T., & Phan, Q. P. T. (2019). Green human resource management: A comprehensive review and future research agenda. *International Journal of Manpower, 41*(7), 845–878.
17. De Souza, C. G., Barbastefano, R. G., & Teixeira, R. C. (2017). Life cycle assessment research in Brazil: Characteristics, interdiciplinarity, and applications. *The International Journal of Life Cycle Assessment, 22*(2), 266–276.
18. Williams, W. M. (2018). Underrepresentation of women in science: International and cross-disciplinary evidence and debate. *Frontiers in Psychology, 8*, 2352.
19. Beeler, W. H., Smith-Doody, K. A., Ha, R., Aiyar, R. S., Schwarzbach, E., Solomon, S. L., & Jagsi, R. (2019). Institutional report cards for gender equality: Lessons learned from benchmarking efforts for women in STEM. *Cell Stem Cell, 25*(3), 306–310.
20. Zhu, J., & Liu, W. (2020). A tale of two databases: The use of Web of Science and Scopus in academic papers. *Scientometrics, 123*(1), 321–335.
21. Agarwal, A., Durairajanayagam, D., Tatagari, S., Esteves, S. C., Harlev, A., Henkel, R., et al. (2016). Bibliometrics: Tracking research impact by selecting the appropriate metrics. *Asian Journal of Andrology, 18*(2), 296–309. https://doi.org/10.4103/1008-682X.171582
22. Haustein, S., & Larivière, V. (2015). The use of bibliometrics for assessing research: Possibilities, limitations and adverse effects. In I. Welpe, J. Wollersheim, S. Ringelhan, & M. Osterloh (Eds.), *Incentives and performance*. Springer. https://doi.org/10.1007/978-3-319-09785-5_8

Innovative Responses to the COVID-19 Pandemic in Primary Healthcare: The Case of the Arte Nova Family Health Unit

José Pedro Machado Antunes, Sofia Senra Furtado, Sara Cristina Santos Rocha, Inês Campos Pinto, Maria Espírito Santo da Cunha, Carolina Teixeira Carlos, and Manuel Au-Yong-Oliveira

Abstract Primary healthcare (PHC) is a fundamental pillar in a health system, and its role has only been enhanced in the COVID-19 pandemic. The emergence of COVID-19 required health systems and, in particular, PHCs to develop a constant capacity to adapt and resist adversity. Therefore, it was considered important to identify innovative measures and thus the experience of a family health unit (USF) in Portugal – USF Arte Nova (USFAN) was reported. After carrying out a systematic review of the literature and interviews with the Presidents of the Clinical and Health Councils of five Health Centre Groupings, the measures identified were gathered and compared to the ones carried out by USFAN. Subsequently, the study identified the measures considered to be truly original executed by the health unit. Between the measures presented, the use of an access prioritization score was highlighted, as were daily briefing regarding the management of resources and equip-

J. P. M. Antunes (✉)
USF Arte Nova, Aveiro, Portugal

Department of Economics, Management, Industrial Engineering and Tourism (DEGEIT), University of Aveiro, Aveiro, Portugal

S. S. Furtado · S. C. S. Rocha · I. C. Pinto · M. E. S. da Cunha
USF Arte Nova, Aveiro, Portugal

C. T. Carlos
Department of Economics, Management, Industrial Engineering and Tourism (DEGEIT), University of Aveiro, Aveiro, Portugal

M. Au-Yong-Oliveira
Department of Economics, Management, Industrial Engineering and Tourism (DEGEIT), University of Aveiro, Aveiro, Portugal

INESC TEC, GOVCOPP, Aveiro, Portugal
e-mail: mao@ua.pt

J. C. de Oliveira Matias et al. (eds.), *Quality Innovation and Sustainability*, Springer Proceedings in Business and Economics,
https://doi.org/10.1007/978-3-031-12914-8_14

ment, qualification of the use of personal protective equipment, the project that aimed to optimize the use of paper; the use of a drive-through method to update the National Vaccination Program; and the daily training and sharing of information about COVID-19. In light of the current pandemic, innovative practices and tools have been created and carried out by the healthcare professionals in response to the growing needs of the population. This shows the resilience of these professionals and constitutes an opportunity to share and implement these tools in other healthcare facilities highlighting the continuing chance of improving.

Keywords Primary care · COVID-19 · Pandemic · General practice · Innovation

1 Introduction

Primary care offers the first point of access to a health system and as such has significant responsibility for the overall delivery of effective care and efficient spending [1]. Primary healthcare (PHC) is a central element of the Portuguese National Health Service (NHS) and assumes, in an integrated perspective and articulation with other services for continuity of care, important roles in health promotion and disease prevention, healthcare provision, and quality monitoring and proximity to the population [2, 3]. The Portuguese PHC core are family health units (USF) with teams with a wide functional autonomy, composed of family doctors, nurses, and clinical secretaries [4]. USFs – family health units are part of a health center, which, in turn, is included in a Grouping of Health Centers (ACeS). ACeS are health services with administrative autonomy, composed of several functional units that integrate one or more health centers. The ACeS bodies are the executive director, the executive board, the community council, and the clinical and health council. The members of the clinical and health council must have technical knowledge in primary healthcare, practice in care quality assurance and auditing processes, as well as mastery of risk management techniques [5].

The COVID-19 pandemic led to the overloading of health services around the world. On March 19, 2020, a call was issued for professionals to be proactive in reorganizing services in General Practice (GP) in the context of the exception of the fight against COVID-19 [6]. In the Portuguese NHS, the family doctor acts as a gatekeeper, having kept this role also during the pandemic, adding to their usual tasks, the need to perform new tasks and daily updates [7]. The innovation of services, within the scope of health, is relevant, while accommodating this innovation with the increasingly pressing technological competitiveness [8–10]. Innovation in health is carried out with the ultimate purpose of producing an improvement in the quality of the service provided and, consequently, an improvement in the quality of life. Therefore, innovation in health can be seen in a triad of dimensions, namely, technology (new devices), procedure (new techniques), and forms of organization (palliative medicine) [11]. As Flessa & Huebner (2021) [11] describe: "The existing world is the result of past innovation processes – and the future world will be the consequence of today's innovation processes. Whether the future will bring a better quality of life, more efficient healthcare delivery and equity among *stakeholders* depends on the innovation we do today, the adoption of processes and the diffusion of innovation."

The USF Arte Nova (USFAN) team belongs to the ACeS Baixo Vouga (ARS Centro) based in Oliveirinha, Aveiro, Portugal, serving a population of approximately 13,000 users. The multiprofessional team consists of 11 physicians (7 specialists and 4 residents), 7 nurses, and 5 clinical secretaries. At the end of 2020, its overall performance index (IDG) was 84/100, having been the best Functional Unit in the Center region and one of the best at its level at a national level. Thus, the results stimulated the in-depth study of the preventive and organizational strategies carried out according to the guidelines of the Directorate General for Health (DGS).

The objectives of the study are as follows:

1. Identify innovative tools in the response to COVID-19 carried out by the USF Arte Nova that may have contributed to its performance in 2020 in primary care
2. Compare/explain the innovative measures present in the literature with the measures reported by the USF Arte nova and, through an interview, by the Presidents of the Clinical and Health Councils

The structure of the chapter is as follows: the following section presents the methodology followed; a literature review was performed as well as qualitative and descriptive empirical work; the researchers hence then present the results of the study, the discussion/conclusion, as well as the references used throughout the research effort.

2 Methodology

A systematic search was conducted on the Scopus® and PubMed® platforms, using the keywords in title and abstract "innovation" AND "covid-19" AND "primary care," accessed on January 17, 2022. The exclusion criteria assumed were the absence of free access, articles related to public health or governmental measures, and repeated articles. The articles were reviewed by three authors, working with a common standardized checklist.

In addition to the database search, semi-structured interviews were proposed to the Presidents of the Clinical and Health Councils (PCCS) of six ACeS of different Regional Health Authorities of continental Portugal. Selection was by convenience and Exclusion of the Chair of the Clinical Board of USFAN's area. The following questions were addressed:

1. Do you know of innovative projects/measures to fight the pandemic COVID-19 carried out within family health units?
2. Can you list and give a brief description?
3. Do you consider the USFAN's measures innovative?

Table 1 Criteria for classifying innovative measures

1. Descriptions exist in the documentary collection or are reported in interviews with USFAN professionals
2. Not described in the literature review
3. Not being described as innovative activities by the PCCS
4. They are considered innovative by all the CSCP respondents

The access to the document collection of USFAN was unanimously approved by the team at a multidisciplinary meeting, ensuring unequivocal confidentiality of users and professionals. Interviews were also requested from the USFAN professionals. For the classification of innovative measures of USFAN, the research team used four criteria (Table 1).

3 Results

A total of 234 articles were obtained. After reading the title and abstract, 197 articles were excluded due to the inadequacy of the content with the objective of the study. Six articles were duplicated, thus 31 articles were included for full reading and identification of innovative measures of response to COVID in primary healthcare.

The digital document collection of USFAN, referring to the years 2019–2021, was analyzed. Free interviews were conducted with the medical element of the technical council and a nurse from the executive support group of USFAN. Interviews were requested with six PCSS from different ACES and four different ARS, who performed their functions in the period under analysis. There was an 83.3% response rate ($N = 5$), and the interviews were conducted between January 27 and 29, 2022. The interviews were semi-structured and standardized.

3.1 Innovative Responses in Primary Healthcare in Portugal and Worldwide: Literature Review

Telemedicine was the most widely described measure in the literature used to facilitate and maintain access to healthcare by avoiding direct contact between users and health professionals. Strategies such as virtual consultations (via telephone or video) and electronic medical prescription were thus implemented or strengthened [13–15]. The implementation of regular meetings with staff in primary healthcare units has been described to combat the frequent updating of standards and guidelines by health entities [12, 16]. The importance of regular adjustments of task distribution within care teams, fostering interdisciplinary collaboration and teamwork in order to optimize the response to the growing needs of the population, was also mentioned [17]. In order to ensure the maintenance of therapeutic adherence, one of the measures mentioned in the literature was the home delivery of medication to patients, simultaneously ensuring that they did not have to travel, thus minimizing the consequent risk of infection [18]. The installation of triage models was also one of the most described strategies. The professionals responsible for it varied according to the items and could be either administrative, medical, or nursing staff. It facilitated, on the one hand, the distinction between nonsuspects and suspects for COVID-19 infection and, in turn, within the latter, those at risk of worsening the condition and subsequent orientation [12, 19]. In this line of thought, the constitution of specific and properly adjusted areas for the observation of patients suspected or infected with COVID-19 safely allowed these patients not to feel segregated and to have appropriate guidance [19–21]. Thus, in order to ensure the safety of patients and health professionals, circulation circuits with specific signage were implemented in several institutions. In addition, new cleaning and disinfection protocols were also instituted to minimize the risks of disease transmission [12]. One of the most commonly reported techniques for disease management and early detection of alarm signals for worsening was the creation of follow-up algorithms based on disease evolution in the light of state-of-the-art knowledge [19, 22–24]. In addition to these algorithms, the constitution of tracing platforms has allowed the management and monitoring of patients from a population point of view by health professionals, coping with the growing and challenging number of cases [25]. The construction of testing chains, in a drive-through model [12], as well as the support of medical students and other health professionals, were strategies pointed out in the literature as auxiliary in fighting the pandemic [12, 16]. The use of oximetry in the home was a useful form of monitoring and risk assessment, according to a UK study [26].

In order to ensure the maintenance of care for the most vulnerable population groups, strategies have been reported for the homeless (in refuge homes) in South Africa [27] and for residents in long-term care institutions in Catalonia [28]. During the course of the COVID-19 pandemic, the increasing amount of information and its rapid dissemination generated the need to develop online health portals where information resources and training sessions aimed at the population were made available in order to provide evidence-based information and clarify any doubts [29]. The level of burnout in healthcare professionals is a concern that preceded the pandemic and has been aggravated by it, and there are articles in the literature with suggestions for strategies to combat it [30]. However, according to Cairns et al., there are still no high-quality, sufficiently well-founded tools and interventions targeting this gap, thus posing a challenge for the future [31].

See Table 2 for a summary of innovative measures.

Table 2 Summary table of innovative measures: bibliographical survey

Area	Category/subarea	Measure	References
Access	Telemedicine virtual consultations	Video and/or telephone consultation	[12–14, 16, 18, 19, 24, 29, 32–38]
	Virtual administrative assistance	Electronic prescription	[15]
Assistance	Respiratory patient management	Reception screening	[12]
		Nursing triage	[12, 19]
		Screening by medical specialists	[12]
		Self-screening	[24]
		Dedicated areas for observation of suspects/positives	[19, 21]
		Tracking/screening algorithms	[19, 22–24]
		Tracing and disease management platform/patient telephone line COVID	[20, 25, 29]
		Home oximetry monitoring	[26]
	Management of the chronically ill	Home delivery of medication	[18, 29, 39]
	Assistance to vulnerable and at-risk groups	Temporary shelter institutions	[27]
		Long-term care institutions	[12, 28]
	Vaccination	Vaccination teams	[40, 41]
Prevention and health promotion	Health literacy	Online health portals	[42]
		Online information resources	[29]
		Online training sessions for the community	[29]
Human resources and attendance	Human resources management	Regular meetings with the team	[12]
		Interdisciplinary collaboration in care	[16, 17]
		Support of medical students/other professionals in the response to the pandemic	[12, 16]
Job satisfaction and security	Infection prevention and control	Specific COVID signage	[12]
		Institution of new cleaning protocols	[12]
		Carrying out tests on the drive-through system	[12]
	Five-year prevention	Preservation of the mental health of health professionals	[30, 31]

3.2 Outcome Innovative Measures in Response to the Pandemic Carried Out by USF Arte Nova (USFAN): Documentary Analysis, Experience Report, and Free Interviews

The identification of innovative measures, described in the documentary repository and/or consistently referred to by the team, was classified into eight areas:

1. Access
2. Procurement and stock management
3. Assistential
4. Training, teaching, and research
5. Community orientation
6. Prevention and health promotion
7. Human resources and attendance
8. Job satisfaction and security

Regarding access (point 1), the team identified as innovative measures the realization of virtual consultations, via telephone, by the medical and/or nursing team with the possibility of conversion into a face-to-face consultation depending on the clinical assessment. In addition, the BRAVU platform (Quick Virtual Customer Service Counter) was redesigned, allowing, through a structured questionnaire addressed to the main reasons for visiting the FHU, faster access to healthcare. A strategic plan – "Mission G" – was identified, which consisted of an increase in the assistance commitment, corresponding to an increase of 1700 users, which aimed to guarantee a family health team to users without PHC coverage in their area of residence, and therefore with limited access to care related to the prevention and management of COVID-19. Still regarding access, at a time when care activities were limited, the team developed a tool – "Score USFAN" – with the objective of identifying users who, having a higher score, required higher priority care.

In the area of supply and stock management (point 2), stocks were shared daily with the team, as well as measures were taken to qualify the use of personal protective equipment (PPE). Within the scope of the management of the users' document archive and in order to prioritize the nonphysical communication of information, the "USF Without Paper" mission was established, using the stimulus of receiving information via telephone, email, and digitalization, thus calling for the maximization of computerization, and also the issuing and sending of prescriptions and credentials by telephone/email.

Regarding care activity (point 3), in the context of preventive medicine (3.1), the user's visit to the FHU was optimized, taking advantage of opportunistic surveillance activities in all contacts of the user with the unit, optimizing resources and reducing the need for future trips. Based on the management of respiratory patients, an adapted office was created, adjacent to the functional unit, dedicated to face-to-face care of patients with suspected or diagnosed infection by SARS-CoV-2. Additionally, the team collaborated with the areas dedicated to COVID of the Aveiro Healthcare Centre, participating in the preparation of procedures, circuits, and articulation manual, as well as sharing USFAN's human resources for assistance activities. Still regarding the management of respiratory patients, USFAN developed a weekly schedule of teams dedicated to the surveillance of users, based on the TRACE-COVID® platform, in order to guarantee their easy access to healthcare. The scale was adapted periodically, allocating the number of professionals according to the volume of users under management and the remaining care activity. The DOMUS-P program aimed at carrying out preventive home visits to patients at increased risk of complications from COVID-19 by avoiding the displacement and potential increase of risk contacts with their visit to the unit. In addition, the drive-through vaccination model allowed the update of the National Vaccination Program (PNV) and other opportunistic vaccines in an outpatient circuit.

The team considered it important to maintain the culture of training, teaching, and research (point 4) among the team, involving medical professionals with specific training in FGM in the various activities. Given the constant update of new data regarding the pandemic, a periodic training model was adopted with updated guidelines that encouraged team discussion, with a view to improving procedures and management of care activity. The activity "COVID in numbers" took place daily in a multiprofessional meeting in order to update the team on the regional and national epidemiological situation. The team considered it important to keep a record of the information shared in meetings and training to review the strategies adopted and for further research purposes. Thus, a summary of the activities carried out during the meeting was made and shared via email by the team, as well as, when applicable, a photographic and videographic record was made.

Another point that the team considered innovative was community orientation (point 5), through the articulation of community intervention agents. In the team's care activity, in addition to the articulation with health professionals from the PHC and secondary healthcare, the community council was created, which aimed to facilitate dynamic and two-way communication with the various local entities (parish councils, pharmacies, users' representatives) in order to optimize the team's decision-making, also taking into account the needs, concerns, and resources available in the population.

Communication training of professionals and increasing the population's level of health literacy have been investments made by the team. The pandemic-related infodemic led the team to develop communication strategies adapted to the pandemic phase and the population it was serving at a given time. The team realized that the longitudinal and close follow-up, characteristic of PHC, represented a reliable source of information for users considering prevention and health promotion (Point 6). The communication of transparent information regarding the update of guidelines related to the pandemic and its reflection on the functioning of the FHU was carried out through exposure inside and outside the FHU, as well as using articles in the local journal and digital platforms. The team considered it important to invest in the standardization of information transmitted to the user as well as in the acquisition of communication skills by professionals. In this context, some users expressed satisfaction and perception of proximity and accessibility to health resources, even if they did not physically go to the unit. In addition to information on the pandemic, infographics related to other topics, including mental health and cardiovascular health, were also disseminated.

One of the challenges overcome by the team was human resources management and attendance (point 7). The concern for infection prevention and anticipation of possible professional absences (unscheduled and scheduled for other work tasks) led to the discussion and adoption of team strategies related to resource management, adapted to current guidelines. In addition to optimizing the inter-substitution systems, well-defined schedules and transitional tasks were established for the various professional categories. At a certain moment, mirror schedules were adopted, with the care of holding a daily multiprofessional meeting with the presence of the largest possible number of professionals, which would come to be considered fundamental for the remaining areas of intervention described.

The pandemic led to critical reflection by the team regarding the innovative development of measures to preserve professional satisfaction and safety (point 8). The USFAN team advocates that one of the means of effective clinical governance is the development, empowerment, and involvement of professionals in the various areas of action for the continuous improvement of care. To this end, 5-year prevention is an indispensable factor. Thus, adaptation strategies were promoted to keep the team healthy and motivated in the fight against the pandemic, while preserving its identity. Besides the daily corridor meetings, several teambuilding activities were carried out, which were evaluated, qualitatively, positively by the professionals. Regarding the prevention and control of infection, the circuits and procedures of the unit were redefined, namely, the acquisition of proper uniforms with professional identification, internal laundry circuits and guidelines to cleaning professionals for the hygiene of spaces/equipment. The service provision locations were also readapted with the allocation, definition, and identification of the spaces according to the activity to be developed there. In the head office building, two distinct wings were defined (curative wing and preventive wing) and for the pole, the provision of activities within the scope of maternal and child health was allocated. In both buildings, the signage of the spaces was optimized, namely, in the isolation areas. In order to improve the organization of spaces/equipment and stock management, in addition to the inventory of each space, Lean Thinking was applied through the 5S tool to standardize the layout and content of the spaces adapted to the activity to be developed, in order to facilitate hygiene and reduce material to the essential. The professionals also contributed to the risk map of the buildings, identifying potential risks and improvement proposals to increase the safety of users and professionals.

3.3 *Innovative Measures in the View of the Presidents of the Clinical and Health Councils (PCCs): Interviews*

All the PCCS accepted being interviewed and answered the three main questions.

When asked about their knowledge of innovative projects/measures to combat the COVID-19 pandemic carried out within family health units, all responded positively. In this sense, the innovative projects/measures were questioned and, together, cataloged in an area and subarea, as represented in Table 3.

The measures considered innovative by USFAN were presented to all the PCCS interviewed and questioned whether they considered the measure innovative. The results are present in Table 4.

Thus, there was consensus in classifying five measures as innovative, fundamentally

1. Score USFAN
2. Daily record and shared with the team the management of resources and equipment and qualification of the use of personal protective equipment
3. Paperless USF project
4. Use of the drive-through method for updating the National Vaccination Program
5. Daily training and information about COVID-19

Table 3 Innovative measures reported by PCCS

Area	Subarea/category	Measure
Access	Virtual administrative user service	Communication platform with the residential structure for the elderly (ERPI)
Access	Virtual administrative user service	Platform service platform
Assistance	Respiratory patient management	Provide auxiliary material for follow-up and monitoring of signs of the disease for use at home
Assistance	Telemedicine: virtual consultations	Telephone consultations
Assistance	Respiratory patient management	Dedicated teams for respiratory patient management
Assistance	Respiratory patient management	Platform for articulation of levels of care to manage access to emergency services
Training, teaching, and research	Formative actions	Refresher training sessions between various functional units
Training education and research	Training	Use of online training platforms
Human resources and attendance	Talent and time management	Permanent schedule adaptation
Job security and satisfaction	Five-year prevention	Burnout prevention activities –URAP
Safety and job satisfaction	Professional assistance	Network and helpline for professionals

Table 4 Results of PCCS classification of USFAN's innovative measures

Area	Subarea/category	Measure	Innovative?	
			Yes	No
Access	Telemedicine virtual consultations	Telephone enquiries	3	2
	Virtual administrative user service	BRAVU® – Fast Virtual Patient Service Counter	4	1
	Broadening the commitment to assistance	Strategic plan – "MIssion G" – increase of 1700 users	2	3
	Prioritization of access	Preparation of the USF AN score	5	0
Procurement and stock management	Resources and equipment management	Daily record and shared with the team the management of resources and equipment and qualification of the use of personal protective equipment	5	0
	Resources and equipment management	USF paperless	5	0
Assistance	Respiratory patient management	Area dedicated to COVID in the Functional Unit	3	2
	Management of the chronically ill	Domus-P program	3	2
	Respiratory patient management	Dedicated patient management teams	4	1
	Preventive medicine	Opportunistic activity	0	5
	Respiratory patient management	Areas dedicated to the Health Center COVID	3	2
	Vaccination	Drive-through vaccination – update of the national vaccination program	5	0
Training education and research	Daily training sessions and pandemic status updates	Periodic training sessions with updated guidelines, COVID in numbers daily	5	0
Community orientation	Articulation of community intervention agents	Creation of a community council	1	4
Prevention and health promotion	Health literacy	Use of platforms (Facebook® and Instagram®) and local newsletters and newspapers	0	5
Human resources and attendance	Talent and time management	Adaptation of schedules	3	2
Job security and satisfaction	Five-year prevention	Burnout prevention activities	3	2
	Prevention and infection control	Internal and autonomous circuit for sanitizing uniforms	2	3
	Prevention and infection control	Risk map	2	3
	Prevention and infection control	Readaptation of the service provision sites	4	1
	Organization of spaces and equipment	Application of the 5s and LEAN methodology	4	1
	Standardization and investment in communication skills	Training	1	4

4 Discussion/Conclusion

The pandemic COVID-19 required health systems and, in particular, PHCs to develop a constant capacity to adapt and resist adversity. In the literature review conducted, the most frequently reported measure was telemedicine, which includes conducting consultations via telephone or video, yet this measure was referred to as innovative by three of the five PCCS. Other measures were also described as innovative by the PCCS that were already described in the literature, such as the delivery of material for monitoring and follow-up of signs of disease for use by users at home [18]. The PCCS considered five measures/projects of USFAN as innovative, which were also not previously described in the authors' literature review. There were, therefore, five measures/projects considered as innovative carried out by USFAN that met the four criteria proposed by the authors (Table 1). Considering the topicality of the subject, the authors point out that there may be other innovative measures that are not yet described in the literature and, given the sample size of the PCCS, are known to other professionals with management and leadership positions. Additionally, high levels of job satisfaction were identified among the members of the USFAN, which may be related to the implementation of innovation measures. Innovation in health, as previously mentioned, is significant and is consistent with the increasingly present reality of technological competitiveness [10], which emphasizes the importance of creating and implementing these measures, particularly in primary healthcare, in order to increase the adaptability of health professionals to the new demands of the reality of the pandemic context. This research aimed to conduct a comprehensive, and equally detailed, look at the innovative responses in the fight against the COVID-19 pandemic, in primary healthcare – with the practical applicability of the context, with a case study, of USF ARTE NOVA. Furthermore, the adaptability and resilience of health professionals when faced with new clinical challenges in the context of a pandemic should be recognized and further explored in the literature. Additionally, the need for innovative practices for the exercise of the profession, in a way consistent with the daily demands in primary healthcare settings, is an important issue that will enable the identification and introduction of such innovative practices. Thus, the relevance of this study lies in the importance of sharing these strategies found among peers, so that they can be reflected upon, improved, and also implemented in other healthcare units.

References

1. Neri, M., Cubi-Molla, P., & Cookson, G. (2022). Approaches to measure efficiency in primary care: A systematic literature review. *Applied Health Economics and Health Policy, 20*(1), 19–33. https://doi.org/10.1007/s40258-021-00669-x
2. *Decree-Law 73/2017.*: Diário da República No. 118/2017, Series I of 2017-06-21.
3. Biscaia, A. R., & Heleno, L. C. V. (2017). The reform of primary health care in Portugal: Portuguese, modern and innovative. *Ciência & Saúde Coletiva, 22*(3), 701–712. https://doi.org/10.1590/1413-81232017223.33152016

4. Biscaia, A. (2015). Unidades de Saúde Familiar: USF - modelo positivo do presente e para o futuro. USF-AN.
5. Decree-Law No. 28/2008, Diário da República No. 38/2008, Series I of 2008-02-22.
6. Familiar, C. de M. G. E. (n.d.). Reorganização de serviços em Medicina Geral e Familiar no contexto de exceção do combate à COVID-19. (2020). Last acessed 2022/03/02. https://ordemdosmedicos.pt/reorganizacao-de-servicos-em-medicina-geral-e-familiar-no-contexto-de-excecao-do-combate-a-covid-19/
7. Baptista, S., Teixeira, A., Castro, L., Cunha, M., Serrão, C., Rodrigues, A., & Duarte, I. (2021). Physician burnout in primary care during the COVID-19 pandemic: A cross-sectional study in Portugal. *Journal of Primary Care & Community Health, 12*, 1–9. https://doi.org/10.1177/21501327211008437
8. Lyng, H. B., & Brun, E. C. (2018). Knowledge transition: A conceptual model of knowledge transfer for cross-industry innovation. *International Journal of Innovation and Technology Management, 15*(5). https://doi.org/10.1142/S0219877018500438
9. Lyng, H. B., Macrae, C., Guise, V., Haraldseid-Driftland, C., Fagerdal, B., Schibevaag, L., Alsvik, J. G., & Wiig, S. (2021). Balancing adaptation and innovation for resilience in healthcare - a metasynthesis of narratives. *BMC Health Services Research, 21*(1), 1–13. https://doi.org/10.1186/S12913-021-06592-0/FIGURES/2
10. Berry, L. L. (2019). Service innovation is urgent in healthcare. *AMS Review, 9*(1–2), 78–92. https://doi.org/10.1007/s13162-019-00135-x
11. Flessa, S., & Huebner, C. (2021). Innovations in health care-a conceptual framework. *International Journal of Environmental Research and Public Health, 18*(19), 10026. https://doi.org/10.3390/ijerph181910026
12. Kippen, R., O'Sullivan, B., Hickson, H., Leach, M., & Wallace, G. (2020). A national survey of COVID-19 challenges, responses and effects in Australian general practice. *Australian Journal of General Practice, 49*(11), 745–751. https://doi.org/10.31128/AJGP-06-20-5465
13. Eberly, L. A., Kallan, M. J., Julien, H. M., Haynes, N., Khatana, S. A. M., Nathan, A. S., Snider, C., Chokshi, N. P., Eneanya, N. D., Takvorian, S. U., Anastos-Wallen, R., Chaiyachati, K., Ambrose, M., O'Quinn, R., Seigerman, M., Goldberg, L. R., Leri, D., Choi, K., Gitelman, Y., Kolansky, D. M., Cappola, T. P., Victor, A. F., Hanson, C. W., Deleener, M. E., & Adusumalli, S. (2020). Patient characteristics associated with telemedicine access for primary and specialty ambulatory care during the COVID-19 pandemic. *JAMA Network Open, 3*(12). https://doi.org/10.1001/JAMANETWORKOPEN.2020.31640
14. Taylor, A., Tieman, J., & Maeder, A. (2021). Digital innovations for aged care: Impacts in the COVID-19 pandemic. *Studies in Health Technology and Informatics, 276*, 51–57. https://doi.org/10.3233/SHTI210010
15. Wilson, G., Windner, Z., Bidwell, S., Currie, O., Dowell, A., Halim, A. A., Toop, L., Savage, R., Ranaweera, U., Beadel, H., & Hudson, B. (2021). "Here to stay": Changes to prescribing medication in general practice during the COVID-19 pandemic in New Zealand. *Journal of Primary Health Care, 13*(3), 222–230. https://doi.org/10.1071/HC21035
16. Doolittle, B. R., Richards, B., Tarabar, A., Ellman, M., & Tobin, D. (2020). The day the residents left: Lessons learnt from COVID-19 for ambulatory clinics. *Family Medicine and Community Health, 8*(3), 513. https://doi.org/10.1136/FMCH-2020-000513
17. Schmidt, S., Gupta, R., Bracey, J., Volerman, A., Henry, T., Jackson, C., Levine, D., Jones, D., Richter, J. M., Schmidt, J., Rotenstein, L., Berger, Z., Peairs, K., Singh, H., Schwartz, M., Bailey, J., & Goodson, J. (2021). Primary care practice transformation in the era of COVID-19 and beyond: Key principles for general internal medicine practitioners. *Journal of General Internal Medicine, 37*, 459–466. https://doi.org/10.1007/S11606-021-07205-6
18. Ray, S., & Mash, R. (2021). Innovation in primary health care responses to COVID-19 in Sub-Saharan Africa. *Primary Health Care Research & Development, 22*(44), 1–11. https://doi.org/10.1017/S1463423621000451
19. Blazey-Martin, D., Barnhart, E., Gillis, J., & Vazquez, G. A. (2020). Primary care population management for COVID-19 patients. *Journal of General Internal Medicine, 35*(10), 3077. https://doi.org/10.1007/S11606-020-05981-1

20. Al-Tawfiq, J. A., Kheir, H., Al-Dakheel, T., Al-Qahtani, S., AlKhadra, H., Sarhan, A., Bu Halaiga, M., & Ibrahim, R. (2021). COVID-19 home monitoring program: Healthcare innovation in developing, maintaining, and impacting the outcome of SARS-CoV-2 infected patients. *Travel Medicine and Infectious Disease, 43*, 102089. https://doi.org/10.1016/j.tmaid.2021.102089
21. Mulcahy, S. S., Fox, R., Mannion, M., Joyce, D., de Brún, A., Glynn, L., Ryan, D., Keane, N., & McAuliffe, E. (2021). A mixed methods protocol to evaluate the effectiveness and acceptability of COVID-19 Community Assessment Hubs. *HRB Open Research, 4*(16). https://doi.org/10.12688/hrbopenres.13217.2
22. Fiorentino, F., Prociuk, D., Gonzalez, A. B. E., Neves, A. L., Husain, L., Ramtale, S. C., Mi, E., Mi, E., Macartney, J., Anand, S. N., Sherlock, J., Saravanakumar, K., Mayer, E., de Lusignan, S., Greenhalgh, T., & Delaney, B. C. (2021). An early warning risk prediction tool (RECAP-V1) for patients diagnosed with COVID-19: Protocol for a statistical analysis plan. *JMIR Research Protocols, 10*(10). https://doi.org/10.2196/30083
23. Greenhalgh, T., Thompson, P., Weiringa, S., Neves, A. L., Husain, L., Dunlop, M., Rushforth, A., Nunan, D., de Lusignan, S., & Delaney, B. (2020). Original research: What items should be included in an early warning score for remote assessment of suspected COVID-19? Qualitative and Delphi study. *BMJ Open, 10*(11). https://doi.org/10.1136/BMJOPEN-2020-042626
24. Judson, T. J., Odisho, A. Y., Neinstein, A. B., Chao, J., Williams, A., Miller, C., Moriarty, T., Gleason, N., Intinarelli, G., & Gonzales, R. (2020). Rapid design and implementation of an integrated patient self-triage and self-scheduling tool for COVID-19. *Journal of the American Medical Informatics Association: JAMIA, 27*(6), 860–866. https://doi.org/10.1093/JAMIA/OCAA051
25. Driver, J. A., Strymish, J., Clement, S., Hayes, B., Craig, K., Cervera, A., Morreale-Karl, M., Linsenmeyer, K., Grudberg, S., Davidson, H., Spencer, J., Kind, A. H. J., & Fantes, T. (2021). Front-line innovation: Rapid implementation of a nurse-driven protocol for care of outpatients with COVID-19. *Journal of Clinical Nursing, 30*(11–12), 1564–1572. https://doi.org/10.1111/jocn.15704
26. Clarke, J., Flott, K., Fernandez Crespo, R., Ashrafian, H., Fontana, G., Benger, J., Darzi, A., & Elkin, S. (2021). Assessing the safety of home oximetry for COVID-19: A multisite retrospective observational study. *BMJ Open, 11*(9), 049235. https://doi.org/10.1136/bmjopen-2021-049235
27. Madela-Mntla, E. N., Renkin, W., Hugo, J. F., Germishuys, P. S., & Madela-Mntla, E. (2021). The role played by family physicians in providing health services for the sheltered homeless populations during COVID-19 lockdown in the Tshwane district. *African Journal of Primary Health Care & Family Medicine, 13*(1), 1–5. https://doi.org/10.4102/PHCFM.V13I1.3060
28. Echeverría, P., Mas Bergas, M. A., Puig, J., Isnard, M., Massot, M., Vedia, C., Peiró, R., Ordorica, Y., Pablo, S., Ulldemolins, M., Iruela, M., Balart, D., Ruiz, J. M., Herms, J., Sala, B. C., & Negredo, E. (2020). COVIDApp as an innovative strategy for the management and follow-up of COVID-19 cases in long-term care facilities in Catalonia: Implementation study. *JMIR Public Health and Surveillance, 6*(3). https://doi.org/10.2196/21163
29. Darko, D. O. (2021). Person-centred innovations for family medicine and COVID-19 in Ghana: A short report. *African Journal of Primary Health Care & Family Medicine, 13*(1), 1–4. https://doi.org/10.4102/phcfm.v13i1.3016
30. Sullivan, E. E., Dwiel, K., Hunt, L. S., Conroy, K., & Gergen, B. K. (2021). Moving the needle on primary care burnout: Using a driver diagram to accelerate impact. *Healthcare, 9*(4), 100595. https://doi.org/10.1016/j.hjdsi.2021.100595
31. Cairns, P., Aitken, G., Pope, L. M., Cecil, J. E., Cunningham, K. B., Ferguson, J., Gibson Smith, K., Gordon, L., Johnston, P., Laidlaw, A., Scanlan, G. M., Tooman, T. R., Wakeling, J., & Walker, K. (2021). Interventions for the well-being of healthcare workers during a pandemic or other crisis: Scoping review. *BMJ Open, 11*(8), 047498. https://doi.org/10.1136/bmjopen-2020-047498

32. Knierim, K., Palmer, C., Kramer, E. S., Rodriguez, R. S., Vanwyk, J., Shmerling, A., Smith, P., Holmstrom, H., Bacak, B. S., Levey, S. M. B., Staton, E. W., & Holtrop, J. S. (2021). Lessons learned during COVID-19 that can move telehealth in primary care forward, *34*, 196–202. https://doi.org/10.3122/jabfm.2021.S1.200419
33. White, S. J., Nguyen, A., Roger, P., Tse, T., Cartmill, J. A., & Willcock, S. M. (2021). Experiences of telehealth in general practice in Australia: Research protocol for a mixed methods study. *BJGP Open*, BJGPO.2021.0187. https://doi.org/10.3399/BJGPO.2021.0187
34. Neves, A. L., Li, E., Serafini, A., Jimenez, G., Lingner, H., Koskela, T. H., Hoffman, R. D., Collins, C., Petek, D., Claveria, A., Tsopra, R., Irving, G., Gusso, G., O'Neill, B. G., Hoedebecke, K., Espitia, S. M., Ungan, M., Nessler, K., Lazic, V., Laranjo, L., Memarian, E., Fernandez, M. J., Ghafur, S., Fontana, G., Majeed, A., Car, J., Darzi, A., & Darzi, A. (2021). Evaluating the impact of COVID-19 on the adoption of virtual care in general practice in 20 countries (inSIGHT): Protocol and rationale study. *JMIR Research Protocols, 10*(8). https://doi.org/10.2196/30099
35. Wanat, M., Hoste, M. E., Gobat, N. H., Anastasaki, M., Böhmer, F., Chlabicz, S., Colliers, A., Farrell, K., Hollerbach, S., Karkana, M.-N., Kinsman, J., Lionis, C., Marcinowicz, L., Reinhardt, K., Skoglund, I., Sundvall, P.-D., Vellinga, A., Goossens, H., Butler, C., van der Velden, A., Anthierens, S., & Tonkin-Crine, S. (2022). Patients' and clinicians' perspectives on the primary care consultations for acute respiratory infections during the first wave of the COVID-19 pandemic: An eight-country qualitative study in Europe. *BJGP Open, BJGPO*. https://doi.org/10.3399/BJGPO.2021.0172
36. Hamadi, H. Y., Zhao, M., Haley, D. R., Dunn, A., Paryani, S., & Spaulding, A. (2021). Medicare and telehealth: The impact of COVID-19 pandemic. *Journal of Evaluation in Clinical Practice, 28*(1), 43–48. https://doi.org/10.1111/JEP.13634
37. DeVoe, J. E., Likumahuwa-Ackman, S. M., Angier, H. E., Huguet, N., Cohen, D. J., Flocke, S. A., Marino, M., & Gold, R. (2020). A practice-based research network (PBRN) roadmap for evaluating COVID-19 in community health centers: A report from the OCHIN PBRN. *Journal of the American Board of Family Medicine: JABFM, 33*(5), 774. https://doi.org/10.3122/JABFM.2020.05.200053
38. Der-Martirosian, C., Wyte-Lake, T., Balut, M., Chu, K., Heyworth, L., Leung, L., Ziaeian, B., Tubbesing, S., Mullur, R., & Dobalian, A. (2021). Implementation of telehealth services at the US Department of Veterans Affairs during the COVID-19 pandemic: Mixed methods study. *JMIR Formative Research, 5*(9). https://doi.org/10.2196/29429
39. Brey, Z., Mash, R., Goliath, C., & Roman, D. (2020). Home delivery of medication during Coronavirus disease 2019, Cape Town, South Africa: Short report. *African Journal of Primary Health Care & Family Medicine, 12*(1), 2449. https://doi.org/10.4102/phcfm.v12i1.2449
40. Agrawal, U., Katikireddi, S. V., McCowan, C., Mulholland, R. H., Azcoaga-Lorenzo, A., Amele, S., Fagbamigbe, A. F., Vasileiou, E., Grange, Z., Shi, T., Kerr, S., Moore, E., Murray, J. L. K., Shah, S. A., Ritchie, L., O'Reilly, D., Stock, S. J., Beggs, J., Chuter, A., Torabi, F., Akbari, A., Bedston, S., McMenamin, J., Wood, R., Tang, R. S. M., De Lusignan, S., Hobbs, F. D. R., Woolhouse, M., Simpson, C. R., Robertson, C., & Sheikh, A. (2021). COVID-19 hospital admissions and deaths after BNT162b2 and ChAdOx1 nCoV-19 vaccinations in 2-57 million people in Scotland (EAVE II): A prospective cohort study. *The Lancet Respiratory Medicine, 9*(12), 1439–1449. https://doi.org/10.1016/S2213-2600(21)00380-5
41. Rocha, T. A. H., Boitrago, G. M., Mônica, R. B., Almeida, D. G., Silva, N. C. D., Silva, D. M., Terabe, S. H., Staton, C., Facchini, L. A., & Vissoci, J. R. N. (2021). National COVID-19 vaccination plan: Using artificial spatial intelligence to overcome challenges in Brazil. *Ciencia & saude coletiva, 26*(5), 1885–1898. https://doi.org/10.1590/1413-81232021265.02312021
42. Vilendrer, S., Amano, A., Brown Johnson, C. G., Favet, M., Safaeinili, N., Villasenor, J., Shaw, J. G., Hertelendy, A. J., Asch, S. M., & Mahoney, M. (2021). An app-based intervention to support first responders and essential workers during the COVID-19 pandemic: Needs assessment and mixed methods implementation study. *Journal of Medical Internet Research, 23*(5). https://doi.org/10.2196/26573

Competitive and Business Intelligence: A Bibliometric Analysis

Bruno de Sousa Lopes, Vanessa Amorim, Manuel Au-Yong-Oliveira, and Orlando Lima Rua

Abstract This study aims to conduct a bibliometric analysis inherent to assessing competitive intelligence and business intelligence concepts using the Scopus database and the Bibliometrix R software. The study's articles were found using precise criteria in the Scopus database. The 42 publications were then examined with Bibliometrix software, which included extensive parameterization for each component under evaluation. The results of this study consisted of establishing the number of existing publications on the topic under analysis between 2017 and 2021 – in this sense, it was possible to identify that the publications are experiencing an annual decrease rate of 22.69%; the trends in terms of publications and collaborations between countries; the most relevant journals in the area; and the interconnections between authors, keywords, and publications. This study has as an added value the possibility to evaluate the relevance attributed by academics to ascertain the most important contributions in terms of authors, articles, and journals. One major limitation in this study could be addressed in future research. The study focused on a limited study field in the context of business, management, and accounting, so it would be very pertinent to understand how this topic has evolved, particularly in the area of computer science.

Keywords Bibliometric analysis · Business intelligence · Competitive intelligence · Scientometrics

---

B. de Sousa Lopes (✉) · V. Amorim
Department of Economics, Management, Industrial Engineering and Tourism (DEGEIT), University of Aveiro, Aveiro, Portugal
e-mail: vanessa.amorim@ua.pt

M. Au-Yong-Oliveira
INESC TEC, GOVCOPP, Department of Economics, Management, Industrial Engineering and Tourism, University of Aveiro, Aveiro, Portugal

O. L. Rua
Center for Organisational and Social Studies (CEOS.PP), Porto School of Accounting and Business (ISCAP), Polytechnic of Porto (P.PORTO), S. Mamede de Infesta, Portugal

J. C. de Oliveira Matias et al. (eds.), *Quality Innovation and Sustainability*, Springer Proceedings in Business and Economics,
https://doi.org/10.1007/978-3-031-12914-8_15

1 Introduction

The current markets where organizations operate are increasingly unstable and uncertain due to technological changes becoming more disruptive [1]. In this sense, the authors [2] suggest that maintaining dynamism in an increasingly competitive context is synonymous with business intelligence. According to this, business intelligence, which is an analytical tool, helps businesses make important decisions while taking their long-term performance into account [3].

Although knowledge is regarded as a valuable resource in an organization because it contributes to the achievement of competitive advantage [4], given the technological context in which we currently live, it is essential to emphasize its integration within the scope of competitive intelligence because this concept is broader and includes the agglomeration of the concepts of data, information, and knowledge to develop strategies and insights that help organizations to strategically make decisions [1]. Competitive intelligence, like business intelligence, is widely regarded as a necessary tool for every organization attempting to survive and, more importantly, remain competitive in the marketplace [5].

Given the importance and recognition defined in the literature, the concepts of business intelligence and competitive intelligence are critical for enhancing organizations' competitiveness and development in an increasingly technological environment where decision-making ability is crucial for an effective strategy. In this scenario, it becomes critical to examine how researchers have contributed to this subject by bibliometric analysis. Following the literature review to identify possible gaps in the literature, it is concluded through the Scopus database search that there are no studies that are a bibliometric literature review researching the concepts of business and competitive intelligence.

After considering the research question, "Which authors, articles, and journals contribute significantly to state of the art in the area of business and competitive intelligence?" it was established that the objectives are based on defining and understanding the trends of publications and collaborations between countries; the most relevant journals in the area; and the interconnections between authors, keywords, and publications. Thus, this paper will resort to a bibliometric analysis, parameterizing the query used in the Scopus database between the years 2017 and 2021.

Between 2017 and 2021, the results demonstrate that 132 authors authored 42 publications on business and competitive intelligence. Regarding the evolution and classification of scientific production, it is clear that the publication of research on business and competitive intelligence has decreased significantly since 2017. The United States has the most publications (19), followed by China (16), Australia (12), Italy (12), and Portugal (12). The United States and Turkey, Portugal, and Spain are the leading countries the authors collaborate with. In terms of publications with the most significant impact on the scientific community, this analysis determined that *Decision Support Systems* and *Industrial Management and Data Systems* journals have the highest total number of citations and H-index. Finally, the authors

[6, 7] summarize the articles that received the most global citations between 2017 and 2021.

The chapter is divided into five sections. Section 1 introduces the concepts of business intelligence and competitive intelligence and their associations, and presents the research question, objectives, and main results from the bibliometric analysis. Section 2 presents the literature review, divided into two subsections, namely, business and competitive intelligence, highlighting the main concepts and their interrelation to organizations. Section 3 presents a detailed analysis of the methodology used and the criteria applied in the Scopus database. Section 4 presents the study's results, subdivided into subsections that specifically address (1) characterization of the data collected; (2) evolution and characterization of scientific production; and (3) bibliometric analysis. Finally, Sect. 5 presents the final considerations of this study, further identifying its limitations and the possibility of future research.

2 Literature Review

2.1 Business Intelligence

Business intelligence, like competitive intelligence, originated in the 1950s due to researchers'efforts to generate and collect massive volumes of data [8]. The concept of business intelligence is a significant mainstream trend in organizations nowadays as it allows unsupported work to be eliminated or reduced, allowing specific decisions to be made based on automated information systems, which in addition provide better management of immediate decisions [2]. In this follow-up, organizations use business intelligence systems to analyze disparate data sources and, as a result, perform predictive analytics that enables the building of real-time responses [3]. Additionally, the authors [9] assert that there is actual evidence that business intelligence is favorably connected with an organization's ability to develop internationally.

Business intelligence tools enable decision support through various elements [3]. Thus, the following components might be considered to help organizations improve their decision-making processes: (1) personalized reports; (2) information overload mitigation; (3) management performance evaluation; (4) business process simulation and scenario planning; (5) add-on features; (6) collaborative business intelligence; (7) data governance and cloud connection; and (8) collaborative business intelligence [3].

Despite the apparent benefits of business intelligence systems, limitations to their use are highlighted, including the following: (1) "its use as a data analytic system for achieving strategies since its typical use is in reporting KPIs by metric measurement"; (2) "its usability limited to few select staff instead of every BI user for establishing decisions"; (3) "drilling down through data layers for extracting information instead of using data supplied into BI tool for analysis"; (4)

"responsibility of the IT team for data integration from disparate sources instead of business managers"; and, finally, (5) "poor data quality maintenance discipline by the organization" [3].

2.2 *Competitive Intelligence*

According to the authors [1], competitive intelligence is "a process that generates actionable information about the firm and its external environment to help firms in making market-related decisions." From the same perspective, the Strategic & Competitive Intelligence Professionals adds that the concept "is a discipline that enables organizations to reduce strategic risk and increase revenue opportunities by having a deep understanding of what has happened, what is happening, and what may happen in their operating environment" [10]. The author [5] notes, on the other side, that the concept is not recent, having been introduced into business reality in the 1960s as part of marketing research.

Considering the volatility and nonstability of the organizations' environments [11], this concept extends far beyond the fact that it attributes competitive advantages to organizations while seeking to ensure business sustainability. Insofar as the new structural resources of organizations are currently represented by data, information, tacit and explicit knowledge, it is essential for organizations in the current century to develop intelligent competitive environments [1].

In terms of competitive intelligence's results in Europe, the authors [12] demonstrate in their study that the application's reach has expanded incrementally in areas connected to the organizations' clients, technology, and markets.

3 Methodology

To gain a comprehensive understanding of business and competitive intelligence in the business context, scientometrics was used to identify the significant contributions to this topic. In 1969, Nalimov used scientometrics to refer to a quantitative technique for assessing science as an information process [13]. The assessment of the quality and influence of research publications and academic journals, the comprehension of scientific citations, and the mapping of scientific areas are significant research concerns [14].

The research was conducted in February 2022, throughout the Scopus database, applying the terms "business intelligence" and "competitive intelligence" within article titles, abstracts, and keywords, as it is possible to observe through the following query – TITLE-ABS-KEY("business intelligence" AND "competitive intelligence") AND (LIMIT-TO (PUBSTAGE, "final")) AND (LIMIT-TO (DOCTYPE,"ar")) AND (LIMIT-TO (SUBJAREA, "BUSI")) AND (LIMIT-TO (EXACTKEYWORD,"Competitive Intelligence") OR LIMIT-TO

Table 1 List of items and search criteria

Items	Criteria
Time horizon	2017–2021
Database	Scopus
Keywords	"Business intelligence" and "competitive intelligence"
Subject area	Business, management, and accounting
Document type	Articles
Type of publications	Journals
Publication stage	Final
Language	English
Software	Bibliometrix R
Documents analyzed	42

(EXACTKEYWORD,"Business Intelligence")) AND (LIMIT-TO (LANGUAGE,"English")) AND (LIMIT-TO (SRCTYPE,"j")) AND (LIMIT-TO (PUBYEAR,2021) OR LIMIT-TO (PUBYEAR,2020) OR LIMIT-TO (PUBYEAR,2019) OR LIMIT-TO (PUBYEAR,2018) OR LIMIT-TO (PUBYEAR,2017)). Scopus's inclusion criteria are listed in Table 1.

4 Results

4.1 Characterization of the Data Collected

Between 2017 and 2021, 132 authors published 42 articles on business and competitive intelligence diffusion, with 2475 references. The literature is highly collaborative since collaboration among authors appears to be a vital component, and as a result, just five authors have written works independently. Table 2 presents a summary of the data collected.

4.2 Evolution and Characterization of Scientific Production

In terms of the temporal evolution of publications, Fig. 1 illustrates researchers' loss of interest in the topics. Accordingly, there has been a significant reduction in research publishing on business and competitive intelligence since 2017.

Additionally, the current analysis attempted to discover which nations publish the most significant amount of publications on the topics under investigation. The relevance of each country in terms of the total number of publications evaluated is depicted in Fig. 2, according to the land of the primary author. As it is possible to

Table 2 Characteristics of the data collected

Description	Results
Main information about data	
Time span	2017:2021
Articles	42
Average years from publication	3.55
Average citations per document	13.02
Average citations per year per doc	2.748
References	2475
Document contents	
Author's keywords (DE)	162
Authors	
Authors	132
Authors of single-authored documents	5
Authors of multi-authored documents	127

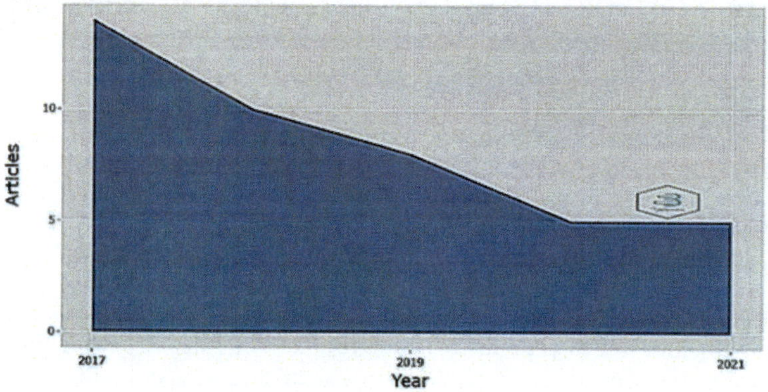

Fig. 1 Evolution of the number of publications by year

verify, the countries with the highest scientific production are the United States with 19 publications, China (16), Australia (12), Italy (12), and Portugal (12).

Furthermore, a map can be used to determine the origins of the contributing authors. As illustrated in Fig. 3, the significant collaborations between authors include the United States of America and and Turkey, and Portugal and Spain.

The H-index is a bibliometric indicator that is widely regarded as extremely valuable. According to the authors [15], the H-index should (1) "reflect elements of reality that are useful for evaluation and meaningful in a statistical sense and ideally have predictive power"; (2) "not lead to undesirable incentives that are detrimental to the progress of science"; (3) "not be too sensitive to small variations in bibliometric records that could be due to random events"; and (4) "be not too difficult to obtain from existing databases."

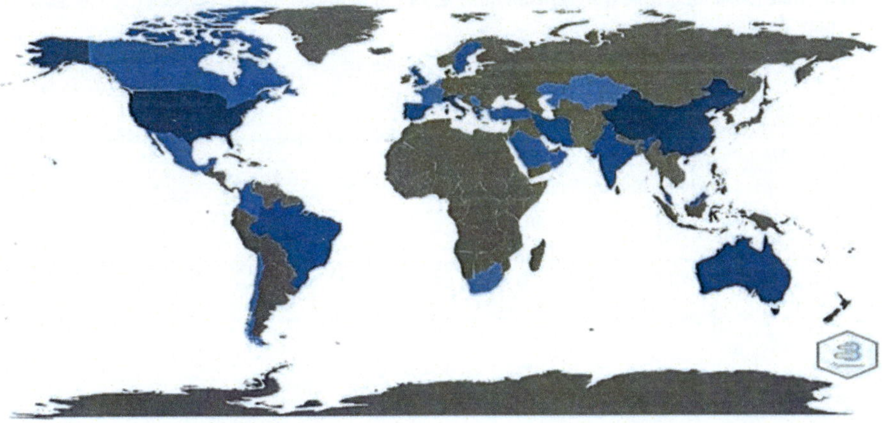

Fig. 2 Publications by country

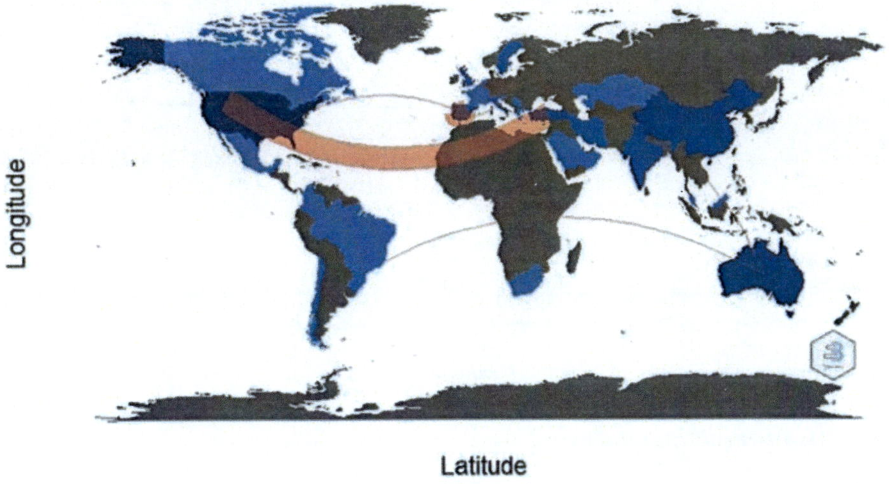

Fig. 3 Country collaboration map

The association between the most prominent journals on the topic of analysis, the number of publications, their H-index value, and the total citations index is shown in Table 3.

The word growth associated with the keywords, as shown in Fig. 4, in terms of cumulative occurrences after 2019 suffered a slight growth, but has remained with a stabilizing trend. On the other hand, it can be seen that the keyword "competitive intelligence" represents about 40% of the occurrences, which justifies the relevance of this topic, and the keyword "information analysis" is the second keyword with more occurrences, gathering about 25% of them.

Table 3 Journals with the most significant number of publications and the respective H-index and total citations

Journal	Number of publications	H-index	Total citations index
Decision Support Systems	6	6	261
Industrial Management and Data Systems	4	3	40
Journal of Cases on Information Technology	3	1	5
International Journal of Business Intelligence and Data Mining	2	–	–
International Journal of Information Management	2	2	22

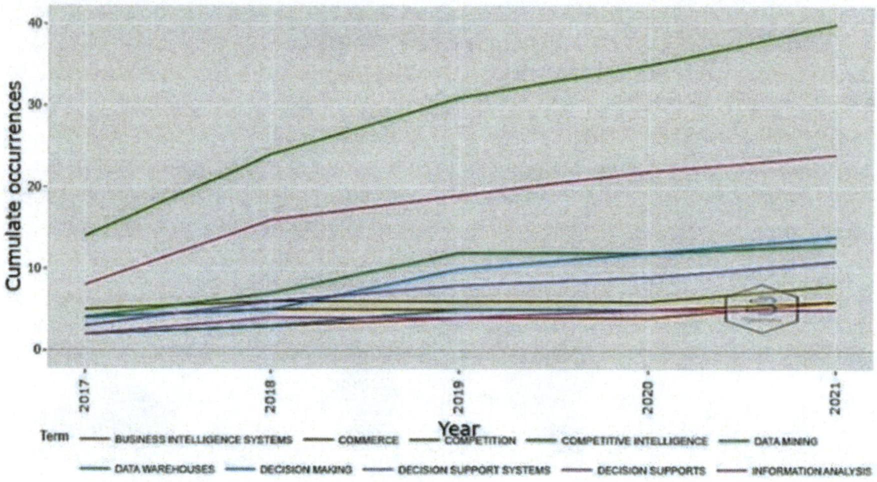

Fig. 4 Word growth in terms of cumulate occurrences

4.3 Bibliometric Analysis

The relationship between the authors, the topics investigated, and the journals in which their results are published can be seen in the three-field chart in Fig. 5. The bigger the colored rectangles are, the more critical a journal, keyword, or author is in the graphical analysis.

The bibliometric analysis also establishes a direct relationship between the researchers with the highest number of publications and the overall production of the principal authors over several years in the context of business and competitive intelligence research (Fig. 6).

Regarding direct citations through Fig. 7, it is possible to analyze the documents, authors, and journals that accumulate the most significant number of global

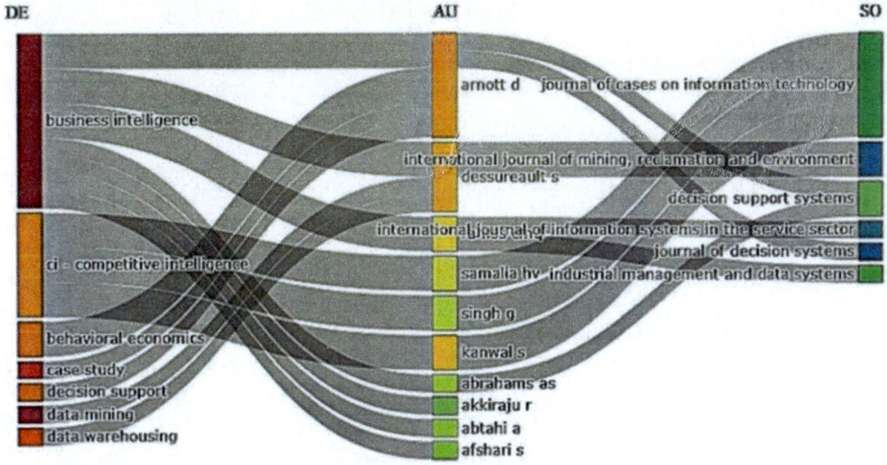

Fig. 5 Collaborative network between keywords (left), authors (center), and journals (right)

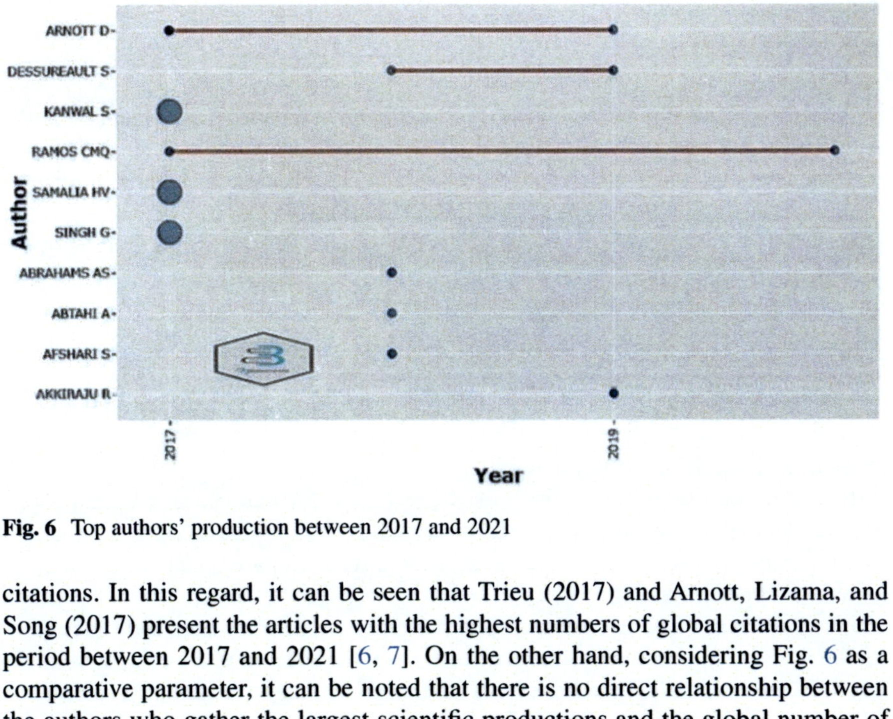

Fig. 6 Top authors' production between 2017 and 2021

citations. In this regard, it can be seen that Trieu (2017) and Arnott, Lizama, and Song (2017) present the articles with the highest numbers of global citations in the period between 2017 and 2021 [6, 7]. On the other hand, considering Fig. 6 as a comparative parameter, it can be noted that there is no direct relationship between the authors who gather the largest scientific productions and the global number of citations.

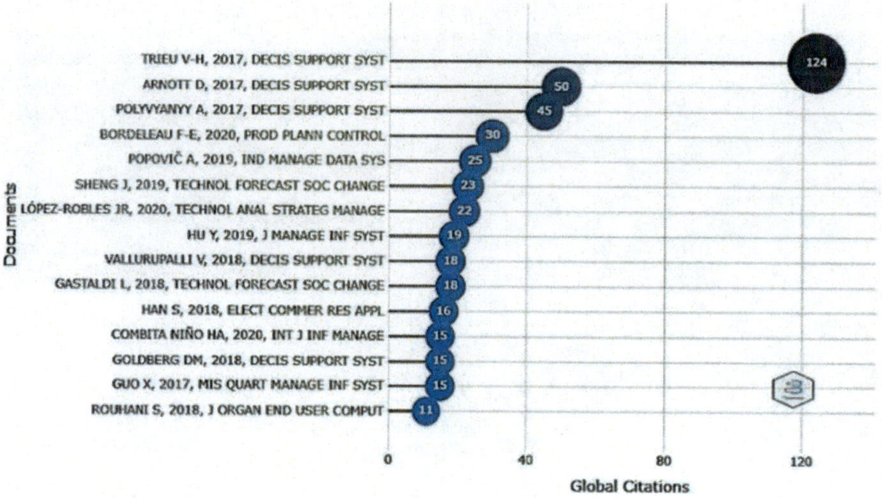

Fig. 7 Most globally cited documents between 2017 and 2021

5 Conclusions

A bibliometric analysis of scientific production on "business intelligence" and "competitive intelligence" available in the Scopus database in the domains of business, management, and accounting was carried out, resulting in the identification of 42 publications in the final stages of publication in these fields.

This analysis was used to characterize the scientific production framework, which included determining which countries contribute the most to scientific publication and the collaboration between countries, the temporal evolution of scientific production, which journals contribute the most to publications on this topic, the impact on terms of the H-index, and the number of publications.

There has been stabilization in scientific production on the topic addressed in recent years. Research peaked in 2017 and demonstrated that the new trend is data mining and decision support systems.

Since the topics of competitive intelligence, business intelligence, data mining, and decision support systems are interconnected, it is expected that the scientific community approaches all the issues to add value through theoretical and practical implications in organizations.

The results also allow us to establish that greater cooperation between the international academic community should be promoted because, in terms of geographic analysis, there is a smaller number of nations that contribute to the scientific production of this theme, given the existence of an international collaboration that is not significant, except the cooperation between the United States and Turkey and Portugal and Spain.

This work has a significant limitation that could be addressed in future research. In addition, because the analysis was limited to a specific study field within the context of business, management, and accounting, it would be very beneficial to understand how this topic has grown through time, particularly in computer science, considering a more extended period for this purpose.

Acknowledgments This work was financed by national funds through the FCT – Foundation for Science and Technology, I. P., under the project with the reference UI/BD/152804/2022.

References

1. Cavallo, A., Sanasi, S., Ghezzi, A., & Rangone, A. (2021). Competitive intelligence and strategy formulation: Connecting the dots. *Competitiveness Review: An International Business Journal, 31*(2), 250–275.
2. Srinivasan, S. M., Shah, P., & Surendra, S. S. (2021). An approach to enhance business intelligence and operations by sentimental analysis. *Journal of System and Management Sciences, 11*(3), 27–40.
3. Mathrani, S. (2021). Critical business intelligence practices to create meta-knowledge. *International Journal of Business Information Systems, 36*(1), 1–20.
4. Hasnain, S. (2015). Organisational knowledge: The Panacea to NGO-operations. *Procedia–Social and Behavioral Sciences, 195*, 197–201.
5. Priporas, C. V. (2019). Competitive intelligence practice in liquor retailing: Evidence from a longitudinal case analysis. *International Journal of Retail & Distribution Management, 47*(9), 997–1010.
6. Trieu, V.-H. (2017). Getting value from Business Intelligence systems: A review and research agenda. *Decision Support Systems, 93*, 111–124.
7. Arnott, D., Lizama, F., & Song, Y. (2017). Patterns of business intelligence systems use in organizations. *Decision Support Systems, 97*, 58–68.
8. Eggert, M., & Alberts, J. (2020). Frontiers of business intelligence and analytics 3.0: A taxonomy-based literature review and research agenda. *Business Research, 13*, 685–739.
9. Cheng, C., Zhong, H., & Cao, L. (2020). Facilitating speed of internationalization: The roles of business intelligence and organizational agility. *Journal of Business Research, 110*, 95–103.
10. SCIP – Strategic & Competitive Intelligence Professionals. https://www.scip.org/page/CI-MI-Basics-Topic-Hub. Last accessed 22 Feb 2022.
11. Tahmasebifard, H. (2018). The role of competitive intelligence and its subtypes on achieving market performance. *Cogent Business & Management, 5*, 1–16.
12. Calof, J., Arcos, R., & Sewdass, N. (2018). Competitive intelligence practices of European firms. *Technology Analysis & Strategic Management, 30*(6), 658–671.
13. Hood, W. W., & Wilson, C. S. (2001). The literature of bibliometrics, scientometrics, and informetrics. *Scientometrics, 52*(2), 291–314.
14. Mingers, J., & Leydesdorff, L. (2015). A review of theory and practice in scientometrics. *European Journal of Operational Research, 246*(1), 1–19.
15. Hirsch, J. E., & Buela-Casal, G. (2014). The meaning of the h-index. *International Journal of Clinical and Health Psychology, 14*(2), 161–164.

Integrated Management System Role-Play Simulation: Training and Development Tool

Kristína Zgodavová, Kristína Lengyelová, Peter Bober, and Tibor Vincze

Abstract This chapter aims to present the history of role-play simulation (RPS), its application to the design of an integrated management system, and show the survey results using IMS-RPS to improve the existing course. The target group discussion methodology was used based on interviews conducted via the Internet during the Covid-19 pandemic. The objective of the group discussion was to innovate the quality management system master course curriculum at the Technical University of Košice (TUKE). The learning outcomes for the study were selected by a team of experts involved in the KEGA 018TUKE-4/2022 and VEGA 1/0633/20 projects. We tested a new approach to education during a pandemic on a sample of master's students using the innovated RPS virtual environment, smart glasses, and WebEx team collaboration tool using Verbal Feedback method. The results contributed to learning, knowledge creation, and knowledge transfer. The limitation of the study is the complexity of the organization's environment and the measurability of achieved real improvement as well as possible empirical verification of the results.

Keywords Integration · Quality · Environment · Occupation healthy · Safety · Management system · Role-play · Simulation

---

K. Zgodavová (✉)
Faculty of Materials, Metallurgy and Recycling, Technical University of Košice, Košice, Slovakia
e-mail: kristina.zgodavova@tuke.sk

K. Lengyelová
Faculty of Mechanical Engineering, Technical University of Košice, Košice, Slovakia

P. Bober · T. Vincze
Faculty of Electrical Engineering and Informatics, Technical University of Košice, Košice, Slovakia

J. C. de Oliveira Matias et al. (eds.), *Quality Innovation and Sustainability*, Springer Proceedings in Business and Economics,
https://doi.org/10.1007/978-3-031-12914-8_16

1 Introduction

Simulation of real situations has been with humankind for a long time. Today, simulation as advisable and purposeful experimenting with a model has found its fixed place in many fields and disciplines [1]. Besides simulations in dynamic socio-eco-technical systems, the simulation could also be used for training people. In a simulated environment, a person acquires new skills and learns how to react to new situations. Such training is cost-effective, and the experience can confirm its fruitfulness [2, 3]. A specific tool for such education is role-play simulation (RPS), an experiential learning method where participants (role-players) improvise the story with learners as part of a simulated scenario. Role-play is widely acknowledged as a powerful technique across multiple avenues of training and education [4, 5].

RPS dates back to the 1960s when William A. Gamson at the University of Michigan developed SIMSOC [6], which primarily aimed to build first-person experience in a safe and supportive environment. Using RPS for medical patient role-playing was invented by Howard Barrows in 1963 at the University of Southern California, and later it was also used for military training [7], law enforcement training [8], business leadership training [9, 10], quality control training [11], management system development [1, 12], and also for overcoming the resistance of change and forecasting [13, 14].

RPS in the industry emerged in the late 1990s, primarily in the United Kingdom, when *Simulation & Gaming Yearbook* began to be published [15]. Many companies specialize in providing role-players for workplace simulations. Practical training can be more expensive the more complex the system. Similarly, education for students without the possibility of internships is inefficient. A complex situation is a situation that can be difficult to define and can change significantly in response to a solution; it may not have a single "correct" answer; has many interconnected causal forces, and many stakeholders [16].

Whenever individual learning involves personal mastery, mental models and team learning require a shared vision and values, a comprehensive understanding of the organization's processes and products and their interconnections are needed, and systems thinking is essential [17, 18]. Systems thinking offers a way to predict future results better and navigate difficult situations. All new situations require some learning, and in some cases, learning can take place in a low-cost environment in the form of simulation and role-playing [19]. Such a situation is also a project for developing an organization's quality management system or an even more complex problem when it is necessary to integrate several management systems [20, 21]. Management system role-play simulation (MS-RPS) methodology and software help overcome this complexity. Further development of this idea is using human activity in a simulated environment to modify it and create a new, better environment. The results are two facts: (1) a new environment and (2) new skills in this environment. New skills acquired this way include overcoming resistance to change, which arises as a natural human response to change and often hampers new practices and tools [13]. By including a person in the situation, a modification of the environment is achieved that reflects the current abilities of the persons involved. This is also advantageous when implementing a new management system created using RPS.

This chapter aims to present the history of role-play simulation (RPS), its application to the design of an integrated management system, and show the Verbal Feedback survey results using IMS-RPS to improve the existing course. The following sections describe the existing MS-RPS methodology, its extension for the integrated management systems projects, and the results of target group discussion, which allowed us to design the sequence of steps needed to achieve learning objectives and learning outcomes for contemporary Industry 4.0.

2 Description of the Management System Role-Play Simulation

The first version of the role-playing simulation, developed at the Technical University in Košice [19], took the form of paper-playing cards with a user manual and focused on the quality management system (QMS) development project and participants' training. Since then, understanding of business models, management systems, and value creation has changed and developed considerably. Markets, industries, as well as society are changeable [22]: from local to global [23]; from product-oriented to process-oriented [24]; from analog to digital [25]; from the supply side to the demand side [26]; from closed to open [27]. Information and communication technologies (ICTs) have also been developing in parallel.

RPS is intended to provide players and potentially interested parties with the opportunity to experiment with the simulated context (that mimics reality) of the organization, organizational structure, process model, roles and responsibilities, inputs and outputs, and internal auditing. The teacher/moderator uses the role-playing methodology to gradually achieve the course's learning objectives during the course. The role-play simulation‘s new, integrated management system model focuses on management systems design using the integration of ISO 9001:2015, ISO 14001:2015, and ISO 45001:2018 standards. The integrated management system role-play simulation (IMS-RPS) aims to contribute to learning, knowledge creation, and knowledge transfer obtained through a web-based role-play simulation environment as a training and development tool [1, 28]. An IMS combines all aspects of an organization's systems, processes, and standards into one interconnected unit. This integration allows a business to streamline its management, save time, and increase efficiency by viewing all components of the management system as a whole [29, 30]. The essence of the innovated role-play simulation is team experimentation with pre-prepared processes and their inputs and outputs within the IMS. The team consists of students or managers, engineers, and employees of any organization. Experimentation concerns the organization's context, policy, strategy, and goals. Players occupying managerial positions then play for process ownership,

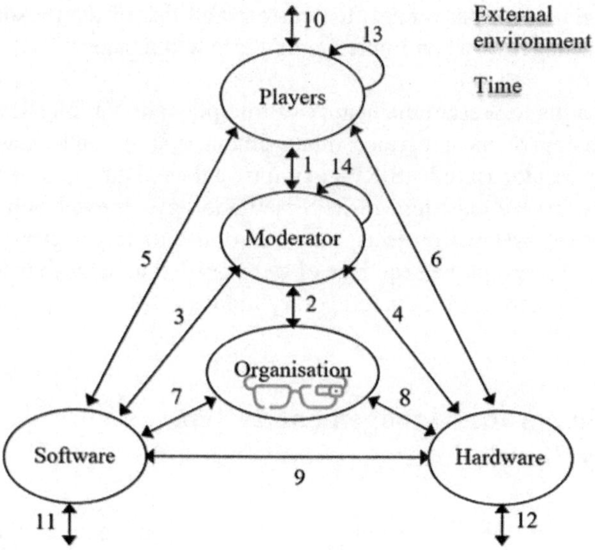

Fig. 1 Linking RPS participants and hardware and software during education

linking inputs and outputs, and internal audit of the management system. The experimentation results are recorded in pre-prepared templates, which form the basis for documenting the information. The links between the RPS participants and hardware and software are shown in Fig. 1.

Elements:

- Moderator: teacher
- Player: student, manager, engineer, employee
- Organization: assistant from the organization
- Software: IMS-RPS, smart glass, process modeling
- Hardware: ICT, Internet, smart glasses

Relations:

- 1, 2, 3, 4, 5, 6 – links between moderator and player, an assistant from the organization, software, and hardware
- 7, 8, 9 – links between assistant and smart glass software and ICT
- 10, 11, 12 – links between players, software, ICT, and external environment during specific the time
- 13 – feedback from the players for RPS or moderator improvement, memory information, and self-assessment
- 14 – moderator feedback for RPS maintenance and improvement

After choosing the management system to be played, the moderator explains to players that they will play the role of the organization's managers. Their task is to create an organization's management system according to one of the ISO management standards or integrated system.

Under the guidance of the moderator, players go through the individual stages of the role-play and perform the assigned tasks. They can help each other, discuss with the moderator, study online documentation, and use additional equipment such as smart glasses. Typical smart glasses applications are listed in [31]. Smart glasses are used to understand a student's experience, abilities, and nature as a teacher. They can be used for video recording and capturing essential points using an image, but mainly a virtual accompaniment of participants through the environment of an entire organization for which a new management system is to be prepared according to predefined requirements.

3 MS-RPS Software

Web-based software was developed in the PHP language. MySQL database was selected for data storage. The database contains

- Main situation types for QM–RPS in an organization from the points of view of the quality policy, efficiency, and effectiveness.
- Main situation types in product quality from the technological, material, production, and service provision points of view.
- Main situation types in production processes quality concerning performance and qualitative capability.
- Main managing activities for the detailed development and valuation of process importance in RPS (planning, organization, leadership, and control).
- Worksheets – "Play cards" for the process definition in quality management role-play simulation in compliance with ISO 9001:2008, 9004:2000, the EFQM Excellence Model, and the MBQA National Quality Award Model. These should be partially completed by processes to be solved.
- Blank graphical management systems models of process definitions for players to take notes.
- Blank matrices of integrated quality management processes and managers participating in role-play simulation.
- Graphical model of main processes in the value chain and relevant activities or groups of activities that have to be performed by managers participating in RPS.

Language versions of role-plays are recorded in the database, enabling various players to participate in the play in various languages simultaneously. As part of the play, there is a monolingual explanatory dictionary and a library of detailed information to facilitate the players' decision-making process.

Examples of a player's graphical interface for phases 1 and 5 are provided in Figs. 2 and 3. The remaining phases' interfaces are provided at http://web.tuke.sk/simpro-ims/index_en.php.

Examples of a moderator's graphical interface are provided in Figs. 4 and 5.

Other interface details are visible after logging the moderator into the system.

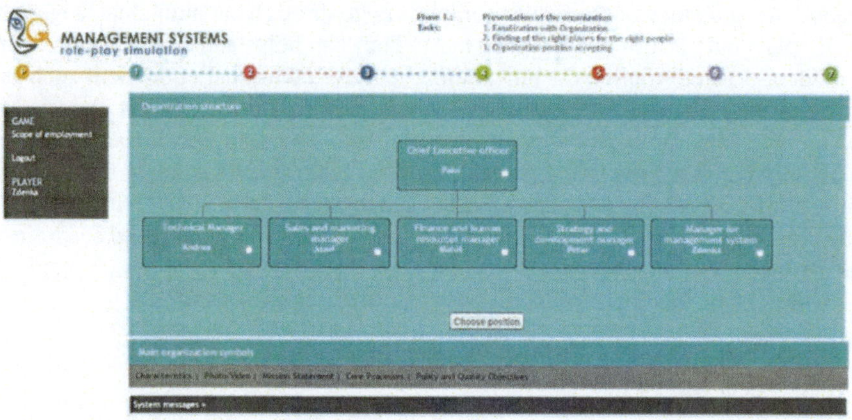

Fig. 2 Phase 1: presentation of the organization

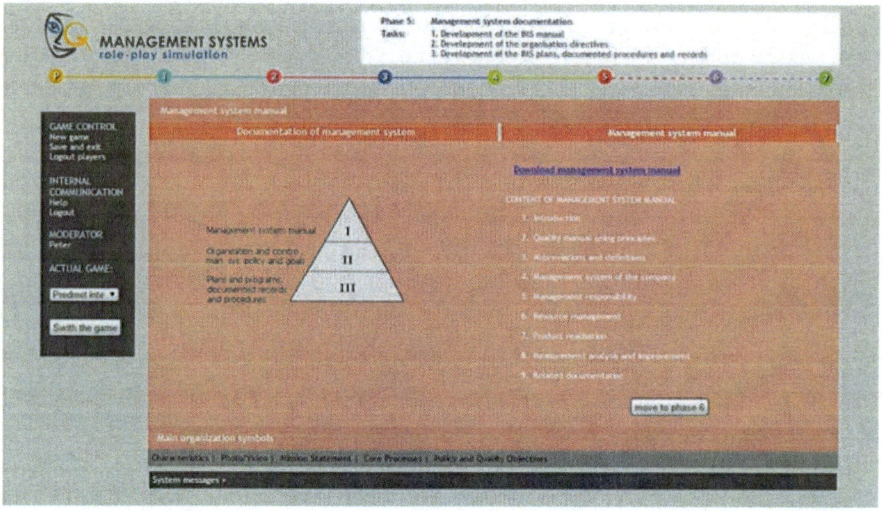

Fig. 3 Phase 5: management system documentation

4 Knowledge, Skills, and Competencies Acquired Using RPS

The research aimed to identify specific knowledge, skills, and competencies necessary for the positions of quality managers, health and safety managers, environmental managers, integrated systems managers, and process engineers. RPS has become a key educational tool for teaching management systems courses for master's degree students at the Technical University of Košice (TUKE).

Learning objectives for the courses have been discussed by an expert team consisting of 5 representatives of industrial organization, 5 teachers, and 10 students.

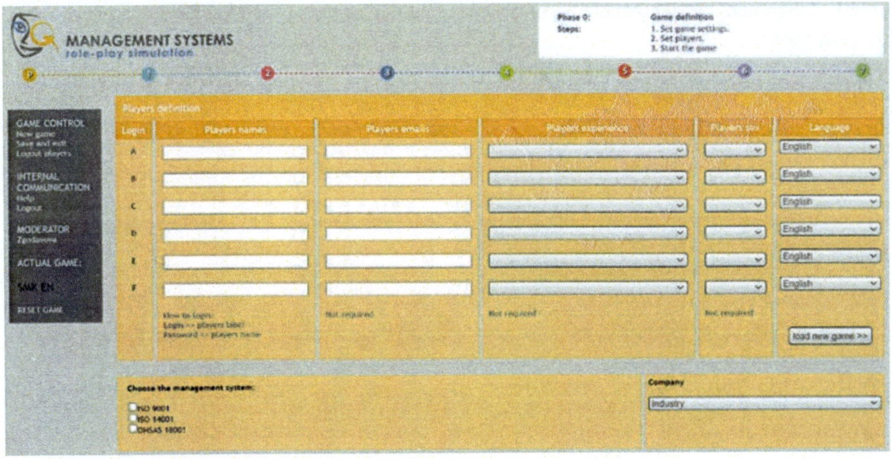

Fig. 4 Following registered players by a moderator

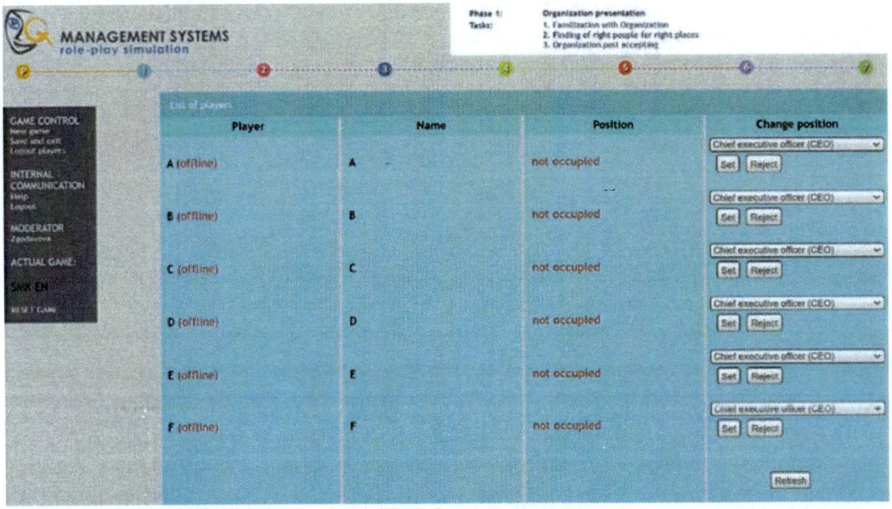

Fig. 5 Monitoring the occupation of job positions by the moderator

The discussion survey began by compiling a list of required learning outcomes in various areas related to managerial and engineering skills, ISO standards, and the expertise needed for today's Industry 4.0, according to [32]. The list was distributed among the respondents, followed by comments and questions, collected and critically assessed to consolidate or discard unnecessary items. Learning objectives observed Bloom's taxonomy [33] to be clear and appropriately structured. The result is six phases of IMS-RPS during which students achieve learning objectives as shown in Table 1.

Table 1 Integrated management system role-play simulation phases and knowledge creation and skills acquisition

Phases	Knowledge and skills
1. Organization presentation Familiarization with the organization Finding of the right people for the right roles Organization post accepting	Knowledge acquisition
2. Experimenting with integrated management system processes Selection from randomly distributed processes Purposely distributed processes acceptance Process responsibility acceptance	Knowledge creation
3. Experimenting with integrated management system process inputs and outputs Selection of inputs for chosen processes Inspection and revision of purposely distributed process inputs Assumption of inputs	Knowledge creation
4. Process definition and modeling Definition of the main (value-adding) process Modeling of these processes Simulation of these processes	Skills acquisition
5. Management system documentation development Development of the management system's manual Development of the organizational directives Development of plans, programs, documented procedures, and records	Skills acquisition
6. Experimenting with IMS support Auditing of partner's processes Creating audit objectives, plans, and report Evaluation of the audit results	Skills acquisition

5 Methodology

A target group discussion methodology was used. The lead researcher brought together two groups of individuals to discuss specific topics based on the participants' complex personal experiences, beliefs, perceptions, and attitudes through moderated interaction [34, 35]. Identifying the main goal and defining the key research objectives of the study is the beginning of the research process [36]. Based on the research objectives, a list of questions was prepared as guidance for each focus group discussion session. Four questions were discussed with teachers and students, and industry representatives, such as satisfaction with the RPS teaching method, achievement of learning objectives, and the ability to comment and make decisions at each step to obtain relevant information about each item. The Verbal Feedback method [37] was used to evaluate the course by moderators/teachers as well as players/students.

6 Discussion and Results

By providing students with realistic scenarios, similar to those in the workplace, students are able to solve problems and consider different perspectives while collaborating with classmates. This is consistent with Herrington's description of "cognitive reality" (Herrington et al., 2009, p. 45), where students can apply new knowledge and develop skills needed for the real world. Role-play is widely acknowledged as a powerful teaching technique in face-to-face, blended, and online settings and has been singled out as an example of good practice, e.g., by the Australian Learning and Teaching Council [5, 38].

After completing the course, we asked teachers and students to evaluate the course. The most interesting findings from the verbal evaluation by teachers are described in Sect. 6.1 and by students in Sect. 6.2.

6.1 Teacher Satisfaction Survey Results

- Graduates have a profound knowledge of terms and definitions related to standardized management systems.
- They can distinguish between different types of organizations' cultures and design an organization's vision and values according to its mission.
- They can design, implement, and improve the quality of processes, products, and systems, make fact-based decisions, and plan and develop management systems projects.
- Based on the knowledge of the management systems requirements according to ISO 9001:2016, ISO 14001:2015, ISO 45001:2018, they understand organizational structures and related roles, can design authorities and responsibilities matrix, and prepare management system manual and related documentation. Can assess the risks associated with processes and products and manage change.
- They have the skills needed to audit management systems based on the recommendations of ISO 19011:2019. They can compile an audit plan, perform an internal audit, and compile an audit report.
- They can present and defend the results of their creative activity, evaluate and make decisions in complex situations, lead larger teams, and prepare and implement extensive projects.

6.2 Student Satisfaction Survey Results

- RPS uses basic terms and definitions according to ISO standards and IMS principles and concepts that are easy to remember.
- RPS links issues with organizations where ISO standards are in place.

- The teacher had a high level of expertise in moderating the RPSs and an excellent overview for answering my questions.
- The first phase: Organization presentation was beneficial as it brought us into the organization environment, and the use of smart glass was especially interesting and new for us.
- The ability to model the organization's processes using a software tool via remote server access was advantageous, especially during the pandemics.
- The ability to audit a process owned by a colleague helped me understand the principle of auditing.

7 Conclusion

The long-term application and development of IMS-RPS have proved its high effectiveness both in educational processes and in implementing single management system according to ISO 9001, ISO 14001, and ISO 4500, or an integrated management system in organizations [39]. The benefits of QM-RPS depend on the following facts: the purposefulness of an entrance QMS model, and competence and motivation of both moderators (teachers/trainers) and role-players (students/managers).

Pros and cons according to [40] and our experiences are as follows:

Pros:
- RPS allows students to use their knowledge and skills in practice.
- RPS outperforms resistance to change.
- RPS breaks the ice both between teacher and students and amongst the students themselves.
- Teacher will learn more about their students.
- Both students and teachers have a good feeling after RPS-based course.

Cons:
- Course may not go according to a plan.
- Students may be skeptical about it.
- What works for one group may not work for another.

The theoretical implication of the presented IMS-RPS reflects on learning in a way that makes tacit knowledge much more explicit and engaging. The practical sense of the RPS is when it is used for the IMS development in an organization. IMS can be developed in less time, role-play will overcome managers' resistance to change, and participants will better understand the objectives, nature, and importance of the management system and documented information management.

The limitation of this study is the complexity of the organization's environment and the measurability of achieved real improvement as well as possible empirical verification of the results.

Acknowledgments This research was funded by the Ministry of Education, Science, Research and Sport of the Slovak Republic, Grant No. KEGA 018TUKE-4/2022 "Creation of new study materials, including an interactive multimedia university textbook for computer-aided engineering activities" and Grant No. VEGA 1/0633/20 "Research of the variability of properties and functions of products made of composite materials produced by additive manufacturing."

References

1. Zgodavová, K., Košč, P., & Kekäle, T. (2001). Learning before doing: Untilising a cooperative role-play for quality management in a virtual organisation. *Journal of Work Place Learning, 13*(3), 113–119. https://doi.org/10.1108/13665620110388415
2. Houshyar, A., & Nuila, V. (1998). Survey of academia and industry on exploitation of simulation. *International Journal of Modelling and Simulation, 18*(1), 45–53. https://doi.org/10.1080/02286203.1998.11760356
3. Cadotte, E. R., & Bruce, H. J. (2003). *The management of strategy in the marketplace*. Knoxville South-Western.
4. Julius, K., & Osman, A. (2015). Role-play technique as an antecedent of performance in English language: Evidence from secondary schools in Wareng District, Uasin Gishu County, Kenya. *Journal of Education and Practice, 6*, 5.
5. CSULX. *Students assume a role to resolve a problem or create an artefact*. The Online Learning Exchange. [Online] Charles Sturt University. [Cited: 16 4 2022.] http://timklapdor.github.io/olx-design/strategy/role-play.html
6. Gamson, W. A. (2013). Games throughout the life cycle. *Simulation & Gaming, 44*(5), 609–623. https://doi.org/10.1177/1046878113503301
7. Bumbuc, S. (2020). Using the role play method in military pedagogy. *Land Forces Academy Review, XXV*(4), 317–324. https://doi.org/10.2478/raft-2020-0038
8. Diener, E., Whitmore, J., & Munn, J. (1976). Role playing in police training. *Police Journal, 49*, 4.
9. Sundararajan, B. (2021). Role-play simulation: Using cases to teach business concepts. In M. Schnurr & A. MacLeod (Eds.), *Simulations and student learning*.
10. Barnabe, F., et al. (2013). The strategic micro-firm: A role-play in management training for dynamic businesses. *The Journal of Workplace Learning, 25*(5), 328–342. https://doi.org/10.1108/JWL-May-2012-0041
11. Bouwmeester, J. Ch., Komisar, V., & Swarup, A. (2017). *Using role-playing simulations to teach quality control in the design of medical devices*. Toronto: University of Toronto. Proceedings of the Canadian engineering education association (CEEA) conference. pp. Paper 062, 1–4.
12. Stratton, E. W. (1992). An effective role-playing exercise for teaching. *Developments in Business Simulation & Experiential Exercises, 19*, 224.
13. Reger, R. K., et al. (1994). Reframing the organisation: Why implementing total quality is easier said than done. *Academy of Management Review, 19*(3), 565–584. https://doi.org/10.2307/258939
14. Armstrong, J. S. (2001). Role playing: A method to forecast decisions. In *Principles of forecasting: A handbook for researchers and practitioners*. Kluwer.
15. Saunders, D., Percival, F. & Vartiainen, M. Games and simulations to enhance quality learning. The Simulation and Gaming Yearbook. IV, 1996.
16. Bennet, A., & Bennet, D. The decision-making process in a complex situation. In *Handbook on decision support systems 1. International handbooks information system* (pp. 3–20). Springer. https://doi.org/10.1007/978-3-540-48713-5_1
17. Senge, P. M. (2006). *The fifth discipline: The art & practice of the learning organization*. 2. Doubleday/Currency: Random House Business.

18. Šolc, M., Markulik, Š., & Sütőová, A. (2014). Quality in contemporary university environments. *Procedia - Social and Behavioral Science, 143*(1), 703–707. https://doi.org/10.1016/j.sbspro.2014.07.467
19. Zgodavová, K. (1998). Management system role-play simulation. In *Slovak: Simulačné projektovanie systémov riadenia kvality*. Q-Projekt Plus.
20. Jørgensen, T. H., Remmen, A., & Mellado, M. D. (2006). Integrated management systems: Three different levels of integration. *Journal of Cleaner Production, 14*, 713–722. https://doi.org/10.1016/J.JCLEPRO.2005.04.005
21. Sá, J. C., et al. (2022). A model of integration ISO 9001 with Lean six sigma and main benefits achieved. *Total Quality Management & Business Excellence, 33*(1–2), 218–242. https://doi.org/10.1080/14783363.2020.1829969
22. Reimann, M., Lunemann, U. F., & Chase, R. B. (2008). Uncertainty avoidance as a moderator of the relationship between perceived service quality and customer. *Journal of Service Research, 11*(1), 63e73.
23. Levitt, T. (1983). *The globalization of markets*.
24. ISO. (2015). *ISO 9000: 2015 quality management systems -- requirements*. ISO.
25. Tapscott, D. (1996). *The digital economy: Promise and peril in the age of networked intelligence* (1st ed.). McGraw-Hill.
26. Combs, K. L., Shapiro, C., & Varian, H. R. (2000). Information rules: A strategic guide to the network economy. *The Journal of Technology Transfer, 25*, 250–252. https://doi.org/10.1023/A:1007897212472
27. Chesbrough, H. (2003). *Open innovation the new imperative for creating and profiting from technology*. 1. (pp. 60–61). Harvard Business School Press.
28. Zgodavová, K., et al. (2012). *Integrated management systems web-based collaborative learning: (Interactive lesson presentation)*. 15th International Conference on Interactive Collaborative Learning (ICL), pp. 1–3. https://doi.org/10.1109/ICL.2012.6402166
29. Price, C. (2022). *What is an integrated management system?* [Online] QMS. [Cited: 25. 2 2022.] https://www.qmsuk.com/
30. Rebelo, M. F., Santos, G., & Silva, R. (2014). A methodology to develop the integration of the environmental management system with other standardized management systems. *Computational Water, Energy, and Environmental Engineering., 3*, 170–181. https://doi.org/10.4236/cweee.2014.34018
31. Kumar, N. M., Krishna, P. R., Pagadala, P. K., Kumar, S. N. M. (2018). *Use of smart glasses in education-a study*. 2nd International conference on I-SMAC (IoT in social, mobile, analytics and cloud), pp. 56–59. https://doi.org/10.1109/I-SMAC.2018.8653666
32. Zgodavová, K., Santos, G., & Lengyelová, K. (2022). *Integrated management systems study program affected by industry 4.0 needs and the Covid-19*. Proceedings, (manuscript).
33. Newton, P. M., Da Silva, A., & Peters, L. G. (2020). A pragmatic master list of action verbs for Bloom's taxonomy. *Frontiers in Education*, 107. https://doi.org/10.3389/feduc.2020.00107
34. Edmunds, H. (2000). *The focus group. Research handbook*. 1. McGraw-Hill.
35. Flores, J. G., & Alonso, C. G. (1995). Using focus groups in educational research: Exploring teachers' perspectives on educational change. *Evaluation Review, 19*, 84–101. https://doi.org/10.1177/0193841X9501900104
36. Nyumba, T. O., et al. (2018). The use of focus group discussion methodology: Insights from two decades of application in conservation. *Methods in Ecology and Evolution, 9*, 20–32. https://doi.org/10.1111/2041-210X.12860
37. Tocco, A. (2021, October 19). *Verbal feedback: An alternative to standard student evaluations*. CETL Blog. [Online]. [Cited: 21 4 2022.]
38. Wills, S., et al. (2009). *Encouraging role-based online learning environments–The BLUE report 2009*. [Online]. https://ro.uow.edu.au/asdpapers/116
39. Zgodavová, K., et al. (2016) Intelligent approaches for an organisation's management system change, *The TQM Journal*, 28, 760–773. https://doi.org/10.1108/TQM-10-2015-0130
40. Kruglova, A. (2020, June 30). *Using role-play in a classroom: Pros and cons. BISA*. [Online]. [Cited: 16 4 2022.] https://www.bisa.ac.uk/articles/using-role-play-classroom-pros-and-cons

Towards a Conceptual Framework for Agroforestry Residual Biomass Sustainable Business Models

Prabalta Rijal, Helena Carvalho, João Matias, Susana Garrido, and Carina Pimentel

Abstract Residual agroforestry valorization reduces the dependency on nonrenewable resources. It is also a source of cheaper, renewable, and greener raw materials. Apart from being used for energy production, it has the potential to be used as feedstock in various other industries if properly valorized. In this study, the potential biomass residues produced in Portugal are identified. An unstructured literature review was performed to identify potential residual biomass streams produced by agroforestry activities. Building on previous findings, a framework to support the identification of factors that can impact valorization efforts is proposed. The framework can help to frame future research in developing sustainable business models to convert those low-value residues into value-added feedstock and products that can support local communities, businesses, and industries.

Keywords Agroforestry residues · Residual biomass valorization · Sustainable practices · Sustainable biomass business models · Biomass supply chain

P. Rijal
Department of Economics, Management, Industrial Engineering and Tourism (DEGEIT), University of Aveiro, Aveiro, Portugal

H. Carvalho (✉)
UNIDEMI, Department of Mechanical and Industrial Engineering, NOVA School of Science and Technology, NOVA University Lisbon, Caparica, Portugal

Laboratório Associado de Sistemas Inteligentes, LASI, Guimarães, Portugal
e-mail: hmlc@fct.unl.pt

J. Matias · C. Pimentel
GOVCOPP R&D Center, Department of Economics, Management, Industrial Engineering and Tourism (DEGEIT), University of Aveiro, Aveiro, Portugal

S. Garrido
CeBER, Faculty of Economics, University Coimbra, Coimbra, Portugal

J. C. de Oliveira Matias et al. (eds.), *Quality Innovation and Sustainability*, Springer Proceedings in Business and Economics,
https://doi.org/10.1007/978-3-031-12914-8_17

1 Introduction

Residual agroforestry biomass valorization is a topic of great interest in the context of Portugal because 90% of the Portuguese territory is occupied by forests, trees, shrubs, and agriculture [1]. Over the past couple of years, there has been a steady rise in the agroforestry sector, especially after 2012. In 2014, 25% of agriculture and forestry in mainland Portugal consisted of agroforestry systems [2]. An agroforestry system is a land-use practice, system, or technology where tree plants are deliberately integrated with crops and/or animals in the same land management unit – in some form of a temporal sequence or spatial arrangement – to improve, diversify, and sustain production for social, economic, and environmental benefits [3].

Forests account for over one-third of the Portuguese territory. Similarly, the agro-food industry in Portugal is one of the largest processing industries, with a turnover of over 15 billion euros in 2017 [4]. These activities result in large quantities of biomass residues that have the potential to become raw materials for other bio-based industries. A corporate study titled 'Portugal's Big Biomass' estimated that the biomass consumption for electricity generation in 2020 had reached 2.8 million tonnes [5]. While another study, 'Biomass resources in Portugal: Current status and prospects', estimated the agricultural and forestry residual biomass only in the Alto Alentejo region to be 4000 dry tonnes per year and 40,000 dry tonnes per year, respectively [6]. This biomass can produce about 158,000 GJ of energy per year [6].

Similarly, Europe produces over 120 million tonnes of crop residues and 40 million tonnes of residues from forestry. These two sectors were responsible for approximately 30% of the total waste [7] produced in the continent. The estimated substantial amounts of residual biomass from agriculture and forestry are primarily used in the energy sector. Therefore, there is a need for more research to find out how to valorize these residues better and to support national policies toward sustaining and expanding local bio-based industries. Apart from its use in the energy sector, residual biomass from agroforestry can produce various products like cosmetics, dietary supplements, polymers for plastics, and even biofuels like ethanol, which can fuel vehicles and other machinery. Most agroforestry residues found in Portugal originate in forest maintenance and farming activities, including agricultural residues, animal manure, and food processing units [6].

To promote the residual agroforestry biomass valorization is essential to understand its supply chain. This will support the development of sustainable business models, hence benefiting local stakeholders and driving economic growth. Further studies are required to identify the various sectors that can create value from residual agroforestry biomass, helping to create new jobs and income streams in rural and urban communities. An estimated 1.2 billion people worldwide are believed to be dependent on agroforestry systems [8], so this study would be useful in Portugal and other countries facing similar challenges of turning agroforestry residues into value-added products.

This study intends to understand residual agroforestry biomass valorization in the Portuguese context. This research was done by doing an unstructured literature review related to the residual agroforestry biomass valorization options and challenges. After, a framework is proposed to contextualize the factors that can impact the residual agroforestry biomass valorization. The findings from this study can be used to find how to promote the residual agroforestry biomass valorization by local farms, businesses, and industries that could.

Following the introduction, the paper delves into residual biomass valorization, including the challenges and national policies that drive it. A proposed framework, future research needs, and the study's conclusion are also discussed.

2 Residual Biomass Valorization

2.1 Data Collection Method

The following steps were followed to collect the most recent and appropriate literature:

- Step 1: The keywords 'Agroforestry residues', 'residual biomass valorization', 'sustainable practices', 'sustainable biomass business models', and 'biomass supply chain' were used in the search query. All kinds of documents were retrieved, including journal articles, book chapters, books, reports, conference proceedings, news reports, independent reports, and studies by government and non-government bodies.
- Step 2: The articles and documents retrieved were then screened by their titles and keywords to ensure that they match the requirements of the study. All documents that were not written in English or Portuguese were excluded. Documents written in Portuguese were machine-translated to English to check for relevance. Duplicate articles and articles that were written before 2009 were removed to ensure that only the most recent studies have been used. A preliminary sample of 110 documents was selected.
- Step 3: The articles were then screened by analysing their abstracts. Only the most relevant articles were used for the review. The final sample consists of 60 documents, which were fully analysed. Only 36 of these texts have been used as references.

2.2 Food Crop Production

Eurostat for 2017 shows that about 54.3% of the agricultural land in Portugal is used for crop production, especially fruits and vegetables. Cereal crops like maize, rice, wheat, barley, and oats are farmed in sizeable amounts in the country. For example, statistics published by INE (Statistics Portugal) show that in 2019, 700,000 tonnes of maize, 160,000 tonnes of rice, 68,000 tonnes of wheat, 60,000 tonnes of barley,

Table 1 Estimated values of food crop production in mainland Portugal between 2019 and 2020

Food crops production quantity (tonnes)						
Cereals	Dried pulses	Potatoes	Cultures for industry	Vegetables	Fodder crops	Fresh fruits
1,044,002	5421	409,641	1,265,439	1,214,815	4,415,984	508,524
Berry species	Subtropical fruits	Citrus fruits	Dried fruits	Vineyard	Olive trees	
43,779	89,646	422,929	79,117	853,384	722,578	

Source INE

and 56,000 tonnes of oats were produced, which means that there is generally an abundance of residues from these food crops. Similarly, tomatoes make up approximately 98% of the industrial crops in Portugal, according to INE. Likewise, vegetables – cabbage, fresh tomatoes, and carrots – and fruits like grapes can also be considered primary food crops in Portugal [12]. Table 1 shows the number of food crops produced in tonnes between 2019 and 2020.

Wheat straw, maize stover, and barley straw have also been identified as the most promising source of feedstock for the bio-economy [9], as they contain a total concentration of lignocellulose of more than 80% of the dry matter. Lignocellulose is a sugar found in plants that can be converted into biofuels that can replace petroleum-based fuels, including jet fuel [10]. In 2019 Portugal produced an estimated 7.04 thousand barrels of biofuel per day from biomass feedstocks [11].

2.3 *Forestry Production*

In Portugal, 5 species of trees are most relevant in production: 3.22 million hectares of forest area are home to the eucalyptus, maritime pine, umbrella pine, cork oak, and oak trees. These trees are found in the country's Northern, Central, and Southern parts. The umbrella pine, evergreen oak, and cork oak are found in the southern regions, while the maritime pine is found in the Northern and Central regions, and the eucalyptus trees are found in all three regions. This makes forestry and its related value chains (cork, hardwood, paper, and pulp) key Portuguese economic sectors.

Private entities own forest ownership in Portugal as 91% of the forests, about 6% by local communities, and the state only 3% of it. According to the latest online data on the Institute of National Statistics Portugal (INE) website, the total value of forest production in 2017 was 1.25 billion euros. According to INE, the Gross Value Added (GVA) by forestry was over 800 million euros between 2015 and 2019, as shown in Table 2.

INE claims that the GVA of forestry production decreased in value in 2017 due to a series of large wildfires that covered about 168,000 hectares of land in 2016 and 502 hectares of land in 2017. However, cork production, which is a crucial export, was not affected by the fires, and there was a 9.1% increase in its price. Portugal is the world's largest producer of cork. Agroforestry residues are mainly used for

Table 2 Portuguese forestry economic accounts. Value of prices in millions (€)

Forestry production	2015	2016	2017	2018	2019
Coniferous timber for industrial uses	149.04	151.62	162.41	165.64	156.62
Coniferous saw logs	128.11	130.67	142.42	145.64	135.44
Coniferous pulp wood	15.72	15.25	13.73	13.89	15.28
Industrial uses: other	5.23	5.7	6.26	6.14	5.83
Non-coniferous timber for industrial uses	309.11	299.62	303.5	288.59	283.29
Non-coniferous saw logs	4.85	4.61	4.61	4.66	5.13
Non-coniferous pulp wood	302.19	293.12	297.03	282.18	276.54
Industrial uses: other	2.07	1.89	1.86	1.74	1.57
Fuel wood	51.87	50.68	53.21	55.35	62.33
Cork	236.46	245.75	230.66	242.88	208.88
Forestry and nursery plants	6.02	5.1	5.62	4.53	4.36
Gross value added at basic prices	929.78	916.82	900.37	869.12	812.79

Source: INE

energy production, and they include almond shell, grape stalk, cherry tree chips, chestnut tree chips, corncob, dried oil millstone, eucalyptus bark, extracted olive pomace, extracted olive pomace pellets, cereal straw, pruning residues from the vineyard, fruit trees, olive, and almond residues [13].

2.4 *Residual Agroforestry Biomass Valorization*

Residual agroforestry biomass valorization may be the key to finding a reliable, renewable source of raw materials or energy for various industries. Any industry needs to have access to raw materials to grow. Bio-based economies can only be sustainable if a justifiable amount of feedstock supports them. Therefore, industries need to work jointly with primary sectors to create and add value to residual biomass such as food, feed, wood, and breweries. Studying further and finding ways to create new value chains that can support different industries and sectors is necessary.

The agroforestry biomass valorization can mitigate the dependency on nonrenewable resources, restore degraded lands, protect crops and livestock from high winds, and improve soil quality and water conservation [14]. Similarly, biomass valorization can convert agricultural waste into energy and the possibility of developing bioenergy plants [15, 16]. Despite all the potential benefits, the residual agroforestry biomass valorization is just beginning to be considered as an alternative option and driver for sustainable economic growth [17–19]. Other value-added products, aside from bioenergy, are still being researched. In Latin America, for example, the use of residues to produce high-value biopolymers and plastics, biofuels, and nano-cellulose fibres from banana waste is of great interest [20].

Even though its importance and benefits have been researched, sustainable business models to promote the residual agroforestry biomass valorization still need to

Table 3 Overview of studies on the valorization of residual agroforestry biomass

References	Residues	Study findings
[7]	Pine tree branches, pine stump chips, tomato fruits, and winery wastewater	Four abundant residues from the Portuguese agroforestry industry showed promising potential within the biorefinery context
[24]	Popular fruit crops like pears, oranges, and apples	Residues from popular fruit crops could be used to extract bioactive compounds that can be potentially used in the nutraceutical and cosmetics industry. Need for further research for deeper characterization of these to obtain safe and reproducible products
[1]	Mediterranean agroforestry vegetation	Shrubs from forests were found to be better sources of energy as they have higher heating value and emit fewer amounts of undesirable elements and ashes than agricultural residues
[25]	Hazelnut and olive tree residues	Premium quality biochar could be obtained from biomass residues. The researchers intended to define a biomass chain

be developed. The development of sustainable business models relies greatly on the supply chain and how it is managed. It is, therefore, necessary to identify the supply, demand drivers, resource availability, and the current government policies that influence them [21].

Since residual agroforestry biomass has little or no value, it becomes essential to understand how it can be turned into value-adding products [22]. Despite the importance of developing innovative and sustainable business models, current studies show that agroforestry biomass supply chains have been facing substantial challenges on the economic, operational, management, logistical, and inventory front [19].

Understanding the supply chain is important in developing and implementing a reliable framework based on strategic planning and efficient decisions to reach business targets. A report [23] pointed out that relatively new businesses that used biomass to produce biofuels and biochemicals faced resistance from farmers, who were unaware of its benefits. For supply chains to work smoothly, producers, distributors, and consumers need to understand the targets, plans, and activities that are incorporated into the whole network. This will lead to a more efficient and competitive supply chain.

Table 3 provides an overview of a few studies on the valorization of residual agroforestry biomass.

3 Challenges

Many traditional agroforestry systems across the European continent were extinct in the twentieth century due to industrialization, commercialization, and mechanization resulting in the loss of traditional knowledge [26]. This caused a lack of knowledge and advisory support in implementing agroforestry systems, which is considered a major barrier to creating profitability through agroforestry systems. Education and information sharing can help farmers choose species that enhance yields without compromising the main crop [27] while also understanding the nature of the biomass residues and how they can benefit from it.

In 2020 Portugal produced 860,000 tonnes of wood pellets. This required approximately 1.8 million tonnes of biomass, mainly from pine trees. There are also claims that much of the pellets that are being produced are not genuine forestry residues [6]. These claims of actual wood being used instead of residual forestry biomass to produce pellets pose a high risk as business competition in this area could lead to deforestation as more and more trees are cut down to meet market demands [28, 29].

As mentioned in the introduction section, the Alto Alentejo region alone produced agriculture and forestry residues of 4000 dry tonnes/year and 40,000 dry tonnes/year, respectively. Although this data is not new, it gives an idea of the quantity of biomass that is produced in the country. A report, 'Mapping Portugal's Bio-based potential' [30], refers to the limitation of data on the exact amount of residual biomass produced. This limitation on data availability can make it challenging to analyse the cost of valorization of agroforestry biomass residues. For example, long-distance transportation of pine tree prunings for energy generation may not be cost-effective at all and could lead to little or no revenue. However, if these prunings were being used to create pellets that can be used to create premium quality biochar, then the distance may not be much of an issue in terms of cost-effectiveness.

The Portuguese government has supported agroforestry biomass valorization in various ways, like promoting investments in agriculture by allowing farmers grants through the Common Agricultural Policy [31]. Similarly, over the years, there have been several policies that have come into existence to regulate and promote the agroforestry sector. Recently, an 11-point recommendation to regulate bioenergy plants from misusing natural forests for their benefit [32] has been formed. Factors like the promotion of resilient ecosystems and systems that allow residual biomass systems to be incorporated into the soil to avoid ecological imbalance and the use of mostly surplus forestry residues for energy production (in contrast to the use of non-residual biomass) have been elaborated in the document.

Other initiatives to develop agroforestry valorization streams by the Portuguese government are as follows: (i) 'The National Plan for promotion of Biorefineries' – which aims to promote biorefineries in the Portuguese territory that can utilize low valued streams of residual biomass from agriculture and forestry sources to produce a variety of feedstock/raw materials [4] – and (ii) 'The Circular Economy Action Plan' – intends to develop methods which can minimize resource extraction, increase and encourage reuse while also developing new business models for ecologically efficient products and services [4].

4 Proposed Framework

Sustainable business models to promote the residual agroforestry biomass valorization still need to be understood. We can refer to a sustainable business model as a business design with a holistic approach that considers not just the economic value of the business but also the social and environmental value it can provide [32]. No industry or business operates in isolation; it exists within an ecosystem. It relies upon a supply chain, making it important to identify the supply, demand drivers, resource availability, and the current government policies that influence them [21].

Figure 1 proposes a conceptual framework to summarize the main elements of a residual agroforestry biomass system. Three factors need to be taken into consideration: (i) collection processing and transportation cost; (ii) residual agroforestry biomass characterization; and (iii) availability and knowledge about valorization options for the biomass suppliers (private forest owners, local communities forest owners, farmers, and small farmers). Knowing this makes it easier for the suppliers to segregate the residues accordingly and sell them for a better price to the different industries and manufacturers wanting to buy these residues. Therefore, it is important for them to be well aware of the various valorization options available for the feedstock they are supplying. The producers need to know the value of their raw materials. They also need to know what kind of value-added feedstock can be obtained from the residues that they produce. Where are they valorized – is it in bioenergy plants or pellet plants, or even biorefineries? Will it be used to produce bioactive compounds, biochar, pellets, or renewable energy? This will make it simpler for them to understand the market demands better, calculate the costs involved in the process like transportation costs, and get returns from these low-value residues [22].

The production of residues depends on several factors like geographic location of production, climatic conditions that may affect supply like drought or wildfires,

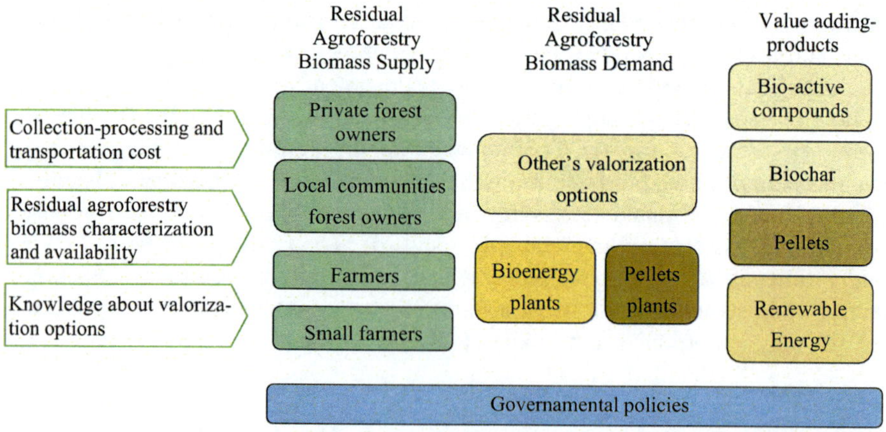

Fig. 1 Framework for a residual agroforestry biomass system

road access for transportation and costs involved, storage, and inventory [33, 34]. It is also important to consider the change in biomass yields from season to season while modelling a supply chain for biofuels or other value-adding feedstock.

However, this can only be possible with the right set of government policies that support residual agroforestry biomass valorization. In short, for residual biomass valorization to become a sustainable practice, it is necessary to identify the supply and demand drivers, the availability of resources, and the current government policies that influence them [1].

A report published in 2015 [23] pointed out that relatively new businesses that used biomass to produce biofuels and biochemicals faced resistance from farmers unaware of its benefits. The existence of good government policies that support the valorization of agroforestry biomass residues can encourage farmers and other stakeholders to invest in valorization initiatives. It is only after farmers and landowners are positively involved in the process can bio-based business streams become sustainable.

5 Conclusion

This paper presents an overview of residual agroforestry biomass valorization. Portugal has an expanding economy, and the number of agricultural products farmed, forestry, manufacturing of food and beverage, sludge, and pulp from the paper manufacturing industry are all strong indicators of the country's high biomass residue production. The country is home to 3.22 million hectares of forest, while the agro-food industry is one of the largest processing industries, with over 15 billion euros in turnover. The biomass residue streams from these two sectors could have immense potential. But so far, few streams of biomass valorization have been promoted apart from the use of woody biomass in the manufacturing of pellets to generate energy. A framework to support the identification of the factors that can impact valorization efforts has been proposed in this study.

Further studies are required to understand better how new feedstock can be generated from residues produced by agroforestry systems in the country and how it can also support other sectors of the economy by recognizing new employment sectors and creating new jobs and income streams in rural and urban communities. Future research in this subject matter can involve an in-depth study into the different sources of residual biomass like food crop production and forestry as well as to better understand how the residual biomass from these activities can be valorized and used to support other industries, local businesses, farmers, and other stakeholders. Similarly, this study found that there is a need for future studies that explore the drivers and barriers of residual biomass valorization and provide methodologies for a better characterization of the supply chains that could operationalize sustainable business models for agroforestry residual biomass valorization.

Acknowledgements This work has been funded by national funds through FCT – Fundação para a Ciência e a Tecnologia, I.P., through national funds and, when applicable, cofinanced by the FEDER under the new partnership agreement PT2020, grant number PCIF/GVB/0083/2019, as well as via the projects UIDB/05037/2020 (CeBER), UIDB/04058/2020 (GOVCOPP) and UIDB/00667/2020 (UNIDEMI).

References

1. Enes, T., Aranha, J., Fonseca, T., Matos, C., Barros, A., & Lousada, J. (2019). Residual agroforestry biomass–thermochemical properties. *Forests, 10*, 1072.
2. Reis, P., Rolo, J., Coelho, I., & Belo, C. C. (2014). Agro forestry systems in the Portuguese mainland. Part I: Economy and agro forest regions. *Revista de Ciências Agrárias, 37*(2), 122–130.
3. Atangana, A., Khasa, D., Chang, S., & Degrande, A. (2013). Definitions and classification of agroforestry systems. In *Tropical agroforestry* (pp. 35–47). Wiley.
4. Biobased Industries Consortium. (2021). *Country report: Mapping Portugal's bio-based potential.*
5. Biofuel Watch, Z. *Big biomass in Portugal – biofuelwatch.* https://www.biofuelwatch.org.uk/2021/biomass-portugal/. Last accessed 28 Feb 2022.
6. Ferreira, S., Monteiro, E., Brito, P., & Vilarinho, C. (2017). Biomass resources in Portugal: Current status and prospects. *Renewable and Sustainable Energy Reviews, 78*, 1221–1235.
7. Gaspar, M. C., Mendes, C. V., Pinela, S. R., Moreira, R., Carvalho, M. G., Quina, M. J., Braga, M. E., & Portugal, A. T. (2019). Assessment of agroforestry residues: Their potential within the biorefinery context. *ACS Sustainable Chemistry & Engineering, 7*, 17154–17165.
8. Sheppard, J. P., Bohn Reckziegel, R., Borrass, L., Chirwa, P. W., Cuaranhua, C. J., Hassler, S. K., Hoffmeister, S., Kestel, F., Maier, R., Mälicke, M., Morhart, C., Ndlovu, N. P., Veste, M., Funk, R., Lang, F., Seifert, T., du Toit, B., & Kahle, H.-P. (2020). Agroforestry: An appropriate and sustainable response to a changing climate in Southern Africa? *Sustainability, 12*, 6796.
9. Thorenz, A., Wietschel, L., Stindt, D., & Tuma, A. (2018). Assessment of agroforestry residue potentials for the bioeconomy in the European Union. *Journal of Cleaner Production, 176*, 348–359.
10. Gutiérrez-Antonio, C., Romero-Izquierdo, A. G., Gómez-Castro, F. I., & Hernández, S. (2021). Production processes from lignocellulosic feedstock. In *Production processes of renewable aviation fuel* (pp. 129–169). Elsevier.
11. Atlas, World Data. *Portugal biofuels production, 1949–2021.* https://knoema.com/atlas/Portugal/topics/Energy/Renewables/Biofuels-production. Last accessed 28 Feb 2022.
12. Estatística, I. N. (2020). *Instituto Nacional de Estatística – Agriculture Statistics.* https://www.ine.pt/xportal/xmain?xpid=INE&xpgid=ine_publicacoes&PUBLICACOESpub_boui=437147278&PUBLICACOESmodo=2. Last accessed 28 Feb 2022.
13. Sirous, R., da Silva, F. J., da Cruz Tarelho, L. A., & Martins, N. A. (2020). Mixed biomass pelleting potential for Portugal, step forward to circular use of biomass residues. *Energy Reports, 6*, 940–945.
14. Rutkowska, B., Schmid, T., Szulc, W., Witters, N., & Sæbø, A. (2018). Intensify production, transform biomass to energy and novel goods and protect soils in Europe—A vision how to mobilize marginal lands. *Science of The Total Environment, 616–617*, 1101–1123.
15. Donner, M., Gohier, R., & de Vries, H. (2020). A new circular business model typology for creating value from agro-waste. *Science of The Total Environment, 716*, 137065.
16. Malico, I., Carrajola, J., Gomes, C. P., & Lima, J. C. (2016). Biomass residues for energy production and habitat preservation. Case study in a Montado area in Southwestern Europe. *Journal of Cleaner Production, 112*, 3676–3683.

17. Torreiro, Y., Pérez, L., Piñeiro, G., Pedras, F., & Rodríguez-Abalde, A. (2020). The role of energy valuation of agroforestry biomass on the circular economy. *Energies, 13*, 2516.
18. Diakosavvas, D., & Freza, C. (2019). Bio-economy and the sustainability of the agriculture and food system. *OECD Food, Agriculture and Fisheries Papers*.
19. van Dam, J. *Sustainable biomass production and use lessons learned from the Netherlands Programme Sustainable Biomass*. https://www.etipbioenergy.eu/images/netherlands-sustainable-biomass.pdf. Last accessed 28 Feb 2022.
20. Redondo-Gómez, C., Rodríguez Quesada, M., Vallejo Astúa, S., Murillo Zamora, J. P., Lopretti, M., & Vega-Baudrit, J. R. (2020). Biorefinery of biomass of agro-industrial banana waste to obtain high-value biopolymers. *Molecules, 25*, 3829.
21. Kardung, M., Cingiz, K., Costenoble, O., Delahaye, R., Heijman, W., Lovrić, M., van Leeuwen, M., M'Barek, R., van Meijl, H., Piotrowski, S., Ronzon, T., Sauer, J., Verhoog, D., Verkerk, P. J., Vrachioli, M., Wesseler, J. H., & Zhu, B. X. (2021). Development of the circular bioeconomy: Drivers and indicators. *Sustainability, 13*, 413.
22. Dashtpeyma, M., & Ghodsi, R. (2021). Forest biomass and bioenergy supply chain resilience: A systematic literature review on the barriers and enablers. *Sustainability, 13*, 6964.
23. Hoye, I. V. *Eip-agri workshop: Building new biomass supply chains for the bio-based economy*. https://ec.europa.eu/eip/agriculture/en/event/eip-agri-workshop-building-new-biomass-supply. Last accessed 26 Feb 2022.
24. Pascoalino, L. A., Reis, F. S., Prieto, M. A., Barreira, J. C., Ferreira, I. C., & Barros, L. (2021). Valorization of bio-residues from the processing of main Portuguese fruit crops: From discarded waste to health-promoting compounds. *Molecules, 26*, 2624.
25. Zambon, I., Colosimo, F., Monarca, D., Cecchini, M., Gallucci, F., Proto, A., Lord, R., & Colantoni, A. (2016). An innovative agroforestry supply chain for residual biomass: Physicochemical characterization of biochar from olive and hazelnut pellets. *Energies, 9*, 526.
26. Quinkenstein, A., Wöllecke, J., Böhm, C., Grünewald, H., Freese, D., Schneider, B. U., & Hüttl, R. F. (2009). Ecological benefits of the alley cropping agroforestry system in sensitive regions of Europe. *Environmental Science & Policy, 12*, 1112–1121.
27. Sollen-Norrlin, M., Ghaley, B. B., & Rintoul, N. L. (2020). Agroforestry benefits and challenges for adoption in Europe and beyond. *Sustainability, 12*, 7001.
28. Orchestrator, O. *Chipping trees in Portugal for energy in the UK*. https://forestdefenders.eu/chipping-trees-in-portugal-for-energy-in-the-uk/
29. Guynup, S. (2021). *COP26: Surging wood pellet industry threatens climate, say, experts*.
30. Commission, E., Investment Bank, E. (2020). *Financial needs in the agriculture and agri-food sectors in Portugal*.
31. Goes, D. B., & Pereira, M. S. *Trends for a public support system to forest biomass plants in Portugal – knowledge*. https://www.csassociados.pt/en/knowledge/flashes/Trends-for-public-support-system-to-forest-biomass-plants-in-Portugal/1598/. Last accessed 26 Feb 2022.
32. Nosratabadi, S., Mosavi, A., Shamshirband, S., Kazimieras Zavadskas, E., Rakotonirainy, A., & Chau, K. W. (2019). Sustainable business models: A review. *Sustainability, 11*, 1663.
33. Sperandio, G., Acampora, A., Civitarese, V., Bajocco, S., & Bascietto, M. (2021). Transport cost estimation model of the agroforestry biomass in a small-scale energy chain. *Forests, 12*, 158.
34. Rentizelas, A. A. (2013). Biomass supply chains. In L. Rosendahl (Ed.), *Biomass combustion science, technology and engineering* (pp. 9–35). Woodhead.

Environmental and Economic Sustainability of Electric Vehicles vs. Combustion Engine Vehicles Fueled with B15 and B30 Blends of Biodiesel

Luis Serrano, Marcelo Gaspar, and Cristina Correia

Abstract This research assesses and analyzes the impacts on energy consumption and the related carbon dioxide (CO_2) emissions resulting from the use of electric vehicles (EVs) when compared with internal combustion (IC) diesel vehicles fueled with biofuels. Enhancing commercial fossil diesel fuel with 15% biodiesel (B15) to be used by a fleet of light-duty diesel vehicles was considered the starting point for this study. Based on data collected from this IC vehicles fleet, it was possible to analyze and compare the energy consumption and the related emissions resulting from the use of EVs with the same work operation. This research showed that the use of EVs presents a few comparative advantages, but also has some significant limitations. On what concerns the global CO_2 emissions and energy efficiency differences of both EVs and biodiesel fueled IC vehicles, these were not very significant. The use of electric motorization implies a 50% reduction in CO_2 emissions compared with the reference fossil diesel fuel and the use of B15 and B30 correspond to a reduction of about 11.8% and 22.4%, respectively, in these same emissions. However, there are additional constraints related to economic factors since the purchase of EVs is significantly more expensive than that of similar IC vehicles. The use and fleet management of EVs is also more complex due to the limitation in terms of vehicle autonomy, requiring more EVs and more people for the same operation. Thus, this type of analysis shows to be essential to support selecting the most sustainable and efficient options for the operation of a light commercial vehicles fleet.

L. Serrano (✉)
School of Technology and Management, Polytechnic Institute of Leiria, Leiria, Portugal

ADAI, Department of Mechanical Engineering, University of Coimbra, Coimbra, Portugal
e-mail: luis.serrano@ipleiria.pt

M. Gaspar
School of Technology and Management, Polytechnic Institute of Leiria, Leiria, Portugal

C. Correia
PRIO Energy, Aveiro, Portugal

J. C. de Oliveira Matias et al. (eds.), *Quality Innovation and Sustainability*, Springer Proceedings in Business and Economics,
https://doi.org/10.1007/978-3-031-12914-8_18

Keywords Sustainability · Energetic efficiency · CO_2 emissions · renewable energy · biofuels · Electric vehicles

1 Introduction

1.1 Energy and the Transport Sector

The transport sector has a very significant impact on greenhouse gases (GHG) emissions and other environmentally harmful pollutants. In fact, in Europe, this sector has an overwhelming dependence on liquid fossil fuels, corresponding in 2017 to 95.3% of the energy sources used by vehicles [1]. Transport effectively represents one of the sectors with the greatest difficulty in changing its growth trend and in promoting an energy transition to sources with reduced environmental impacts. Automotive vehicles present high energy intensity, significant impacts on GHG and air pollution, with the main incidence in urban areas [2]. Compared to the year 2020, a doubling of the number of vehicles worldwide is expected to occur before 2050 [3]. In the European Union, about 25% of CO_2 emissions are the responsibility of the transport sector, considering that the road transport component contributes about 2/3 of these emissions and transport also represents about one-third of the total energy consumption [4, 5].

The charts presented in Fig. 1 show that some of the non-fossil energy solutions that have been used in 2017 still have very low values. These non-fossil green alternatives to conventional fuels include biofuels (biodiesel and bioethanol) and electricity. It is understood that any of these solutions may present greater weight in the energy supply to the transport sector with significant benefits for the environment, both locally and on a global scale.

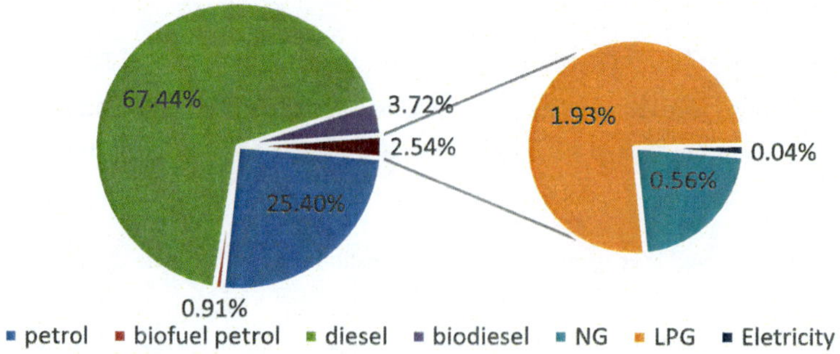

Fig. 1 Road transport energy consumption in Europe in 2017. (Adapted from Statistical Office of the European Union (Eurostat) [6])

Biodiesel is an organic substance gathered from the fatty matter of vegetable or animal products through a transesterification reaction. It is being seen as a solution for today's fleets and almost all countries are preparing a policy for the production and use of biodiesel in their transportation sector [7, 8].

Generally, EVs are considered zero-emission vehicles (0 g CO_2/km). However, even though the use of electricity in vehicle propulsion represents an effective decrease in CO_2 emissions, these emissions still must be considered and assessed. The determination of these emissions is dependent on some uncertainty based on three factors: The clear definition of the effective market penetration, the energy mix that is used in the production of electricity and, finally, the impacts on infrastructure, whether in the production of batteries and vehicles or in the very need to change the electricity distribution network [9, 10].

1.2 Main Goals of the Research

To discuss the sustainability and economic impacts of the two energetic alternative options to be used as a replacement for fossil fuels in actual vehicles, namely, electricity and biodiesel, a real case study based on a fleet of commercial IC vehicles was considered. Selecting a fleet of light-duty vehicles that perform a day-by-day commercial trip that uses diesel fuel, a comparative analysis was carried out to discuss the impacts on energy consumption and carbon dioxide emissions considering the use of biodiesel and electricity. Biodiesel has as one of the main advantages the characteristic of not requiring the change of the current vehicles' technology, once it can be used in IC engines like those operating in actual fleets. Nevertheless, it also has the disadvantage of still having some pollutant emissions, even though it allows lowering the pollution problem with a significant CO_2 reduction, the remaining pollutants can only be partially reduced, and therefore not fully eliminated [11, 12]. Nevertheless, electric energy also presents several critical issues, namely, the range anxiety of EV drivers due to its limited autonomy range and the fact that changes in the energetic consumption depend significantly on the driving patterns [13].

The main objective of this work is to reduce some of the uncertainties related to several assumptions and considerations usually presented in the current literature available both for biofuels and for electric energy used in traditional vehicles. To further support this research, a real case study based on an actual fleet of light-duty commercial vehicles was considered. To assess the main advantages and limitations of these green alternatives to fossil fuels, a blend of 15% of biodiesel in diesel fuel (B15) and a blend of 30% of biodiesel in diesel fuel (B30) was considered as a starting point to analyze and discuss the environmental and economic sustainability of the same vehicle using IC engines and electric motorization.

2 Methodology

2.1 *Fleet Characterization*

For over 2 years, the monitoring of a fleet of light-duty vehicles equipped with an IC engine with a cylinder capacity of 1400 cm^3 with direct injection and compression ignition system has been carried out. In the analysis of this fleet, the records of 13 vehicles were considered.

For the present work, two analogous periods were considered in consecutive years. In the first period (February 2019 to September 2019) the vehicles were fueled with commercial diesel (containing up to 7% biodiesel), while in the second period (February 2020 to September 2020), the same vehicles were fueled with a blend of 15% biodiesel in diesel fuel (B15).

Considering the problem of food supply, biofuel must be produced from feedstocks that should not have been directly or indirectly used for food, but should consider other sources, mainly from biomass. The most straightforward way to accomplish this is by considering the production of biofuels from biomass that is currently treated as either waste or that is a co-product of existing production processes with very low or negative current economic value. Some examples that can be considered to obtain biofuels are municipal garbage, crop residues, wood and forest residues, and used vegetable oils from restaurants and people's homes.

The biodiesel employed in the present project considered the process of transesterification of used vegetable oils. This has in fact two main advantages. One, it is the energetic exploitation of a residue instead of using other useful sources which can represent an effective way to reduce carbon emissions; the other is that it gives a value to a resource that represents a problem to the wastewater treatment if it is delivered to the sewage conducts.

The characterization regarding the use of the vehicles, namely, considering the daily average distance travelled by each vehicle and its average speed is presented in Fig. 2. Observing these results, it is possible to verify that the largest distance travelled each day (on average) by the vehicles is lower than 250 km, with some of them travelling less than 50 km per day. On what concerns the average travelling speed of the vehicles in the observed fleet, it ranged between 20 and 40 km/h.

To characterize alternative fuel consumptions and related impacts, this study considered as base results the data collected during two consecutive years from a fleet of light-duty vehicles using commercial diesel and also B15 biodiesel. It is important to mention that the biodiesel used is obtained from used cooking oils, which are subject to a transesterification process and a quality control program to ensure compliance with the EN14124 standard and consequent use in internal combustion engines.

Based on the defined parameters, it was possible to outline different scenarios using green alternatives in order to evaluate, based on this real situation, what are the expected impacts considering either the use for the same fleet of vehicles renewal or the possibility of maintaining the fleet, but increasing the amount of biodiesel incorporation up to 30% (B30).

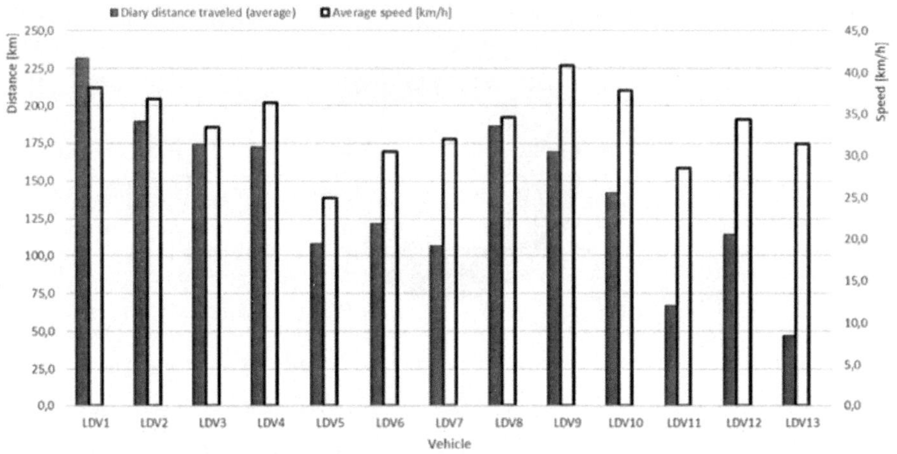

Fig. 2 Fleet of light-duty vehicles' daily average travel speed and distance

2.2 CO_2 Assessment Method

The CO_2 emissions into the atmosphere have a significant impact on global warming. This effect has worsened over the last century, mainly since combustion represents one of the main means of producing energy in the form of work and heat. In the combustion process, the oxidation reaction of hydrocarbons that constitute fuels occurs, with the release of water and carbon dioxide when this reaction occurs in an ideal way and the release of other pollutant compounds such as CO, NOx, and carbonaceous particles that come from disturbances in the reaction process.

Nevertheless, the reduction of CO_2 emissions by using biofuels does not eliminate these pollutant emissions as, to produce energy in the form of work, it is necessary to perform a combustion reaction. This process will correspond to an amount of CO_2 emissions that depends directly on the amount of carbon existing in the fuel. However, the use of biofuels, such as biodiesel, allows capturing CO_2 from the atmosphere through the growth of plants from which oil is extracted, which then allow for the production of biodiesel. Therefore, the use of biodiesel makes it possible for a greater amount of CO_2 not to be accumulated in the atmosphere, as in a short period it extracts the CO_2 molecules that will be emitted after combustion. In a study carried out by the US Departments of Agriculture and Energy [14], it was possible to quantify the CO_2 emission's reduction resulting from the use of B100, which, when compared to fossil fuel, was 78.45%. The same study shows that the benefit in CO_2 reduction with the use of B20 was 15.66%. These values will be used to assess the reduction of CO_2 emissions with the use of B15 and B30. The emission levels associated with biodiesel fuel can be regarded as a carbon credit related to the production of biodiesel using photosynthesis [11].

The determination of the CO_2 emissions related to the use of EVs is not direct or a simple process. Neither can it be done lightly. If not done properly, such

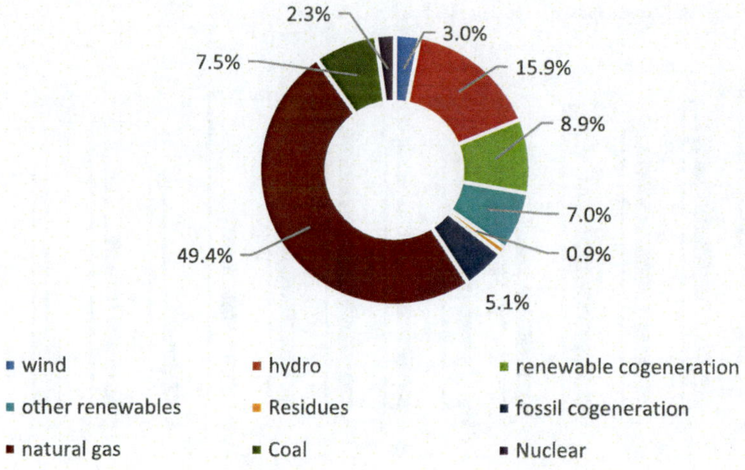

Fig. 3 Energy mix for electric energy produced in Portugal (3° trimester 2021) [10]

assessment may lead to completely different and even conflicting values [9]. On the one hand, the impacts to produce batteries and the EVs themselves have different impacts and requires for distinct resources and processes when compared to traditional IC vehicles. On the other hand, travelling with the EVs requires using electric energy, which can be fully renewable or completely produced in a thermal power plant using the combustion reaction of coal or natural gas. In this sense, it is important to use what is called the local energy mix. Such mix results from different ways and sources of producing electric energy. It conveys, in average terms, the resources used to produce the energy used to charge the EVs' batteries, either by using renewable sources or by using fossil resources. This energy mix globally corresponds to a related amount of CO_2 emissions. In the case of Portugal, considering the third quarter of 2021 as a representative example, it is possible to identify the different sources used to produce the energy distributed in the electric grid, as shown in Fig. 3, which corresponds to a CO_2 emission of 316.74 g/kWh [15].

The results considered for the above-presented energy mix will be used to assess the amount of CO_2 emissions from the use of EVs in Portugal, considering the equivalent routes travelled by the light-duty fleet of IC vehicles.

3 Results

3.1 Measured Values: Diesel and B15

The main characteristics of the current case study fleet of light-duty IC vehicles is presented in Table 1.

The average daily distance traveled by each of the vehicles and the related fuel consumptions are expressed in Table 2. These results consider the monitoring of the

Table 1 Vehicle fleet characteristics

Year	2015–2017
Engine	1396 cm^3
Max power	90 hp
Max torque	220 nm
Max speed	169 km/h
Fuel	Diesel
Weight	1249 kg
Emission homologation type	Euro 6
CO_2 emissions (homologation)	98 g/100 km
Fuel consumption (homologation)	3.8 lts/100 km

Table 2 Vehicle fleet measured results

		Fuel consumption		
	Daily average travel distance	B15	Commercial diesel (B7)	Difference
Vehicle	km	liters/ 100 km	liters/100 km	%
LDV1	231.9	5.89	5.57	6%
LDV2	189.6	5.22	5.36	−3%
LDV3	174.1	5.51	5.19	6%
LDV4	172.6	5.53	5.30	4%
LDV5	107.7	5.47	5.76	−5%
LDV6	121.6	5.97	5.59	7%
LDV7	106.3	5.62	5.73	−2%
LDV8	186.3	6.01	5.64	7%
LDV9	168.9	5.43	5.58	−3%
LDV10	142.2	5.11	5.66	−10%
LDV11	67.0	5.31	5.08	5%
LDV12	114.3	8.13	8.26	−2%
LDV13	47.2	5.43	5.32	2%
Average consumption		**5.74**	**5.70**	
Difference	**liters/100 km**	**0.05**		
	%	**0.01**		

fleet during the two consecutive periods in which the vehicles were fueled by commercial diesel and B15.

Analyzing the results presented in Table 2 allows concluding that the use of B15 as an alternative to commercial diesel in the vehicles that integrate this fleet has no significant difference in fuel consumption. It shows an increase in efficiency with the use of B15. Thus, a decrease in the energetic content of the fuel, which could result in an increase in fuel consumption, was not observed. Nevertheless, some small fluctuations in the fleet's IC vehicles' consumption took place. However, these are expected, as a wide range of parameters can affect the vehicles' fuel consumption. When considering such fluctuations throughout the whole fleet of vehicles, these consumption variations cancel each other out.

Considering the results of Table 2, it is important to further discuss the vehicles' fuel consumption. In the homologation tests, the results for the fuel consumption of these vehicles – and announced to the public – was 3.8 lts/100 km. However, when considering the actual results measured for the fleet of vehicles, it was possible to observe that the homologation claimed value is significantly below the real fuel consumption of such vehicles in normal operation. In fact, the average measured fuel consumption for the fleet of vehicles was 5.7 lts/100 km. This corresponds to an increase of 50% when faced with the expected fuel consumption, both for commercial diesel and B15. This implies that the CO_2 emissions could not be 98 g/km as expressed in the homologation, but should also rise 50%, corresponding to the real value of 147 g CO_2/km for diesel consumption.

Following the same pattern, considering that the use of B15 can reduce the CO_2 emissions by 11.74%, it is possible to conclude that in this case, the change from commercial diesel to B15 biodiesel can result in the lowering of the CO_2 emissions to 129.7 g CO_2/km.

3.2 *Use of B30*

In the next few years, due to the need to reduce the CO_2 emissions of the transport sector, considering the impacts and costs of immediately replacing the fleet of existing vehicles and completely transforming the current electric distribution grid, it is imperative to increase the biodiesel blend in existing vehicles. The use of B30 is a promising alternative since such biofuel blend does not significantly affect the vehicles' fuel consumption and could further contribute to an increased and effective reduction in CO_2 emissions.

Considering that, due to the reduced energetic content of B30 face to diesel, it is possible to expect an increase in fuel consumption of 2%, even if that will not happen, as noticed for the B15 situation, but considering this more conservative scenario, the fuel consumption should rise 1.5% when compared with diesel fuel.

In this scenario, the vehicles' fuel consumption could increase up to 5.81 lts/100 km. The related CO_2 emissions would then lower to 114 g CO_2/km. It represents a reduction of 22.5% of CO_2 emissions when faced with the current use of commercial diesel.

3.3 *Electric Vehicles*

To allow analyzing and discussing the comparative performance of the actual fleet of light-duty IC diesel vehicles fueled with biofuels, comparative EVs had to be selected. To such end, EVs with similar capabilities to those of this case study fleet of IC vehicles were considered. These EVs would allow for travelling the same distances and result in similar operating conditions. It was previously defined that the range of the vehicle should be at least 250 km/h per day. This implies that the EVs should present an autonomy that allowed travelling such distances without the need for recharging.

The characteristics of the selected EVs are presented in Table 3. Such an option considered the same OEM and a wide range of similar characteristics, such as engine power and vehicle size. Again, these EVs should allow for fulfilling identical work requirements.

The first important consideration to extract from the analysis of the data presented in Table 3 is that it is claimed that electric vehicles emit 0 CO_2/100 km, considering the homologation results. This is interesting considering that the energetic consumption defined also in the homologation is 15.3 kWh/100 km. This could be true if the electricity production would, in fact, be zero CO_2 emissions. However, it was previously shown that the energy mix in Portugal for the third trimester of 2021 (Fig. 3) corresponds to CO_2 emissions of 316.74 g/kWh.

Another important aspect highlighted when comparing the two types of vehicles is that for similar engine power of both vehicles the torque for EVs is significantly higher. This means that it is expected for EVs to have better performance in urban traffic. However, another important difference, when comparing the two alternative vehicle types, is that the weight of the electric vehicle is almost 2/3 higher than that of the IC engine vehicles. This implies an additional loss of efficiency, as more energy is required to move the EVs' increased mass.

Getting the same procedure used before, considering that the use of the vehicle in normal condition will require a 50% increase in the energetic source, considering the value of 15.3 kWh/100 km obtained in the homologation tests. It represents an energy consumption of 22.95 kWh required for each 100 km.

Considering the energy mix and the energetic consumption of the EVs, it is possible to compute the overall CO_2 emission related to the EVs operation, with 72.7 g CO_2/km. This value is significantly lower than those of the alternative IC vehicles

Table 3 Comparative EVs fleet characteristics

Year	2021
Engine	Electric
Max power	136 hp
Max torque	395 nm
Max speed	155 km/h
Fuel	Electricity
Weight	2080 kg
Emission homologation type	–
CO_2 emissions (homologation)	0 g/100 km
Energy consumption (homologation)	15.3 kWh/100 km
Autonomy range	400 km

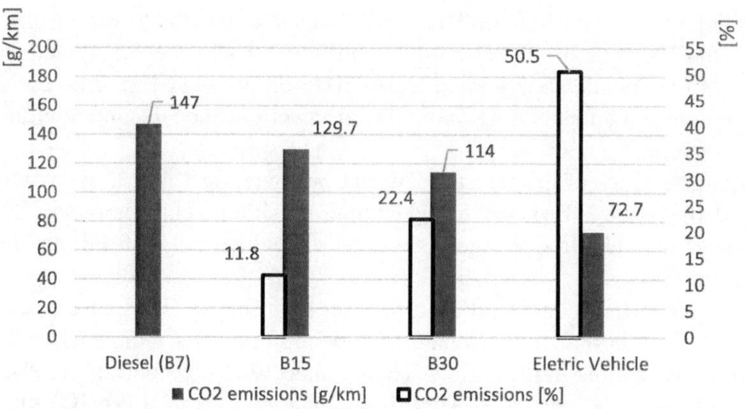

Fig. 4 CO_2 emissions for this study's alternative scenarios (B15, B30, and EVs)

fueled with commercial diesel option and even both alternative biofuel options (B15 and B30), as shown in Fig. 4. Nevertheless, the actual 72.7 g CO_2/km computed for the EVs is significantly higher than the announced value of 0 CO_2/100 km emissions claimed in the EV's homologation.

In fact, there is possible space for the two mentioned solutions. As referred by [16], all European countries still purse both energetic solutions, the use of biofuels, pointing to the products obtained through advanced materials, and implementing a strategy to get an electric and hybrid vehicles dissemination.

3.4 Economic Analysis

From the economic point of view, replacing the existing fleet of IC light-duty diesel vehicles with similar EVs is a significant investment. Thus, from the end-user's point of view, the option of selecting EVs represents an increasingly higher investment than to replace commercial diesel with the use of biofuels. The alternative of changing fueling to B15 or B30 does not require renewing the whole fleet of vehicles once the lifetime of existing vehicles can be extended in a sustainable way and at a low investment cost. The option of replacing existing vehicles with EVs require a significant investment cost to acquire new vehicles, which typically represents a figure of 1.5 to 1.75 times the cost of an IC vehicle.

Another important investment related to the option for EVs to replace existing IC vehicles is the need to improve the existing electric supply lines. The need to recharge every night the 13-vehicle fleet demands for the increased reinforcing of the local distribution grid and the required power supply characteristics for at least more 100 kW, thus allowing recharging simultaneously the vehicles in a 7.2 kW for the regular 6 h charging periods.

It is also expected the need to increase the actual number of vehicles of the existing fleet with two additional vehicles, which will allow reconfiguring the distribution of the vehicles, thus reducing the average distance travelled by each vehicle and allowing minimizing the usual problems related with EVs' limited autonomy-related anxiety of their users. This solution also requires two additional drivers, which will contribute to the increased increase in operational costs related to the use of EVs.

In a different perspective, these investments could be compensated by the cost of the electricity that is lower than the biodiesel price. The cost of fleet operation, considering the two different scenarios and disregarding the maintenance costs, differs significantly and depends highly on the fuels' market prices and the way how the vehicles were recharged.

On what concerns the cost of the biodiesel, during the year 2021, an average price of 1.482 €/liter of fuel was considered. Furthermore, for the price of electricity, the lowest cost scenario to recharge the EVs by using the company tariff for electric energy, an average price of 0.145 €/kWh was considered based on the 2021 costs of energy [17].

This implies that considering both the above scenarios, it is possible to estimate the operational cost of the fleet running on B15 by 3424.2 €/month. For the option of using B30, this could be slightly higher, reaching approximately 3500 €/month. These values can be compared to the lower costs of electric energy required to run the fleet of EVs for the same operation, which would have an average cost of 1339.5 €/month.

Evaluating the implication of the expected increase in energy prices, which will be more significant in electric energy due to the political pressure to eliminate more polluting and cheaper sources such as coal, betting on more sustainable but also more expensive energy sources, there will be some tendency for a decrease in the difference between the price of the two solutions. However, it will not be very significant because the increase in the price of energy will always have an impact on the production and corresponding price of biofuels and because a price increase in electric energy, even if proportionally higher than the predictable increase in the cost of biofuels, will keep the monthly cost below the corresponding one considering the use of biofuels. In this sense, this rise in electric energy price will approximate the operational costs of the two solutions, but there will also be a lower environmental impact due to the use of electric energy with lower CO_2 emissions.

4 Conclusions

The present work allowed acknowledging the following main findings:

1. The values of fuel consumption and CO_2 emissions obtained from the homologation cycles present significant differences when compared with the actual values measured in real road travels.

2. On what concerns the impact on global warming, the expected variations in CO_2 emissions ranged between 53 and 313 g CO_2eq/kWh of battery capacity [18]. This analysis revealed that even considering such a broad range, the EVs' overall life cycle corresponds to lower CO_2 emissions than the ones resulting from the diesel vehicles.

   It became clear that the CO_2 emission factor of battery production for BEV significantly affects the results of the total life-cycle CO_2 emissions [19]. However, the operation of the vehicle implies an effective reduction, but not as significant as theoretically expected. In the present case study, the CO_2 emissions related to the use of EVs, when compared to the IC engine vehicles fueled with diesel, corresponds to a reduction of 50% on CO_2 emissions. However, one must take into account that to produce batteries and manufacture the related vehicles, the CO_2 emissions is significantly higher for EVs than for traditional IC vehicles. Thus, the global benefits of EVs on what concerns the CO_2 emissions is significantly lower than for analyzing merely their daily operation.
3. The use of biodiesel offers the possibility to reduce CO_2 emissions without the need for increased investments once it can represent a simple and effective solution for existing fleets of IC vehicles. The use of B30, considering the overall life cycle analysis, accounts for a reduction of approximately 25% in CO_2 emissions when compared to commercial diesel. Considering the combined CO_2 emissions related to EVs production, the difference in global CO_2 emissions for B30 and the alternative electric vehicles is not as evident as initially foreseen.

Considering this exploratory research, it is now important to measure and analyze the real impacts related to the use of B30 and the operation of alternative EVs to validate the estimated results. A complementary LCA study would also be required to further support the discussion, collecting real data related to the global operation of the proposed alternative solutions. This should also be combined with a deeper economic analysis, thus allowing discussing a more robust scenario that could be presented to governance and decision-makers, to fleet owners and also to the general population to further support the discussion related to sustainable solutions for the current and future mobility. This is reinforced by the study presented by [20], where some critical aspects should be considered, mainly focusing the special electricity tariffs that can be used for car charging and the electricity generation capacity and correspondent average emissions intensity. Also considering the increase percentage of biofuel incorporation should require the assurance for a real CO_2 emission reduction and effective decrease in fuel consumption.

The final results should allow for supporting the evaluation of the alternative options of selecting EVs or biofuels as a solution for the mitigation of the impacts and costs of the transport sector and to minimize the effects of such transportation on its contribution to global warming.

References

1. Energy consumption in transport; Energy statistics – Energy balances provided by Statistical Office of the European Union (Eurostat).
2. IEA. (2009). *World energy outlook*. IEA and Organizations for Economic Co-operation and Development.
3. Meyer, I., Leimbach, M., & Jaeger, C. (2007). International passenger transport and climate change: A sector analysis in car demand and associated CO_2 emissions from 2000 to 2050. *Energy Policy, 35*(12), 6332–6345.
4. EEA, European Environmental Agency. (2012). European Environmental Agency. http://www.eea.europa.eu/themes/air/intro. Accessed 02.07.13.
5. EC. (2014, May). Climate actions. European Commission. http://ec.europa.eu/clima/policies/transport/index_en.htm. Accessed 04.14.
6. Statistical Office of the European Union (Eurostat). (2019). Energy consumption in transport. *Energy consumption in transport*.
7. Shahir, V. K., Jawahar, C. P., & Suresh, P. R. (2015). Comparative study of diesel and biodiesel on CI engine with emphasis to emissions - a review. *Renewable and Sustainable Energy Reviews, 45*, 686–697. https://doi.org/10.1016/j.rser.2015.02.0
8. Abed, K. A., Gad, M. S., El Morsi, A. K., Sayed, M. M., & Elyazeed, S. A. (2019). Effect of biodiesel fuels on diesel engine emissions. *Egyptian Journal of Petroleum, 28*(2), 183–188. https://doi.org/10.1016/j.ejpe.2019.03.001
9. Jochem, P., Babrowski, S., & Fichtner, W. (2015). Assessing CO_2 emissions of electric vehicles in Germany in 2030. *Transportation Research Part A: Policy and Practice, 78*(2015), 68–83. https://doi.org/10.1016/j.tra.2015.05.007
10. Xu, L., Yilmaz, H. Ü., Wang, Z., Poganietz, W. R., & Jochem, P. (2020). Greenhouse gas emissions of electric vehicles in Europe considering different charging strategies. *Transportation Research Part D: Transport and Environment, 87*(September), 1–14. https://doi.org/10.1016/j.trd.2020.102534
11. Coronado, C. R., de Carvalho, J. A., & Silveira, J. L. (2009). Biodiesel CO2 emissions: A comparison with the main fuels in the Brazilian market. *Fuel Processing Technology, 90*(2), 204–211. https://doi.org/10.1016/j.fuproc.2008.09.006
12. Lopes, M., et al. (2013). Emissions characterization from EURO 5 diesel/biodiesel passenger car operating under the new European driving cycle. *Atmospheric Environment, 84*. https://doi.org/10.1016/j.atmosenv.2013.11.071
13. Sagaama, I., Kchiche, A., Trojet, W., & Kamoun, F. (2020). Impact of road gradient on electric vehicle energy consumption in real-world driving. *Advances in Intelligent Systems and Computing, 1151 AISC*, 393–404. https://doi.org/10.1007/978-3-030-44041-1_36
14. Sheehan, J., Camobreco, V., Duffield, J., Graboski, M., & Shapouri, H. (1998). An overview of biodiesel and petroleum diesel life cycles. *U.S.Department Agric. U.S Dep. Energy*, no. May, pp. 1–60, [Online]. Available: http://www.nrel.gov/docs/legosti/fy98/24772.pdf
15. ERSE. mix base do sistema de energia eletrica e fatores de emissão. Accessed 28 Jan 2022.
16. Danielis, R., Scorrano, M., & Giansoldati, M. (2022). Decarbonising transport in Europe: Trends, goals, policies and passenger car scenarios. *Research in Transportation Economics, 91*. https://doi.org/10.1016/j.retrec.2021.101068
17. DGEG. (2021). *Energy prices in Portugal*. https://precoscombustiveis.dgeg.gov.pt/estatistica/preco-medio-diario/. Accessed 28.01.22.
18. Temporelli, A., Carvalho, M. L., & Girardi, P. (2020). Life cycle assessment of electric vehicle batteries: An overview of recent literature. *Energies, 13*(11). https://doi.org/10.3390/en13112864
19. Kawamoto, R., et al. (2019). Estimation of CO_2 emissions of internal combustion engine vehicle and battery electric vehicle using LCA. *Sustainability, 11*(9). https://doi.org/10.3390/su11092690
20. Glensor, K., & María Rosa Muñoz, B. (2019). Life-cycle assessment of Brazilian transport biofuel and electrification pathways. *Sustainability, 11*, 6332. https://doi.org/10.3390/su11226332

The Influence of Consumer Optimism and Pessimism on Purchasing Intention of Eco-Friendly Clothing by Generation Z: Model Proposal

Sofia Antunes, Susana Garrido, and Cristela Bairrada

Abstract Overconsumption has caused a significant reduction in the world's natural resources, negatively influencing the sustainability of individual companies and the sustainable development of the planet. The textile/apparel industry has negatively impacted the ecosystem with the carbon footprint of production units and landfills from non-ethical consumption and waste. This study aims to propose a conceptual model to understand the influence of the psychological mood of consumers (optimism and pessimism) over the intention to purchase eco-friendly clothing by generation Z exploring a novel set of antecedents. Considering these relationships, the proposed study also intends to study the mediating effect of environmental concern and perceived consumer efficiency/effectiveness. An unstructured literature review was used to identify the main factors influencing the intention to purchase eco-friendly clothing by generation Z and suggest a model based on it. The suggested model can help to support future research about the buying behavior of generation Z regarding eco-friendly clothing. This study is the first to suggest the influence of optimism and pessimism of generation Z on the intention to purchase eco-friendly clothing by using as intermediate factors the environmental concerns and the perceived consumer efficiency and effectiveness.

Keywords Purchase intention · Optimism · Pessimism · Environmental concern · Perceived consumer efficiency and effectiveness · Eco-friendly clothing · Generation Z

S. Antunes
University of Coimbra, Faculty of Economics, Coimbra, Portugal

S. Garrido (✉) · C. Bairrada
University of Coimbra, CeBER, Faculty of Economics, Coimbra, Portugal
e-mail: garrido.susana@fe.uc.pt; cristela.bairrada@fe.uc.pt

J. C. de Oliveira Matias et al. (eds.), *Quality Innovation and Sustainability*, Springer Proceedings in Business and Economics,
https://doi.org/10.1007/978-3-031-12914-8_19

1 Introduction

Sustainability has become a central issue nowadays, being part of the discussion at conferences and political meetings where measures are discussed and taken to stop or mitigate the climate change that we are observing. One of the major themes corresponds to individual consumption, which, having grown exponentially in recent years, is gaining special attention in the political debate as it is a highly impacting activity on the environment [1].

In particular, the fast fashion industry has grown significantly, with the number of clothing purchases increasing by 30% from 2012 until 2016 and each individual purchasing, on average, 30 kg of clothing per year [2]. This rate of consumption causes harmful effects on the environment not only because of the resources it consumes but also because of the pollution that its waste causes, with the textile industry being one of the most destructive industries for the environment [3]. In this sense, the slow fashion industry emerges, as opposed to the existing one, which presents clothing focused on quality with a production rhythm that respects environmental resources and with lower waste, which offers the consumer a more resistant product and changes the focus from quantity to quality [2].

This investigation focuses on generation Z as this is the emerging generation of consumers that will dominate and influence the textile market [1]. This is the generation that has lived most intensely through topics such as climate change, energy, and the water crisis. It is also the generation, among the existing ones, that "are most interested in monitoring sustainability into their activities" [4]. With this in mind, Generation Z is changing their consumption behavior by focusing their increasing attention on green products [5].

This study focuses on the environmental aspect of the concept of sustainability to understand the influence of consumer optimism and consumer pessimism on the environmental concerns of generation Z and over the perceived consumer effectiveness. Finally, it is important to understand how these variables affect generation Z's intention to purchase eco-friendly clothing.

This paper aims to propose a conceptual model to understand the influence of the psychological mood of consumers (optimism and pessimism) over the intention to purchase eco-friendly clothing by generation Z exploring a novel set of antecedents.

This study is divided into five sections. After the introduction, a literature review is performed to clarify the variables that form the proposed research model: Intention to Purchase Eco-Friendly Clothing, Optimism, Pessimism, Environmental concern, and Perceived consumer efficiency/effectiveness. Section 3 presents the conceptual model proposed for this study and Sect. 4 describes the methodology that will serve as the basis for all this research work. Finally, in Sect. 5, the main conclusions of this academic work are drawn.

2 Theoretical Framework

2.1 Intention to Purchase Eco-Friendly Clothing

Since its beginning, consumption has been part of the human being, becoming increasingly present in today's society. The reality is that we are now facing over-consumption by society, supported by the linear economy paradigm. Indeed, "in the past fifty years, many advanced societies have gradually become 'consumer societies' in which consumption performs a major role in the stimulation of economic growth" [6]. It becomes a matter of concern when we witness uncontrolled consumption highly harmful to the environment.

Firstly, for the present work, it is important to define the concept of "sustainability." The concept of sustainability "involves complex and changing environmental dynamics that affect human livelihoods and well-being, with intersecting ecological, economic, and sociopolitical dimensions, both globally and locally" [7]. The four components that will motivate a company to have an ecological response to the outside world are as follows: (1) legislation, which is updated according to global environmental needs; (2) pressure from stakeholders, such as customers or local communities; (3) economic opportunities, such as reducing waste, which impacts positively the company economically; and (4) ethical motivations [8]. Consequently, we see that there are many significant aspects that might encourage an organization to become more environmentally friendly in the present time.

In the same line of thought, corporate sustainability is achieved by the intersection of the following three principles: environmental integrity (through environmental management of enterprises), social equity (through corporate social responsibility), and economic prosperity (through value creation) [9]. This concept is also known as the "Triple Bottom Line," explaining the three dimensions on which a company must rely to achieve sustainability. This means "that in full cost accounting of business activities, costs of environmental pollution also have to be included. Only then is the success of green developments and equal achievement of economic, social and ecological goals" [10].

It is evident that the concern with the dimension of sustainability is positive for the company. Not only for economic reasons but also for the new consumers' profile that prefers companies that are environmentally friendly and adopt measures accordingly. In this line, a study concluded that 70% of consumers would be willing to pay up to 5% above the normal price of the product if the packaging were sustainable [11]. Moreover, in the fashion industry, 61% of consumers would pay 20% more than the normal price if the fashion item were eco-friendly [12].

In recent years, there was a proliferation of "green products" in almost every store. The concept is defined as being a product that is ecological or environmentally friendly [13]. The concept, together with "environmental product," is commonly used "to describe those who strive to protect or enhance the natural environment by conserving energy and/or resources and reducing or eliminating the use of toxic agents, pollution and waste" [14]. In addition to the concept, the authors also argue that although no product can have a zero impact on the environment, companies must try to minimize it as much as possible.

In addition, in the fashion industry, the term "slow fashion," in opposition to "fast fashion," was presented by Fletcher [15] and it was highly important for the industry to change its mindset. The author presented a new paradigm that advocates greater awareness of the impact that the industry has on workers, communities, and the ecosystem. It is necessary to create a shift from quantity to quality, with a planned production instead of excessive production in record time that has serious impacts on workers and on the environment [15]. To clarify the differences between the concepts of fast fashion and slow fashion, Mukherjee [16] considers that fast fashion creates small fashion cycles, with a high focus on profit and little focus on quality, which causes poor use of resources, while slow fashion focuses on quality with lower production, reducing the waste of resources [17]. The slow fashion concept argues that consumers think about their fashion purchases as investments in high-quality clothing allowing them to enlarge items' life cycle with no need to keep buying them regularly.

Individual consumption has an outstanding relevance in affecting in a positive way the environment, therefore, the consumption of green products can prevent or even reduce damage to the environment [18]. In that case, it is important to understand which factors influence the purchase of green products.

For this reason, researchers focus on the values, attitudes, and intentions of consumers which leads to two following theories as the ones supporting most studies: Theory of Reasoned Action (TRA), developed by Ajzen and Fishbein in 1980, and Theory of Planned Behavior (TPB), created by Ajzen in 1985 [18].

The Theory of Reasoned Action explains that individual behavior depends on two main factors: individual attitude and social norms. The Theory of Planned Behavior beyond these two factors considers also the perceived behavioral control, which is the resources and opportunities that consumers believe they have at the time of purchase.

Nevertheless, researchers continue to struggle with a weak relationship between a positive attitude of consumers towards the purchase of green products and the actual purchase of these same products. Bray [19] mentioned for the first time the "30:3 phenomenon," which consists in the fact that 30% of consumers are concerned about more ethical consumption, but only 3% of consumption reflects those concerns. Since this discrepancy has been noticed in several studies, other authors have called it the "Ethical Purchasing Gap" or "Attitude-Behaviour Gap." The "Attitude-Behavior Gap" indicates that a consumer who has a positive attitude towards green products does not always transfer this to action – buying, in fact, green products [18].

Consequently, later, Guagnamo in 1995 introduced the Attitude-Behavior-Context model explaining that the purchase behavior of green products is not explained only by the consumer's attitude but also by its context [18]. For a better understanding, some examples of contextual factors are "personal influences, personal values and feelings, abilities, publicity, expectations, physical environment, institutional factors, and temporal perspectives" [20].

In fashion, expectations of the company with environmentally friendly behavior play an important role in purchasing fashion items as much as a social influence by consumers groups [12]. Price, on the other hand, can cause an adverse effect if it is too inflated, but, in general, consumers are willing to pay more for sustainable clothing [12]. Brands need to ensure that slow fashion has better quality, durability, and wearability since that is what distinguishes slow fashion from fast fashion [21].

2.2 *Antecedents of the Intention to Purchase Eco-Friendly Clothing*

Eco-friendly clothing has, like any other product, factors that influence the intention to purchase them. In this section, we address optimism and pessimism as intrinsic characteristics of each consumer's personality that unintentionally affect consumers' intention to purchase eco-friendly clothing. At the end, we also address the concepts of environmental concern and perceived consumer efficiency/effectiveness.

2.2.1 Optimism and Pessimism

Optimism and pessimism refer to a two-dimensional model, as opposed to unidimensionality, which only considers optimism as the only dimension [22].

An optimistic person, after exploring the possibility of something happening, feels a positive reinforcement and a force that something can be done and, finally, proceeds to the action [23]. Optimistic people always believe in the good side of events, believing that everything will go according to their expectations and in a positive way [24]. Positive people are more oriented towards more favorable socialization, which contributes to an overall positive attitude and favorable cognitive processes to the acceptance of circumstances [25]. The motivation and confidence that come from positivism connect consumers to causes and problems in general, as they feel more confident and motivated to overcome the issues that arise in life [23].

On the other hand, negativism results in a negative view of the world, with stress and tension present, even if it is not justified [26]. In addition, pessimistic people "are less likely to have the necessary social and psychological resources, and the negative effect does not motivate them to be engaged in an environment-friendly behavior" [22].

2.2.2 Environmental Concern and Perceived Consumer Efficiency/Effectiveness

In addition to optimism and pessimism, the present investigation also intends to investigate environmental concerns and perceived consumer efficiency/effectiveness as antecedents of the Intention to Purchase Eco-friendly Clothing.

Environmental concerns relate to consumers' awareness of environmental problems and the corresponding willingness to resolve them [27]. Effectively, consumers nowadays are more aware of the impact that their purchases have on the environment, and they know that these also influence the behavior of companies that are increasingly striving to advertise their sustainable campaigns, also because consumers are willing to pay more for a product that they consider more beneficial or less damaging for the environment [12].

Perceived Consumer Effectiveness refers to the consumer's belief that their efforts contribute to the solution of the problem [28]. In the topic of consumption of eco-friendly clothing, this variable gains importance as it is one of the determinants of purchase intention. Consumers must believe that by opting for an eco-friendly product, they are cooperating to reduce their impact on the environment. The aim of the eco-friendly clothing industry will be to develop among consumers a sense of success through their experience or of others, in the impact of their actions.

3 Methodology

This study aims at synthesizing the existing theoretical and empirical findings concerning the environmental aspect of the concept of sustainability to understand the influence of consumer optimism and consumer pessimism on the environmental concerns of generation Z and over the perceived consumer effectiveness.

The synthesis takes the form of a conceptual model that presents the perceived interplay between the key variables affecting the Intention to purchase Eco-friendly clothing in students from the University of Coimbra and belonging to Generation Z.

Miles and Huberman [29] defined a conceptual model as a visual or written product that explains either in a graphical or in a narrative form the key factors, concepts, or variables that comprise it. Developing a conceptual model serves also to present the presumed relationships that exist between the framework's elements. In order to develop the conceptual model, a multi-step research procedure was performed. After defining the theoretical basis for the study, the analysis of literature on the intention to purchase eco-friendly clothing and antecedents of the intention to purchase eco-friendly clothing was performed. It was focused on compiling a list of key variables from these two constructs. It classifies the variables affecting the intention to purchase Eco-Friendly Clothing into "Optimism," Pessimism," Environmental Concerns," and "Perceived consumer Efficiency/Effectiveness." Thus, as a result of applying the research procedure and inductive reasoning, a conceptual model was constructed bringing together the various identified categories specific to affecting the Intention of Gen Z to purchase Eco-friendly clothing. It maps out the various concepts such as Consumer Optimism, Consumer Pessimism, Environmental Concern, Perceived Consumer Efficiency/Effectiveness, and Intention to Purchase Eco-Friendly Clothing. The model builds upon the findings from various fields of study and forms a construct that requires further verification through follow-up investigations that will test the framework.

The first variable to be studied, Consumer Optimism, and the second one, Consumer Pessimism, are measured based on the work of Sadiq, Paul, and Bharti [22]. In order to measure the variables Environmental Concern and Perceived Consumer Efficiency/Effectiveness, the investigation was based on the work of Lee, Kim, Kim, e Choi [30]. Finally, to measure the variable Intention to Purchase Eco-Friendly Clothing, the research was based on the work of Mostafa [31]. The research variables are described in Table 1.

4 Conceptual Model Proposal

The problem that underlies the present work is the need for a drastic change in consumption habits that significantly impact the environment and result in the climate changes that we all witness. Among the various areas that need paradigm shifts, the Youth4Climate group, which was an integral part of the 2021 Climate Summit known as COP26, elaborates that the fashion industry has to be based on materials with low environmental impact, implement a decarbonization strategy, transport materials with low-carbon vehicles, ensure decent working conditions, and explore the circular economy [32]. Not only do companies need to work towards more sustainable production but consumers also need to change their consumption habits. In this sense, it is important to understand the influence of a set of variables related to the psychological mood of consumers both on environmental concerns and on the consumer's perception of efficiency and effectiveness and, consequently, on the purchase intention of sustainable clothing.

In this way, it is possible to know what forms a set of variables related to the purchase intention of sustainable clothing, which is a predecessor of purchase behavior. In this sense, it is important to understand the influence of consumer optimism and consumer pessimism on the environmental concerns of generation Z and perceived consumer effectiveness. Finally, it is important to understand how these variables affect generation Z's intention to purchase eco-friendly clothing. As a result, the present work intends to answer the following research question: There is an influence of the psychological mood of generation Z consumers (optimism and pessimism), on both the environmental concerns and perceived consumer efficiency and effectiveness, and consequently, over the intention to purchase eco-friendly clothing?

Optimism, one of the Intention to Purchase Eco-friendly Clothing antecedents presented, corresponds to a dispositional trait that unconsciously influences the way a consumer evaluates and executes his purchase. The concept is based on the

Table 1 Scales by research variables

Consumer Optimism
In uncertain times, I usually expect the best.
I always look on the bright side of things.
I'm always optimistic about my future.
When I undertake something new, I expect to succeed.
Where there's a will, there's a way.
In general, things turn out all right in the end.
Consumer Pessimism
It is best not to get your hopes too high since you will probably be disappointed.
Rarely do I expect good things to happen.
If something can go wrong for me, it will.
I hardly ever expect things to go my way.
Things never work out the way I want them to.
If I make a decision on my own, I can pretty much count on the fact that it will turn out to be a poor one.
I rarely count on good things happening to me.
Better to expect defeat: then it doesn't hit so hard when it comes.
Give me 50/50 odds and I will choose the wrong answer every time.
Environmental Concern
Humans are severely abusing the environment.
If things continue on their present course, we will soon experience a major ecological catastrophe.
The balance of nature is very delicate and easily upset.
Perceived Consumer Efficiency/Effectiveness
I can protect the environment by buying products that are friendly to the environment.
I feel I can help solve natural resource problems by conserving water and energy.
I feel capable of helping solve environmental problems.
Each person's behavior can have a positive effect on society by signing a petition in support of promoting the environment.
Intention to Purchase Eco-Friendly Clothing
Over the next month, I will consider buying Eco-Friendly Clothing.
Over the next month, I will consider switching to other clothes brands for ecological reasons.
Over the next month, I plan to switch to a green version of clothes.

consumer's focus on positive rather than negative expectations for the future, even if there is no indication of it [33]. According to the authors, optimistic consumers get involved in causes and strive to work on them since they believe that the result will be positive. When barriers arise, positive consumers tend to continue efforts, unlike pessimistic consumers. Moreover, when optimists imagine the future, they tend to look at it positively, which causes them to be very hopeful. Hence, the following hypothesis is suggested:

H1: Optimism has a direct and positive influence on the Intention to Purchase Eco-friendly Clothing.

H2: Optimism has a direct and positive influence on environmental concerns.
H3: Optimism has a direct and positive influence on Perceived Consumer Effectiveness.

Pessimism is another antecedent of the Intention to Purchase Eco-friendly Clothing variable of the conceptual model, and, in contrast to positivism, it has a negative impact on the way consumers face life and see obstacles that arise in it. The pessimistic consumer sees the world in the opposite way to the positive ones, having unfavorable expectations for the future [22]. Pessimistic consumers have little motivation to engage in tasks and react poorly to obstacles, preferring the path of demoralization, and giving up most of the time. According to also Coelho et al. [34] "a tendency to portray the world through a negative lens is likely to contaminate an individual's view about the environment itself. More specifically, it is likely that high Negative effects will lead individuals to focus their attention on negative information about the environment, thus raising concerns about it." Hence, the following hypotheses are suggested:

H4: Pessimism has a direct and negative influence on the Intention to Purchase Eco-friendly Clothing.
H5: Pessimism has a direct and positive influence on environmental concerns.
H6: Pessimism has a direct and negative influence on Perceived Consumer Effectiveness.

As previously presented, environmental concerns relate to consumers' awareness of environmental problems and the corresponding willingness to resolve them [27]. Taking this into account, environmental concerns are directly linked to green purchase behavior [35], indicating that purchase intention is related to consumers' environmental concerns. Hence, the following hypothesis is suggested:

H7: Environmental concerns have a direct and positive influence on the Intention to Purchase Eco-friendly Clothing.

As explained previously, perceived consumer effectiveness refers to the consumer's belief that their efforts contribute to the solution of the problem [28]. A sense of contribution to solving the problem motivates their desire to continue to contribute to the solution of the problem, in this specific case, the reduction of the environmental impact of the textile industry. Bearing in mind that the combination of intention with perceived consumer effectiveness leads to action [36], consumers who believe that the purchase of eco-friendly clothing will have a favorable impact on the environment will proceed to purchase through the intention to purchase eco-friendly clothing. Hence, the following hypothesis is suggested:

H8: Perceived consumer effectiveness has a direct and positive influence on the Intention to Purchase Eco-friendly Clothing.

Given the hypotheses previously defined, the conceptual model described in Fig. 1 is suggested.

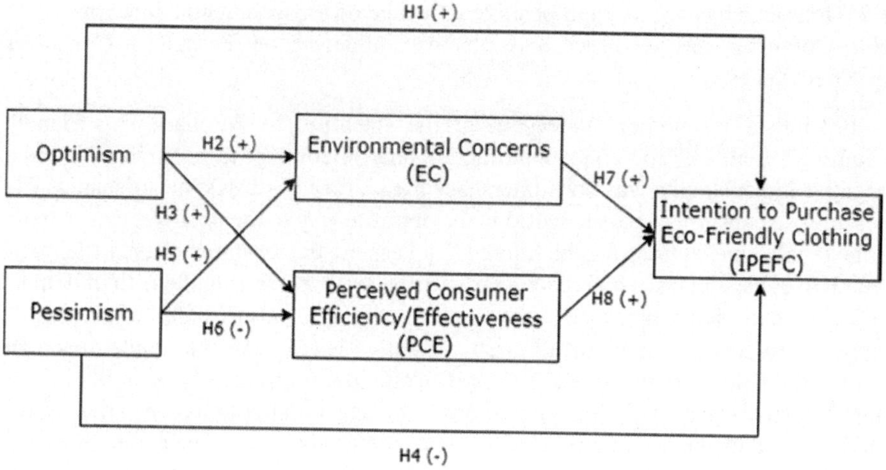

Fig. 1 Conceptual model

5 Conclusion

With the growing importance of sustainability, it is extremely important to understand how the psychological mood of generation Z influences their environmental concern and perceived consumer efficiency/effectiveness over their intention to purchase eco-friendly clothing. The present study is the first to explore the influence of optimism and pessimism of generation Z on the intention to purchase eco-friendly clothing by using the environmental concerns and perceived consumer efficiency/effectiveness as intermediate factors. Saying this, it is clear that this research presents contributions to the fashion industry in general and marketers in particular giving some insight on the Intention to Purchase Eco-Friendly Clothing by generation Z attending to their psychological mood, environmental concern, and perceived consumer efficiency/effectiveness.

This article has an important managerial contribution to enterprises and organizations with an environmentally conscious culture. Specifically, for firms who promote their sustainability initiatives and products and want to connect with today's and tomorrow's customers, generation Z. Along with industry, non-profit environmental organizations receive critical information about generation Z and the impact of dispositional features on their capacity to define environmental concerns.

According to Markowitz [37], personality traits play a significant role in pro-environmental behavior; consequently, the current research will give some insights for advertising firms to tailor their messages in order to maximize the purchase intention of green products. Along with advertising campaigns, products label may evoke certain emotions in order to increase purchase intent. Finally, this study serves as an excellent springboard for future research into the effect of personality factors on pro-environmental behavior.

Acknowledgments This work has been funded by national funds through FCT – Fundação para a Ciência e a Tecnologia, I.P., Project UIDB/05037/2020.

References

1. Chaturvedi, P., Kulshreshtha, K., & Tripathi, V. (2020). Investigating the determinants of behavioral intentions of generation Z for recycled clothing: An evidence from a developing economy. *Young Consumers, 21*(4), 403–417.
2. Jung, S., & Jin, B. (2016). From quantity to quality: Understanding slow fashion consumers for sustainability and consumer education. *International Journal of Consumer Studies, 40*(4), 410–421.
3. Moorhouse, D., & Moorhouse, D. (2017). Sustainable design: Circular economy in fashion and textiles. *The Design Journal, 20*(1), 1948–1959.
4. Dabija, D., Brandusa, B., & Puscas, C. (2020). A qualitative approach to the sustainable orientation of generation Z in retail: The case of Romania. *Journal of Risk and Financial Management, 13*(7), 152.
5. Finantial Times. https://www.ft.com/content/8b08bf4c-e5a0-11e7-8b99-0191e45377ec. Last accessed 2022/03/06.
6. Crane, D. (2010). Environmental change and the future of consumption: Implications for consumer identity. *Anuario Filosofico, 43*, 353–379.
7. Joy, A., Sherry, J. F., Jr., Venkatesh, A., Wang, J., & Chan, R. (2012). Fast fashion, sustainability, and ethical appeal of luxury brands. *Fashion Theory, 16*(3), 273–296.
8. Bansal, P., & Roth, K. (2000). Why companies go green: A model of ecological responsiveness. *The Academy of Management Journal, 43*(4), 717–736.
9. Bansal, P. (2005). Evolving sustainably: A longitudinal study of corporate sustainable development. *Strategic Management Journal, 26*(3), 197–218.
10. Friedrich, D. (2021). What makes bioplastics innovative for fashion retailers? An in-depth analysis according to the Triple Bottom Line Principle. *Journal of Cleaner Production, 316*, 128257.
11. Martinho, G., Pires, A., Portela, G., & Fonseca, M. (2015). Factors affecting consumers' choices concerning sustainable packaging during product purchase and recycling. *Resources, Conservation and Recycling, 103*, 58–68.
12. Ciasullo, M. V., Maione, G., Torre, C., & Troisi, O. (2017). What about sustainability? An empirical analysis of consumers' purchasing behavior in fashion context. *Sustainability, 9*(9), 1617.
13. Tan, B. C., & Lau, T. (2010). Attitude towards the environment and green products: Consumers' perspective. *Management Science and Engineering, 4*(2), 27–39.
14. Ottman, J., Stafford, E., & Hartman, C. (2006). Avoiding green marketing myopia: Ways to improve consumer appeal for environmentally preferable products. *Environment: Science and Policy for Sustainable Development, 48*(5), 22–36.
15. The Ecologist. https://theecologist.org/2007/jun/01/slow-fashion. Last accessed 2022/03/06.
16. Mukherjee, S. (2015). Environmental and social impact of fashion: Towards an eco-friendly, ethical fashion. *International Journal of Interdisciplinary and Multidisciplinary Studies, 2*(3), 22–35.
17. Cline, E. L. (2012). *Overdressed: The shockingly high cost of cheap fashion* (1st ed.). The Penguin Goup.
18. Joshi, Y., & Rahman, Z. (2015). Factors affecting green purchase behaviour and future research directions. *International Strategic Management Review, 3*(1–2), 128–142.
19. Bray, J., Johns, N., & Kilburn, D. (2011). An exploratory study into the factors impeding ethical consumption. *Journal of Business Ethics, 98*(4), 597–608.

20. Yadav, R., Balaji, M. S., & Jebarajakirthy, C. (2019). How psychological and contextual factors contribute to travelers' propensity to choose green hotels? *International Journal of Hospitality Management, 77*, 385–395.
21. Lundblad, L., & Davies, I. A. (2015). The values and motivations behind sustainable fashion consumption. *Journal of Consumer Behaviour, 15*(2), 149–162.
22. Sadiq, M., Paul, J., & Bharti, K. (2020). Dispositional traits and organic food consumption. *Journal of Cleaner Production, 266*, 121961.
23. Peter, P. C., & Honea, H. (2012). Targeting social messages with emotions of change: The call for optimism. *Journal of Public Policy & Marketing, 31*(2), 269–283.
24. Scheier, M., & Carver, C. (1985). Optimism, coping, and health: Assessment and implications of generalized outcome expectancies. *Health Psychology, 4*(3), 219–247.
25. Kolokotroni, P., Anagnostopoulos, F., & Hantzi, A. (2018). The role of optimism, social constraints, coping, and cognitive processing in psychosocial adjustment among breast cancer survivors. *Journal of Clinical Psychology in Medical Settings, 25*(7), 452–462.
26. Judge, T. A., Erez, A., & Thoresen, C. J. (2000). Why negative affectivity (and self-deception) should be included in job stress research: Bathing the baby with the Bath water. *Journal of Organizational Behavior, 21*(1), 101–111.
27. Singh, A. K. S., & Bansal, M. (2012). Green marketing: A study of consumer attitude and environmental concern. *The Indian Journal of Commerce, 65*(2), 273–283.
28. Ellen, P., & Cobb-Walgren, C. (1991). The role of perceived consumer effectiveness in motivation environmentally conscious behaviors. *Journal of Public Policy & Marketing, 10*(2), 102–117.
29. Miles, M. B., & Huberman, A. M. (1994). *Qualitative data analysis: An expanded sourcebook* (2nd ed.). Sage Publications.
30. Lee, Y. K., Kim, S., Kim, M. S., & Choi, J. G. (2014). Antecedents and interrelationships of three types of pro-environmental behavior. *Journal of Business Research, 67*(10), 2097–2105.
31. Mostafa, M. M. (2007). A hierarchical analysis of the green consciousness of the Egyptian consumer. *Psychology & Marketing, 24*(5), 445–473.
32. Youth4Climate. https://unfccc.int/sites/default/files/resource/Youth4Climate-Manifesto.pdf. Last accessed 2022/03/06.
33. Carver, C., & Scheier, M. F. (2014). Dispositional optimism. *Trends in Cognitive Sciences, 18*(6), 1–7.
34. Coelho, F., Pereira, M. C., Cruz, L., Simões, P., & Barata, E. (2017). Affect and the adoption of pro-environmental behaviour: A structural model. *Journal of Environmental Psychology, 54*, 127–138.
35. Albayrak, T., Aksoy, S., & Caber, M. (2012). The effect of environmental concern and scepticism on green purchase behavior. *Marketing Intelligence & Planning, 31*(1), 27–39.
36. Ajzen, I. (1991). The theory of planned behavior. *Organizational Behavior and Human Decision Processes, 50*, 179–211.
37. Markowitz, E. M., Goldberg, L. R., Ashton, M. C., & Lee, K. (2012). Profiling the "pro-environmental individual": A personality perspective. *Journal of Personality, 80*(1), 81–111.

Persistence in Innovation. Do Low-Tech Sectors Differ Much from the High-Tech?

Joana Costa and Nasimeh Tashakori

Abstract Disentangling innovation from growth is unrealistic in the present times. Also, anticipating the future behavior of innovative firms is relevant to the entire innovation ecosystem; and assessing the persistence of innovation and appraising the role of factors affecting ongoing innovation activities in firms is essential. This chapter discusses a very important subject related to the concept of innovation persistence in relation to structural innovation characteristics of firms, with a focus on technological regimes, to better understand if there is change in innivation continuity accordingly to the technological intensity embedded in the sector. The empirical research is based on data from CIS database, comprising 3237 firms which present in the 2014 and 2018 waves. We analyze the innovative persistence behavior of these firms regarding proxies like firm dimension, innovation activities, types of innovation, government funding, and more importantly, technological regimes. To do this, we applied binary logistic regression for developing a model which can forecast the drivers of innovation persistency propensity. The presented study shows that some very important results are achieved. Besides demonstrating innovative persistency in 75% of science-based firms, the findings confirm that firms in high-tech and science-based industries are more prone to continue innovating and, as a result, this consistency in innovation will generate virtuous cycles of innovation. Furthermore, our data shows that complex innovators are more likely to persist than single innovators, proving the existence of complementarities among the innovation types.

J. Costa
DECEIT–Department of Economics, Management, Industrial Engineering and Tourism, Campus Universitario de Santiago, University of Aveiro, Aveiro, Portugal

GOVCOPP–Research Unit on Governance, Competitiveness and Public Policies, Campus Universitário de Santiago, Aveiro, Portugal

N. Tashakori (✉)
DECEIT–Department of Economics, Management, Industrial Engineering and Tourism, Campus Universitario de Santiago, University of Aveiro, Aveiro, Portugal
e-mail: nasimeh.tashakori@ua.pt

J. C. de Oliveira Matias et al. (eds.), *Quality Innovation and Sustainability*, Springer Proceedings in Business and Economics,
https://doi.org/10.1007/978-3-031-12914-8_20

Keywords Innovation persistence · Innovation activity · Technological regimes

1 Introduction

The growing value of knowledge as a productive factor in modern societies needs to shift how people think about innovation in both general and specific terms, like technological innovation, product innovation, and organizational innovation [1]. And, due to the centrality of technology innovation as a factor in a firm's competitive advantage, resilience, and survival, continuity of innovation and innovation persistence has become a core topic in the agenda of Academics, practitioners, and policymakers.

The term "innovation persistence" refers to whether or not a company's innovation activity or performance continues over time. At first, innovation persistence studies such as [2] were focused on firms, leading to further research to specifying the factors of innovation persistence and their heterogeneity in all levels such as technologies, industrial sectors, and countries. In most countries, data for innovation studies is collected through cross-sectional surveys and there aren't many studies on the longitudinal aspect of innovation persistence. However, some empirical studies have been conducted on the continuity of innovation in different countries with various viewpoints [2–9].

Nevertheless, research on this topic focusing on the Portuguese case encompassed data from 2004 to 2010 [10]; the focus of this paper is on the role of innovation activities' continuity at the firm level by identifying effective determinants that are playing a role on innovation persistence such as size, types of innovation, and technological regimes with the latest data. Some questions can be raised in this vein; however, in this article, the main debate relies on: "is persistence associated with innovative activities, size of firms, innovation types, technological regimes or public supports?" what is the relation between them? and "to what extent firms are persistent in different technological regimes?" As few studies addressed the connection between technological regimes and innovation persistence. In this paper, we analyze a group of 3237 firms present in two waves of the CIS database (2014 and 2018) to infer the connection between persistence in innovation and firms' structural characteristics with a core emphasis on technological regimes.

The structure of this paper is as follows: Section 2 introduces empirical studies on this topic and defines the main determinants. Section 3 describes the data and the methodology in the analysis. In Sect. 4, findings are discussed. And finally, in Sect. 5, some suggestions and recommendations for policymakers are presented to help draft policies for the enhancement of the continuity of innovative firms.

2 Literature Review

2.1 Persistence in Innovation

Persistence means "continuing to occur over time." This concept is coined by Schumpeter's [11] cumulation creation process. The fact that technical change is related to large firms' existence in oligopolistic markets (where the accumulation process is activated by Research and Development [R & D] and innovation for making firms continue their performances) brought up three approaches for innovation persistence: path dependence, virtuous cycles of accumulation, and market power dynamics [12]. For instance, the virtuous accumulation cycles approach is used by Nelson and Winter [13]. They showed that persistence is because of the feedback between previous innovations, current investments, and future innovations. Furthermore, they claimed that the process of deciding to innovate can be repeated and become routine in case of being successful [13]. What's more, Malerba and Orsenigo [14] said that persistence is generally about the contribution of knowledge accumulation to innovation's steadiness [14]. In a nutshell, the term "innovation persistence" is introduced in some papers mostly related to the accumulation of knowledge and technological competencies [2, 5, 15]. Moreover, Antonelli [16], Le-Bas, and Latham [12] brought up the time span. Regarding these definitions, we can describe innovation persistence as generating innovation in all possible periods and continuity of innovative activity over time.

To gain a competitive advantage in the market, companies must produce innovations continuously, and this necessitates the existence of variables that have a beneficial impact on the persistence of innovation. Given that innovation persistence is important for continuity of a firm's activity, reasons for developing the persistence of innovation by companies and also intending to create a systematic approach within them are as follows: (1) a better chance for survival: if a company focuses on medium and long-term innovation, it has a better chance of surviving; (2) reducing the chance of being duplicated by competitors: rivals might be unable to copy innovations because the company generates skills that are hard to emulate; (3) less negative impacts on the company: process automation and substitutional human resources could result from thinking of discontinuity of innovation [17]. About the persistence of innovation, four frameworks are presented: market power and innovation [11], success-breeds-success [18], sunk costs [19], and evolutionary innovation theory [13]. These approaches still coexist and are used in some studies [12, 20]. According to the 'market power and innovation' approach [11], innovators are temporary monopolists who benefit from unusual profits. As it is a transient status, the firm will operate the next innovation, increasing the propensity to persist. As new entrants decrease monopoly profits, incumbents will persist in innovating to obstruct entry [20]. The concept of innovative persistence can be related to the Schumpeterian intuition that critical market feedbacks link R & D investment, technological performance, and profitability in its most basic economic interpretation. Firms' evolution is expected to benefit from this strategic behavior, which is strongly linked to the industry life-cycle. The entrepreneurial technological regime favors entry into the industry in the first part of the life cycle. As a result, small businesses innovate, but mortality is high and selection is harsh [21]. A routinized technological regime predominates in the mature phase of the life cycle. In the entrepreneurial regime, innovation occurs in the R & D departments of large oligopolistic firms, which have a longer economic life than small firms. We can expect a small company to innovate in a shorter period than a large company [10]. Indeed, persistence may simply be the result of 'success-breeds-success' processes as used in Nelson and Winter's model [13]: innovative success leads to profits that can be reinvested in R & D, increasing the likelihood of renewed innovation. The persistence of innovation may also be due to organizational factors at the firm level. For example, persistence could result from the establishment of fixed-cost R & D facilities that generate a relatively stable flow of innovations. More generally, firm technological capacities that can only be developed continuously over time, defining what a firm can do now and what the achievements in the future would be, are likely to be the cause of persistence. From this perspective, the qualitative heterogeneity in the population of innovators is most likely connected to the persistence of innovative activities [10, 22, 23].

2.2 Technological Regimes and Taxonomies

The notion of a technological regime is concerned with the technology upon which firms rely in their problem-solving activities, given a broadly defined way of doing things [13]. It is identified by a unique mix of technological opportunity circumstances, appropriability conditions, learning cumulativeness, and knowledge-based nature [24, 25]. Similarly, a technological regime describes the specific knowledge environment in which business problem-solving operations are conducted [21]. Technological regimes are important because they limit the pattern of innovation that emerges in the sector and highlight common aspects of innovative processes across different types of operations [26]. Taxonomy is described as the science of classification and categorization. Divergent patterns of innovation exist in different sets of production activities, according to taxonomic studies of firm innovative activities. These taxonomies frequently overlap with industrial categories, although they frequently group production activities that are not in the same industry [26]. Several definitions and attributes for classifying firms according to their technology intensity have been presented by researchers. Pavitt [27], Hatzichronoglou [28], De Jong and Marsili [29], and Rizzoni [30] are the most known categories. Pavitt [27] identified innovative firms' characteristics in science-based (electrical/electronic and chemicals), specialized suppliers (non-electrical machinery, instruments, and specialty chemicals), supplier dominated (paper and textiles), and scale intensive (food, vehicles, and metals). Although Pavitt's taxonomy does not cite directly to technological regime's definition, it can be related to the conditions of technological opportunity, the threat of technology-based entry, and appropriability [31]. Pavitt's taxonomy has been tested empirically using innovation surveys by others [29–31]. Malerba and Orsenigo [25] identified two distinct types of technological innovation. In nonelectrical machinery, instruments, and old technologies, the Schumpeter Mark I pattern of innovation is characterized by a scattered and tumultuous pattern of innovative activities. A focused and persistent structure of innovative activities distinguishes the Schumpeter Mark II pattern of innovation, which is typical of chemical and electrical–electronic technology. Differences in technological regimes can be cited for these disparate patterns of innovation across industries or technologies.

The fundamental model of the technological regime is suggested by Nelson and Winter as "science-based" and "cumulative" technology. A broad knowledge base, which comes from externally performed intense R & D activity identifies science-based technology, and a rather narrow one, which is cumulatively built through the learning processes inside the firm distinguishes cumulative technology. This model has been used to establish a relationship between technologies, patterns of innovation, and industrial competition. OECD has developed a new taxonomy of sectors based on their R & D intensity level – the ratio of R & D to value-added within each industry. Firms are classified into five groups (high, medium-high, medium, medium-low, and low R & D intensity industries) [33]. Furthermore, Bogliacino and Pianta [34] introduced a revised Pavitt taxonomy to identify specific technological change patterns using CIS2 and CIS3 to assess the relation of innovation and employment for manufacturing and services [34]. Here, we also rely on this taxonomy to assess the persistence of innovation in groups with different technological regimes.

2.3 *Other Determinants of Innovation Persistence*

Some empirical studies (e.g., [2, 3, 32]) have risen about the subject of innovation persistence and different characteristics have been evaluated. Geroski et al. [2] researched a quantitative methodology for assessing firms' innovation persistence [2]. They made a distinction between occasional and persistent innovators. They examine two datasets based on patenting activity to examine firms' innovation activities for over 20 years, finding that firms that produce patents or major innovations regularly are few; furthermore, they proposed that the number of patents granted at the beginning of the period of innovating can be a good indicator of how long that period will last [2].

On the one hand, other studies [9, 33, 34] stressed the critical importance of formal R & D conducted within a company and show that when firms invest consistently in formal R & D programs, they achieve innovative persistence. For example, Leuven [35] illustrates that when innovation is measured on the output side, performance over time is lower than when measured on the input side (R & D). Also, Song et al. express that internal R & D showed higher persistence than joint/external R & D [9]. On the other hand, some of these studies have found low persistence in the innovation activity of firms [2, 4, 14]. Cefis and Orsenigo [4] classify firms based on their patenting activity, such as non-innovators, firms that apply for a small number of patents, and firms that are great innovators applying for a large number of patents, and examine whether firms stay within their initial categories over time. They find that there is little persistence at the firm level across countries over the period 1978–1993. Firms, on the other hand, have a tendency to stay in their original states, and great innovators account for a large proportion of innovation activities [4]. Some researchers [4, 22, 32] consider the size of a firm as a determinant for studying innovation persistence and assess this relation. According to Malerba and his colleague, firm size is an important factor, but it is not directly related to firms' innovating, but rather to the consistency of their innovation activities. Small firms are more likely than large firms to stop innovating; in industries with many small firms and a few large ones, innovation will tend to concentrate over time among the larger firms [22]. Lhuillery [37] examines the impact of different types of innovation, such as marketing activities, on innovation success, particularly persistent innovation success by using three waves of the French CIS innovation survey from 2002 to 2008. The findings appear to show that marketing innovation does not have a positive impact on low-tech industries' long-term innovation success [37]. When it comes to incremental innovation, innovation marketing has a positive impact, but it has a negative impact when it comes to radical innovation [37]. Aligned with mainstream literature, Le Bas and Négassi [38] investigate the relationship between innovation persistence and innovation activity; however, in other cases, the measurement is narrower – a patent. The main point of this paper is that innovative firm persistence is the main phenomenon that affects technological innovativeness across sectors (the sectoral intensity of innovation) and countries (international technological specialization) [38]. Clausen et al. [6] pursue a different approach to the phenomenon, focusing on causality. These authors assess the relation of factors such as type of innovation, firm size, R & D activity, Tech intensity, and innovation strategy with continuity of innovation. They identify various combinations of activities among Norwegian businesses from 1995 to 2004 and confirm persistence in science-based and market-oriented businesses. The results show that innovation strategies are an essential part of the continuity of innovation and as the technological regime does matter, product/process innovation and the low/high-tech sector have different levels of persistence [6]. Raymond et al. [5] examined innovation continuity separately for manufacturing firms in different technological regimes (high and low-tech firms). They discovered that high-tech companies innovated consistently, whereas low-tech companies did not [5]. The relation between types of innovation and innovation persistence is another important factor that is analyzed in most empirical studies [8, 9, 32, 36]. Tavassoli, Karlsson [39] show that the degree of persistency is not equal among various types of innovation, among which product innovators turn out to be the strongest persistent innovators [39], and it is supported by other findings [9].

Following the previous empirical literature on innovation persistence, proxies such as innovation activities, types of innovation, public support, higher education employment, size of firms, and technological regimes are suggested. We will assess its effects (especially technological regimes) to see whether they play any role in innovation persistence among Portuguese firms.

3 Data and Methodology

The empirical research is based on data from two waves (2014 and 2018) of the Portuguese Community Innovation Survey (CIS) database. This database contains information about the innovative activity of enterprises, being conducted every 2 years. To assess the innovation persistence among Portuguese firms, only 3237 firms were kept and analyzed as the selection criteria were having participated in both waves of the survey. A detailed description of the variables in use is provided in Table 1, addressing the description and measurement scale. To consider whether a firm is performing innovation or not, we define it as if the firm introduced one type of innovation (product, process, organizational, marketing) or conducted any type of R & D.

Summary descriptive statistics of the main variables in use are shown in Table 2, presenting the details about the frequency and percentage distribution of the groups.[1] Results evidence that 39% of these firms continue their innovation activity (persistence innovator), and 28% still do not perform any innovation behavior (non-innovator). On the contrary, 23% stop innovating (past innovator) while 10% are new in this field (early innovator). Additionally, almost 75% of the enterprises that are in science base sectors (high-tech firms that rely on R & D from both in-house sources and university research, including industries such as pharmaceuticals and electronics) show persistence in innovation. This result means they perform innovation from 2012 to 2014 and continue it in 2016–2018. A Homogeneity test[2] is performed and showed differences of continuity in innovation among science-based firms (high-tech) and others.

Correlation among the variables is introduced in Table 3. There has been a significant positive correlation between most factors, except the relation between marketing innovation and technological regime, these results point towards the identification of the exogenous variables as enhancers of the innovative strategies.

The econometric estimations were run using binary logistics with all defined variables (as Model 1). Since the dependent variable (persistency) is binary, the logistic binary regression is used to estimate the factors which influence persistency in innovation. This model is demonstrated in Table 4.

The results from Model 1 relate to firms encompassing innovative persistence, following the similar proxies from the extant literature. The overall model is

[1] The focus of the analysis is innovation persistence, consequently, being persistent in innovation implies having performed innovation in both periods of analysis; the predictors of the probability to persist in 2018 will consequently be collected from the 2018 time frame.

[2] By this test, we could answer the question in which persistence in innovation differs in different technological regimes.

Table 1 Variables descriptions

Variables		Description	Measurement
Persistency		Whether the firm continues innovation activity or not	Binary
Innovative behavior category (IBC)	Persistent innovator	A firm in which being innovator in two waves	3
	Early innovator	A firm that starts innovating in the former wave	2
	Past innovator	A firm that stops innovating in the former wave	1
	Non-innovator	A firm in which not perform innovation activity at none of the waves	0
Technological regimes (TECHREGIME)	Supplier dominated (SD)	Firms are divided into four categories, based on Boliaciano and Pianta	1
	Scale-intensive (SII)		2
	Specialized suppliers (SS)		3
	Science base (SB)		4
Type of innovation (TOP)	Product innovation (PI)	A firm performed product innovation	Binary
	Process innovation (PRI)	A firm performed process innovation	Binary
	Organizational innovation (OI)	A firm performed organizational innovation	Binary
	Marketing innovation (MI)	A firm performed marketing innovation	Binary
	Multiple Innovation (MUI)	A firm performed at least two types of innovation	Binary
R & D activity (RDAC)	Intramural R & D(IRD)	A firm performed inbound R & D	Binary
	Extramural R & D(ERD)	A firm performed outbound R & D	Binary
Higher education employment (HEE)	0	Percentage of high education people working in the firm	1–7 degree
	1 to 5		
	5 to 10		
	10 to 25		
	25 to 50		
	50 to 75		
	75 to 100		
Size of the firm (SIZE)		Firm dimension	1–3 degree
Public fund support or tax incentives(PFSTI)		If the firm has received public support or any tax incentives	Binary

Table 2 Descriptive statistics

Variables (Status in the latest observation)		Persistency 0 *N*	% Within category	1 *N*	% Within category	total *N*	Sum	Mean	SD
TECHREGIME	1	1118	66.10	574	33.90	1692	5644	1.74	0.91
	2	503	59.00	349	41.00	852			
	3	314	59.90	210	40.10	524			
	4	43	25.40	126	74.60	169			
	Total	1978	61.10	1259	38.90	3237			
PI	0	1757	87.60	249	12.40	2006	1231	0.38	0.49
	1	221	18.00	1010	82.00	1231			
	Total	1978	61.10	1259	38.90	3237			
PRI	0	1737	89.70	199	10.30	1936	1301	0.4	0.49
	1	241	18.50	1060	81.50	1301			
	Total	1978	61.10	1259	38.90	3237			
OI	0	1823	77.70	524	22.30	2347	890	0.27	0.45
	1	155	17.40	735	82.60	890			
	Total	1978	61.10	1259	38.90	3237			
MI	0	1870	71.90	730	28.10	2600	637	0.2	0.40
	1	108	17.00	529	83.00	637			
	Total	1978	61.10	1259	38.90	3237			
RDAC	0	1867	78.6	508	21.4	2375	862	0.27	0.44
	1	111	12.9	751	87.1	862			
	Total	1978	61.1	1259	38.9	3237			
HEE	1	280	94.30	17	5.70	297	11,779	3.64	1.75
	2	582	72.90	216	27.10	798			
	3	286	60.30	188	39.70	474			
	4	340	48.40	363	51.60	703			
	5	211	51.60	198	48.40	409			
	6	144	50.50	141	49.50	285			
	7	135	49.80	136	50.20	271			
	Total	1978	61.10	1259	38.90	3237			
SIZE	1	1256	75.00	418	25.00	1674	5152	1.59	0.68
	2	593	49.00	618	51.00	1211			
	3	129	36.60	223	63.40	352			
	Total	1978	61.10	1259	38.90	3237			
MUI	0	1673	96.40	63	3.60	1736	1501	0.46	0.50
	1	305	20.30	1196	79.70	1501			
	Total	1978	61.10	1259	38.90	3237			
PFSTI	0	1726	70.30	730	29.70	2456	781	0.24	0.43
	1	252	32.30	529	67.70	781			
	Total	1978	61.10	1259	38.90	3237			

Table 3 Correlation among variables

	(1)	(2)	(3)	(4)	(5)	(6)	(7)	(8)	(9)	(10)	(11)
Persistency (1)	1.00										
TECHREGIME (2)	0.129**	1.00									
PI (3)	0.693**	0.130**	1.00								
PRI (4)	0.716**	0.095**	0.713**	1.00							
OI (5)	0.552**	0.086**	0.541**	0.715**	1.00						
MI (6)	0.448**	0.026	0.496**	0.567**	0.553**	1.00					
MUI (7)	0.778**	0.116**	0.842**	0.882**	0.662**	0.532**	1.00				
RDAC (8)	0.541**	0.146**	0.378**	0.353**	0.274**	0.232**	0.372**	1.00			
HEE (9)	0.257**	0.327**	0.216**	0.212**	0.200**	0.164**	0.226**	0.268**	1.00		
SIZE (10)	0.304**	0.054**	0.264**	0.260**	0.206**	0.159**	0.263**	0.320**	0.189**	1.00	
PFSTI (11)	0.334**	0.077**	0.314**	0.314**	0.246**	0.186**	0.301**	0.275**	0.131**	0.258**	1.00

**Correlation is significant at the 0.01 level (2-tailed)

Table 4 Regression models 1 and 2

	Model 1								Model 2							
	B	S.E	Wald	df	Sig	Exp(*B*)	95% C.I. for Exp(*B*)		*B*	S.E	Wald	df	Sig	Exp(*B*)	95% C.I.for Exp(*B*)	
TECHREGIME			10.83	3	0.01		Lower	Upper			12.80	3	0.01		Lower	Upper
SD	−0.94	0.36	6.81	1	0.01	0.39	0.19	0.79	−1.11	0.34	10.55	1	0.00	0.33	0.17	0.65
SII	−1.13	0.36	9.68	1	0.00	0.32	0.16	0.66	−1.24	0.35	12.62	1	0.00	0.29	0.15	0.57
SS	−1.10	0.37	8.82	1	0.00	0.33	0.16	0.69	−1.17	0.36	10.43	1	0.00	0.31	0.15	0.63
PFSTI	1.31	0.48	7.60	1	0.01	3.71	1.46	9.43	1.29	0.46	7.81	1	0.01	3.64	1.47	9.01
PI	0.38	0.17	4.86	1	0.03	1.46	1.04	2.05	0.39	0.17	5.48	1	0.02	1.48	1.07	2.05
PRI	0.38	0.21	3.23	1	0.07	1.46	0.97	2.19	0.50	0.19	7.32	1	0.01	1.66	1.15	2.38
RDAC	1.85	0.16	127.48	1	0.00	6.34	4.60	8.74	2.11	0.16	181.84	1	0.00	8.26	6.08	11.23
MUI	3.48	0.27	166.46	1	0.00	32.53	19.17	55.22	3.48	0.26	173.48	1	0.00	32.45	19.34	54.47
OI	0.14	0.16	0.68	1	0.41	1.15	0.83	1.58								
MI	0.02	0.16	0.01	1	0.92	1.02	0.74	1.40								
HEE			23.02	6	0.00											
HEE (1)	−1.35	0.38	12.70	1	0.00	0.26	0.12	0.55								
HEE (2)	−0.34	0.26	1.76	1	0.18	0.71	0.43	1.18								
HEE (3)	−0.17	0.28	0.36	1	0.55	0.85	0.49	1.46								
HEE (4)	0.04	0.26	0.03	1	0.37	1.05	0.63	1.74								
HEE (5)	0.16	0.28	0.31	1	0.58	1.17	0.68	2.03								
HEE (6)	0.01	0.29	0.00	1	0.99	1.01	0.57	1.79								
SIZE			11.94	2	0.00											
SIZE (1)	−0.43	0.23	3.41	1	0.07	0.65	0.42	1.03								
SIZE (2)	0.04	0.23	0.02	1	0.88	1.04	0.66	1.62								
Constant	−2.20	0.42	26.82	1	0	0.11			−2.50	0.34	53.95	1	0	0.08		

significant at the 0.01 level according to the Model chi-square statistic, and the model correctly predicts 88.5% of the responses. The results evidence that persistence in innovators is strongly tied to technological innovation types. However, contributions of variables like the size of a firm and higher education employment are not significant in model 1. So, we run a new regression model (model 2) and remove these two variables from the analysis. The result of the test is presented in Table 4 which gives the results of fitting the logistic regression model to data and showing coefficients, which will be used in the equation for making the classifications. The prediction of this model is almost like model 1 (88.4%), which shows a high correct response. In this model, 94% of occurrences and 85% of non-occurrences were correctly predicted. This result is quite unexpected but very interesting, as it proves that continuing in innovation is not connected to firm size as well as the availability of human capital, which is common norm for the allocation of public grants. The model in the analysis can be summarized by the following conceptual model:

$$\text{Persistence in innovation} = f(\text{TECHREGIME}, \text{PFSTI}, \text{PI}, \text{PRI}, \text{RDAC}, \text{MUI}) \quad (1)$$

For evaluating the model, we use the area under the ROC curve. The area under the curve of this model is 0.948 with a 95% confidence interval (0.941, 0.955), which indicates that this model performs significantly better than by chance.

4 Findings

Our results go along with previous findings, which showed that the willingness to persist in performing innovation activity is expected to be related to the sectors and technological regime [5, 6, 40], innovation in product or process, R & D activity, doing more than one type of innovation [7], and using public support or tax incentives [16]. An important distinction emerges between the relative importance of conducting more than one type of innovation in the firm in explaining innovation persistence, and the relatively lesser importance of the size of the firm. In parallel with the case of Norway [6], the firm dimension has less impact on the propensity to persist than the innovation types (product and process). As our main focus in this study is assessing the relation between technological regime and persistence in innovation. Firstly, we observe that like some studies [40], the proportion of persistent innovators is significantly higher in high-tech industries. Also, almost 75% of high-tech firms are persistent in innovation, while less than half of other technological regimes (40%) continue their innovation in their firms. Secondly, firms in high-tech sectors and science-based industries have more incentives to innovate, and hence they tend to innovate persistently. This result is consistent with other findings on Dutch and [5] Norwegian firms [6]. This could be because of the innovation agencies in these two countries and the similarity of their missions with Portuguese's innovation strategy, all support innovation activities across the economy without a focus on practical sectors or technology levels. Findings suggest that those who are in science-based sectors are almost respectively 3, 3.4, and 3.2 times more likely to be persistent innovators than those who are in supplier-dominated (SD), scale-intensive (SII), and specialized suppliers (SS) categories. Furthermore, we can see that as much as firms use public support and tax incentives, the probability of persistence in innovation would increase. In the same vein as Le Bas and Poussing [7], both product and process innovators are more prone to be persistent than innovators performing a single type of innovation, our findings also reflect the nature of the significance of performing more than one type of innovation on the increasing of probabilities of continuity of innovation [7].

5 Recommendations and Limitations

This study evidences that the most relevant drivers of persistence of innovation are those that are related to the innovative strategy, such aspersing more than one type of innovation, also, supporting firms financially with public subsidies is also an important determinant of continuity, and surprisingly, that continuity is less related to firm size. But innovation policies should continue with supporting firms for their innovation activity and innovation strategies should be considered to conduct a variety of innovation types. Furthermore, larger firms are not more prone to persist in innovation, so this proxy is not an essential part of the continuity of innovation, and policymakers should not just focus on the firm dimension and positively discriminate large companies, given that their smaller counterparts are equally prone to continue in innovation.

One of the limitations of this study is that, sampling criteria forced to remove firms that do not present in both waves, disregarding the underlying reason, nor the stopping motivation as well as the igniting force. This might lead to an attrition bias, especially if the exit of firms from samples is correlated with not continuing innovation in innovation performance. The control for such a bias cannot be made; however, future research may minimize this problem by adding additional CIS waves. Second, here, the present appraisal used comparative statics and a memory-based endogenous variable despite the cross-sectional estimation to identify the correlation between proxies and their role in firms' continuity of innovation. Further research may use with panel data analysis and provide additional insights.

Moreover, some limitations are associated with the survey: since these two questionnaires were somehow different, some variables could not be examined, subjectivity might cause not/over-reporting by firms, also a large part of the economy is omitted.

Our main contribution to the literature is that it has extended prior research on innovation persistence with the argument that firms in different technological regimes show different persistent innovative behavior. Furthermore, we contribute to the literature by demonstrating that complex innovators are more likely to persist than single innovators, reinforcing the complementarities among the innovation cycle and the multi-layer facets of the innovative behavior. Future research could also try to better understand why and how firms continue their innovative actions in different spans of time. Moreover, in the innovation ecosystem, openness and collaboration are playing a key role in doing innovation and hence, can be further research subjects for assessing their impact on the continuity of innovation.

Acknowledgments This work was financially supported by the research unit on Governance, Competitiveness and Public Policy (UIDB/04058/2020) + (UIDP/04058/2020), funded by national funds through FCT – Fundação para a Ciência e a Tecnologia.

References

1. Nonaka, I. (1994). A dynamic theory of organizational knowledge creation. *Organization Science, 5*(1), 14–37.
2. Geroski, P. A., Van Reenen, J., & Walters, C. F. (1997). How persistently do firms innovate? *Research Policy, 26*(1), 33–48.
3. Malerba, F., & Orsenigo, L. (1997). Technological regimes and sectoral patterns of innovative activities. *Industrial and Corporate Change, 6*(1), 83–118.
4. Cefis, E., & Orsenigo, L. (2001). The persistence of innovative activities A cross-countries and cross-sector comparative analysis. *Research Policy, 30*(7), 1139–1158.
5. Raymond, W., Mohnen, P., Palm, F., & Van Der Loeff, S. S. (2010). Persistence of innovation in Dutch manufacturing: Is it spurious? *The Review of Economics and Statistics, 92*(3), 495–504.
6. Clausen, T., Pohjola, M., Sapprasert, K., & Verspagen, B. (2012). Innovation strategies as a source of persistent innovation. *Industrial and Corporate Change, 21*(3), 553–585.
7. Le Bas, C., & Poussing, N. (2014). Are complex innovators more persistent than single innovators? An empirical analysis of innovation persistence drivers. *Innovation Management, 18*(1), 1450008.
8. Altuzarra, A. (2017). Are there differences in persistence across different innovation measures? *Innovation, Management, Policy, and Practice, 19*(3), 353–371.
9. Song, C., Lee, J., & Jang, P. (2021). An exploratory study on the characteristics of technology innovation persistence of Korean firms. *The Korea Society for Innovation Management & Economics, 29*(3), 1–31.
10. Costa, J., Botelho, A., & Teixeira, A. A. *Persistence in innovation and innovative behavior in unstable environments.*
11. Schumpeter, J. (1942). *Creative destruction, capitalism, socialism, and democracy* (pp. 82–85).
12. Bas, C. L., & Latham, W. (2006). Persistence in innovation: Definitions and current development of the field. In *The economics of persistent innovation: An evolutionary view* (pp. 1–18). Springer.
13. Winter, S. G., & Nelson, R. R. (1982). An evolutionary theory of economic change, the University of Illinois at Urbana-Champaign's Academy for Entrepreneurial Leadership Historical Research Reference in Entrepreneurship.
14. Malerba, F., & Orsenigo, L. (1999). Technological entry, exit, and survival: An empirical analysis of patent data. *Research Policy, 28*(6), 643–660.
15. Colombelli, A., & von Tunzelmann, N. (2011). The persistence of innovation and path dependence the persistence of innovation and path dependence. In *Handbook on the economic complexity of technological change*. Edward Elgar Publishing
16. Antonelli, C., Crespi, F., & Scellato, G. (2012). Inside innovation persistence: New evidence from Italian micro-data. *Structural Change and Economic Dynamics, 23*(4), 341–353.
17. Arenas, J. J., Gómez, J. E., Ortiz, E., Paz, F., & Parra, C. (2020). Elements of the persistence in innovation: Systematic literature review. *Administrative Sciences, 10*(4), 83.

18. Stoneman, P. (1983). *The economic analysis of technological change*. Oxford University Press.
19. Sutton, J. (1991). *Sunk costs and market structure: Price competition, advertising, and the evolution of concentration*. MIT Press.
20. Le Bas, C., & Scellato, G. (2014). Firm innovation persistence: A fresh look at the frameworks of analysis. *Economics of Innovation and New Technology, 23*(5–6), 423–446.
21. Winter, S. G. (1984). Schumpeterian competition in alternative technological regimes. *Journal of Economic Behavior & Organization, 5*(3–4), 287–320.
22. Malerba, F., Orsenigo, L., & Peretto, P. (1997). Persistence of innovative activities, sectoral patterns of innovation and international technological specialization. *International Journal of Industrial Organization, 15*(6), 801–826.
23. Suarez, D. (2014). Persistence of innovation in unstable environments: Continuity and change in the firm's innovative behavior. *Research Policy, 43*(4), 726–736.
24. Dosi, G. (1982). Technological paradigms and technological trajectories A suggested interpretation of the determinants and directions of technical change. *Research Policy, 11*(3), 147–162.
25. Malerba, F., & Orsenigo, L. (1996). Technological regimes and firm behavior. In *Organization and strategy in the evolution of the enterprise* (pp. 42–71). Palgrave Macmillan.
26. Marsili, O. (1999). *Technological regimes: Theory and evidence*. DYNACOM working.
27. Pavitt, K. (1984). Sectoral patterns of technical change: Towards a taxonomy and a theory. *Research Policy, 13*(6), 343–373.
28. Hatzichronoglou, T. (1997). *Revision of the high-tech sector and product classification*.
29. De Jong, J. P., & Marsili, O. (2006). The fruit flies of innovations: A taxonomy of innovative small firms. *Research Policy, 35*(2), 213–229.
30. Rizzoni, A. (1991). Technological innovation and small firms: A taxonomy. *International Small Business Journal, 9*(3), 31–42.
31. Pavitt, K., Robson, M., & Townsend, J. (1989). Technological accumulation, diversification and organization in UK companies. *Management Science, 35*(1), 81–99.
32. Marsili, O., & Verspagen, B. (2001). *Technological regimes and innovation: Looking for regularities in Dutch manufacturing*. ECIS-Eindhoven University of Technology.
33. OECD homepage. https://www.oecd-ilibrary.org/science-and-technology
34. Bogliacino, F., & Pianta, M. (2010). Innovation and employment: A reinvestigation using revised Pavitt classes. *Research Policy, 39*(6), 799–809.
35. Van Leeuwen, G. (2002). *Linking innovation to productivity growth using two waves of the community innovation survey*.
36. Johansson, B., & Lööf, H. (2010). CESIS Electronic Working Paper Series Innovation Strategy and Firm Performance What is the long-run impact of persistent R & D? Innovation Strategy and Firm Performance What is the long-run impact of persistent R & D?
37. Lhuillery, S. (2014). Marketing and persistent innovation success. *Economics of Innovation and New Technology, 23*(5–6), 517–543.
38. Le Bas, C., & Négassi, S. (2000). Les structures des activités d'innovation en France et comparaison avec celles des principaux partenaires commerciaux. *Convention d'étude, 19*, 72.
39. Tavassoli, S., & Karlsson, C. (2015). Persistence of various types of innovation analyzed and explained. *Research Policy, 44*(10), 1887–1901.
40. Duguet, E., & Monjon, S. (2002). *Creative destruction and the innovative core: Is innovation persistent at the firm level?*

Industrial Symbiosis Applied to Oil Refineries: Drivers and Barriers

Ângela Neves, Hugo Ferreira, Francisco J. Lopes, Radu Godina, and João C. O. Matias

Abstract Oil refineries play an important role in the economy, mainly in the transport and energy sectors; however, they are energy-intensive industries and responsible for a significant amount of waste and carbon dioxide emissions. Industrial symbiosis can contribute significantly to increasing the sustainability of this sector, as it allows entities that are traditionally separate, to cooperate with each other in sharing resources and allows waste valorization. Thus, this article aims to analyze the main drivers, barriers, and challenges that this industry faces when implementing and developing industrial symbiosis. To this end, an exhaustive survey of published case studies of industrial symbiosis with refineries was conducted to analyze the economic activities involved, the type of flow and the role of refineries as source or sink. The study showed that despite the challenges that this sector has to face for the implementation and development of industrial symbiosis, there are numerous possibilities for its realization.

Keywords Industrial symbiosis · sustainability · oil refinery · energy-intensive industry

Â. Neves (✉) · H. Ferreira · F. J. Lopes
Department of Mechanical Engineering, Polytechnic Institute of Viseu, Viseu, Portugal
e-mail: aneves@estgv.ipv.pt

R. Godina
UNIDEMI, Faculty of Science and Technology (FCT), Universidade NOVA de Lisboa, Caparica, Portugal

J. C. O. Matias
GOVCOPP, Unidade de Investigação em Governança, Competitividade e Políticas Públicas, Campus Universitário de Santiago, Universidade de Aveiro, Aveiro, Portugal

J. C. de Oliveira Matias et al. (eds.), *Quality Innovation and Sustainability*, Springer Proceedings in Business and Economics,
https://doi.org/10.1007/978-3-031-12914-8_21

1 Introduction

Oil refineries play an important role in the economy, mainly in the transport and energy sector, but also in the production of goods such as plastics, paints, fertilizers, which have refined petroleum products as raw materials, and by-products such as sulfur. However, they are energy-intensive industries and the manufacture of coke and refined petroleum products sector in 2020 was responsible for 5.2% of the total carbon dioxide emissions in the EU-27, relative to all economic activities [1] and, in 2018, was responsible for the production of 6,080,000 tonnes of waste (hazardous and non-hazardous) [2]. Furthermore, it is heavily dependent on oil price fluctuations, which makes it more susceptible to these fluctuations. Thus, it is essential to find solutions that make this industry more efficient and more sustainable.

Industrial symbiosis enables a shift from a linear to a circular paradigm, in which different entities associate in a mutually beneficial relationship to exchange resources, energy, water, infrastructure, and public services, and where waste is valued as a resource [3, 4]. Driven by various reasons, such as cost savings with raw materials, with the treatment and disposal of waste in landfills, with the compliance with environmental requirements, among others [5], the industrial symbiosis has spread a little around the world with a myriad of economic, environmental, and social benefits [6]. In the scientific literature, there are several articles that compile different case studies of industrial symbiosis [5, 7–9], and research on this practice related to refineries [3, 10, 11]. However, there was a gap in the research regarding the compilation and analysis of industrial symbiosis case studies focused on refineries and their specificities.

Thus, this paper aims to analyze the main drivers for the implementation and development of industrial symbiosis (IS) in refineries, and the barriers and challenges that this sector faces for this practice to take hold. To this end, a literature review was conducted to collect case studies of industrial symbiosis involving refineries to provide a comprehensive overview of the activity of this practice and provide information about the economic activities involved, the types of flows and the role of refineries as sources and sinks. The study is organized as follows: in Sect. 2, the methods used in the review are presented; in Sect. 3, the case studies of industrial symbiosis where refineries are part of are described and analyzed; in Sect. 4, the main drivers, barriers, and challenges to the implementation and development of industrial symbiosis with the involvement of refineries are discussed; and finally, in Sect. 5, the main conclusions are presented.

2 Methodology

To meet the proposed objectives, a methodology developed in different phases was followed: the in-depth collection of existing literature, the screening of relevant publications and their content analysis. To perform a comprehensive search, the academic database Scopus, and the publishers with more publications in this area, such as Elsevier, Wiley Online Library, Springer, MDPI, Inderscience, Taylor & Francis, ACS Publications, SAGE Journals, Emerald Insight, IEEE Xplore, and Annual Reviews were used. Using the keyword combination "industrial symbiosis" AND Refineries, 28 publications were obtained in Scopus and 1052 in the publishers. The combination "industrial symbiosis" AND Case studies was also used to check whether any relevant cases had been excluded. The search resulted in 439 articles from Scopus and 4308 from the publishers.

With the resulting database, and after eliminating repeated publications and articles not written in English, a first selection was made. Thus, only research articles, conference articles, and book chapters were included, and no time interval was imposed. In the next screening, only articles that studied industrial symbiosis were included. Thus, titles, keywords, and abstracts were read to select the relevant publications for the study. Subsequently the whole article text was examined in order to include only articles referring to case studies. All cases of industrial symbiosis studied in a real context, but whose relationships had not yet been materialized, were excluded. Finally, the entities involved in the case studies of industrial symbiosis were verified and only those with refineries were included. This selection resulted in 30 articles that were subject to a more detailed content analysis.

3 Analysis of IS Case Studies Involving Refineries

Refineries are part of very diversified industrial symbiosis networks, in terms of location, size, economic activities involved, and types of flows, as illustrated in Table 1, which compiles the case studies. The results of the bibliographic research show a variable distribution of case studies involving refineries over the years, as illustrated in Fig. 1. Regarding the geographical distribution, there is a predominance of case studies referring to Europe, with seven different cases, studied in 17 articles. Of these, nine concern the Kalundborg case, one of the first and most cited in scientific publications [5]. Asia comes next with five cases studied in six articles and North America with three different cases studied in five articles. Lastly, Oceania with four publications referring to a single case.

In terms of regions (listed in the legend of Fig. 2), the UK has the highest number of cases, three, followed by the USA and South Korea, with two. The remaining countries have only one case, even though this is sometimes studied in more than one publication.

Regarding the economic activities present in the industrial symbiotic networks involving refineries, there is a great diversity, as illustrated in Fig. 3, which appear grouped according to the International Standard Industrial Classification of All Economic Activities, Revision 4 (ISIC, Rev.4).

In the case studies analyzed, there are 38 industrial symbiotic relationships involving refineries. Of these, in 63.2%, the refineries are the ones supplying the waste or utilities and only in 36.8% are the receivers. Regarding the diversity of economic activities, refineries provide waste and utilities to 19 different industries,

Table 1 Published case studies of industrial symbiosis with refinery intervention

Region, country	Economic activities	Type of stream	Industries that generate waste (source)	Industries that use waste (sink)	References
Kalundborg, Denmark	Coal-fired power plant, oil refinery, biotech and pharmaceutical company, liquid fertilizer company, plasterboard manufacturer, soil remediation and recovery company, public wastewater treatment, waste treatment company, the municipality, freshwater provider, district heating distribution, fish farms, neighbouring farms, pig farms, and cement and road aggregate producers, and some material recycling companies	(a) Gas (b) Steam, and boiler water (c) Sulfur (d) Wastewater, and cooling water	(a) Oil refinery (b) Power plant (c) Oil refinery (d) Oil refinery	(a) Plasterboard manufacturer (b) Oil refinery (c) Liquid fertilizer company (d) Power plant	[17, 21–28]
Grangemouth, United Kingdom	Biotechnology company, active ingredient manufacture, refinery, gas supplier, water effluent treatment, de-watering/drying company, combined heat and power plant, petrochemical company, plastics and rubber industry, chemical industry, and plastics-chemical industry	(a) Steam and electricity (b) Compressed air and N_2 (c) Bio-treatment sludge	(a) Combined heat and power plant (b) Gas supplier company (c) Refinery	(a) Refinery (b) Refinery (c) De-watering/drying company	[29]
Forth Valley, United Kingdom	Energy company, refinery, chemical industry, paper mill, oil and gas company, cement manufacturer, and waste processor	(a) Ethane (b) Steam	(a) Refinery (b) Chemical industry	(a) Chemical industry (b) Refinery	[29]
Humber, United Kingdom	Biodiesel production, plaster board manufacturer, chemical industry, refineries, water treatment chemicals, food and fish processing, wastewater treatment, local farms, pet food, and furniture production	Hydrogen	Chemical industry	Refineries	[19]
Rotterdam, Netherlands	Refinery, greenhouse companies, residential area, port, chemical company, truck cleaning company, power plant, and shrimp farm	(a) Waste heat (b) CO_2 emissions	(a) Refinery (b) Refinery	(a) City's district heating system (b) Greenhouse companies	[13, 14]

Region, country	Economic activities	Type of stream	Industries that generate waste (source)	Industries that use waste (sink)	References
Taranto, Italy	Cement factory, steelworks factory, power plants, refinery, construction materials producing companies, distilleries, agricultural firms, and wine companies	(a) Electrical energy, and demineralized water and vapour (b) Fuel gas and fuel oil, sea and well water, and condensing vapour (c) Vapour and demineralized water, and electrical energy	(a) Refinery (b) Refinery (c) Power plant	(a) Steelworks factory (b) Power plant (c) Refinery	[20, 30]
Lysekil, Sweden	Land-based seaweed cultivation plant, and oil refinery	CO_2 obtained from flue gas	Oil refinery	Land-based seaweed cultivation plant	[31]
Guayama, Puerto Rico, USA	Coal-fired power plant, refinery, pharmaceutical firms, industrial landfills, road construction, public wastewater treatment plant, and waste stabilization	(a) Steam (b) Condensate	(a) Coal-fired power plant (b) Refinery	(a) Refinery (b) Coal-fired power plant	[15, 16]
Honolulu, USA	Power plant, cogeneration plant, oil refineries, cement company, quarry, biosolids beneficiation company, local golf course, construction and demolition waste landfill, wastewater treatment plant, oil and tire recovery company, city water agency and recycling company	(a) Steam (b) Make-up water and Granular activated carbon	(a) Power plant (b) Refinery (c) Water Recycling Facility	(a) Refinery (b) Power plant (c) Refinery	[32, 33]
		(c) Reverse osmosis recycled water			

(continued)

Table 1 (continued)

Region, country	Economic activities	Type of stream	Industries that generate waste (source)	Industries that use waste (sink)	References
Sarnia-Lambton, Canada	Fertilizer company, greenhouse operator, gas specialist company, power plant, medium-sized fine-particle manufacturer, oil refinery, chemical company, integrated energy company, and cattle farmers	(a) Residual oil feedstock (b) CO_2 (c) Dried distiller grains	(a) Oil refinery (b) Oil refinery (c) Oil refinery	(a) Fine-particle manufacturer (b) Industrial gases company (c) Cattle farmers	[18]
Ulsan, South Korea	Chemical, petroleum, and petrochemical company; industrial waste treatment and disposal company; chemical companies; tank terminal business; copper smelter and refinery; non-ferrous metal smelting company; paper mill company; wastewater treatment facilities; specialty chemicals and life science products company; integrated water management enterprise; and Ulsan Metropolitan City	(a) Pure water (b) Steam	(a) Refinery (b) Industrial waste treatment and disposal company	(a) Industrial waste treatment and disposal company (b) Refinery	[12, 34]
Yeosu, South Korea	Refineries, power plants, and petrochemical companies	Waste heat			[35]
Tianjin, China	Water treatment plant, industrial, commercial and residential users, wastewater treatment plant, construction companies, cogeneration plant, new water source company, desalination plant, resource recovery company, cast iron company, auto die makers, automatic transmission company, aluminium smelting, resource management company, stemless steel pipe maker, steel scrap contractors, refineries, chemical companies, lead recycling company, cement mill, rubber company, batteries company, and various lead acid battery users		Refinery	Chemical company	[28]

Region, country	Economic activities	Type of stream	Industries that generate waste (source)	Industries that use waste (sink)	References
Osaka, Japan	Gas company, petrochemical plant, refinery, municipalities, and ports				[36]
Jurong Island, Singapore	Chemical plants, multi-utility service provider, storage and terminaling service providers, electricity pool, co-generation plant, refinery, gasification plant, and air separation plant	(a) Hydrogen (b) Electricity	(a) Gasification plant (b) Electricity pool	(a) Refinery (b) Refinery	[37]
Kwinana, Australia	Alumina refinery, fused alumina and zircon producer, worm farm, cement manufacturing, industrial chemical and fertilizer producer, industrial gas producer, coal-fired power station, titanium mineral processing company, turf farm, construction company, cement mill, Blokpave producer, zirconia powder producer, water supply and treatment company, coal mine, insulation plant, chlor alkali plant, cement and lime producer, fertilizer producer, mineral processing plant, industrial gas producer, inorganic chemical producer, pig iron plant, nickel mine, composting facility, oil refinery, synthetic rutile plant, titanium dioxide producer, nickel refinery, gas fired power station, co-generation plant, distributor and producer of LPG	(a) Methyl diethyl amine (b) Steam, electricity, and wastewater (c) Spent CPU catalysts (d) Spent catalysts (e) Sulfur and hydrogen (back-up supply) (f) Sulfur (g) Hydrogen (h) Wastewater (i) Fuel gas	(a) Industrial, chemical and fertilization products (b) Cogeneration plant (c) Oil Refinery (d) Oil Refinery (e) Oil Refinery (f) Oil Refinery (g) Oil Refinery (h) Oil Refinery (i) Oil Refinery	(a) Oil Refinery (b) Oil Refinery (c) Composting facility (d) Cement and lime producer (e) Nickel refinery (f) Inorganic chemical producer (g) Industrial gas producer (h) Industrial chemical and fertilizer producer (i) Cogeneration plant	[38–41]

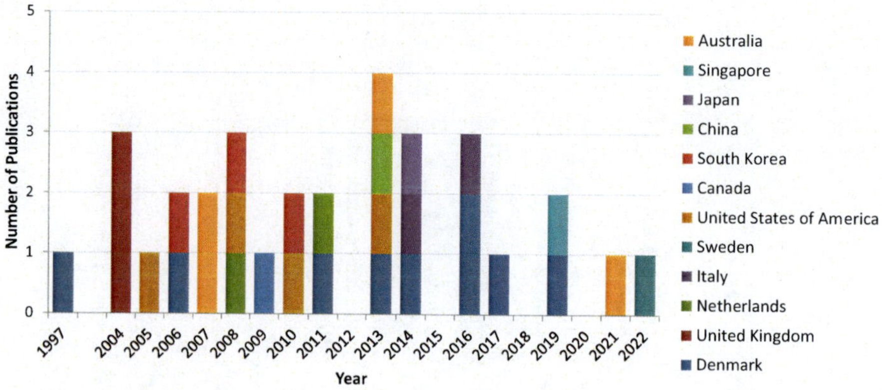

Fig. 1 Distribution of the number of case study publications by year and location

Fig. 2 Mapping of published case studies of industrial symbiosis with refineries as participants

while they are sinks of only nine. Electricity, gas, steam, and air conditioning supply are those that contribute most to the supply of waste to refineries, representing 64.3% of the total. This is followed by manufacturing and water supply; sewerage, waste management, and remediation activities, which account for 21.4% and 14.3%, respectively. The diversity of economic activities that are the sources of the waste produced by refineries is greater than that of sinks. Of the four receiving sections, manufacturing has the greatest weight and diversity, accounting for 54.5% of the total. This fact can be explained by the capacity to absorb waste from other industries, both the same and different, and to incorporate them as raw material. This is followed by Electricity, gas, steam, and air conditioning supply with 22.7%. Finally, Agriculture, forestry and fishing, and water supply; sewerage, waste management, and remediation activities represent 13.6% and 9.1%, respectively. Cities can also be sinks of waste from refineries, namely, waste heat, and thus reduce the environmental and economic impact of energy use in cities.

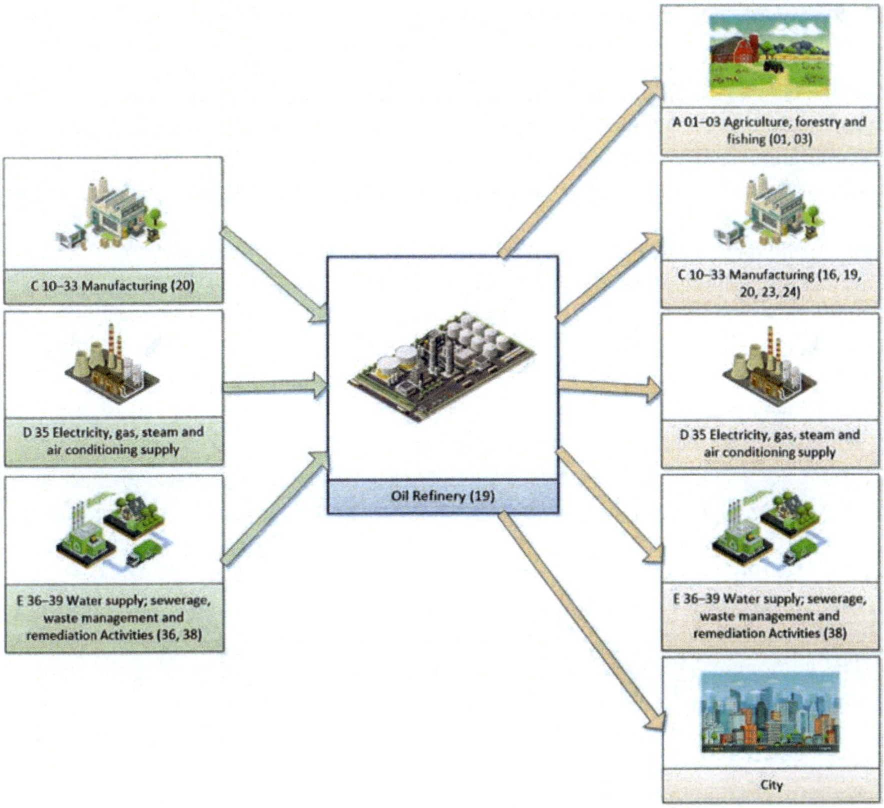

Fig. 3 Economic activities involved in industrial symbiosis with refineries, reported in the published case studies. (The numbers refer to the economic activities present in the industrial symbiosis networks within the respective division defined according to the ISIC)

The type of flow is also very diversified, with a higher incidence on the exchange of steam, gas, CO_2, and water.

As reported in previous studies, also in these cases, the industrial symbiosis activity allowed to achieve several benefits. In the case of Rotterdam, industrial symbiosis allowed greenhouse companies to reduce in 2007, 170,000 tons of CO_2 emissions by saving 95 million m^3 of natural gas [13, 14]. In Guayama, Puerto Rico, the flow exchanged between the power plant and the refinery, steam, allowed a 99.5% reduction of sulfur dioxide (SO_2) emissions, a 95% reduction of particulate matter smaller than 10 mm (PM10) and 84% reduction of nitrous oxide (NOx) [15, 16].

4 Drivers, Barriers, and Challenges

The diversity of economic activities and the flows exchanged in the cases of IS collected in this article reveal the potential of refineries to establish this type of relationship and thereby obtain economic and environmental benefits. However, there are several barriers that constrain this development. One of them is related to the complexity of the production process of petroleum products that require many inputs of materials and energy and therefore makes them more dependent [17]. For this reason, it makes this type of industries very vulnerable to disruptive events, which requires a very high level of trust in the companies involved in the symbiosis network to ensure the quantity and quality of flow required. Furthermore, if there are more companies in a chain, the non-supply of a by-product can compromise the whole IS network [18].

Another major challenge that refineries face is the constraints in the legislation regarding the use of waste. Often there are barriers to consider a waste product as a by-product, which prevents it from being used as a raw material by another entity. Furthermore, the characteristic of the waste may also condition the achievement of industrial symbiosis. One of the examples is the use of waste heat, where infrastructures are required to direct the waste from the source to the receiver, which involves large amounts of money that companies are not always willing to assume.

The companies' initiative is a determining factor for the success of the implementation of industrial symbiosis. However, the lack of knowledge of this practice, the reluctance to disclose waste data and the lack of trust condition its implementation.

Another challenge facing the development of industrial symbiosis is its expansion and continuity over the years, because, as demonstrated in several studies, even if it occurs spontaneously, there must be mechanisms to drive it [19, 20].

4.1 Strategies to Overcome Barriers to Industrial Symbiosis in Refineries

As mentioned earlier, there are many barriers and challenges to the implementation and development of industrial symbiosis involving refineries, so it is essential to find strategies to overcome these barriers. One of the strategies is to implement legislation that facilitates the creation of industrial symbiotic relationships and speeds up the process that allows waste to be considered as a by-product. Financial support from government entities may also help to overcome barriers when creating synergies, namely, support for initial investment to create infrastructures, as is the case of those necessary for the use of waste heat.

The existence of entities that can overcome existing barriers may be essential in some cases, both in building bridges of trust between the various companies and in disseminating the various economic and environmental advantages of this practice.

One of the examples was the programme developed at the Rotterdam Port and Industrial Complex where several initiatives emerged due to the dissemination of the results obtained by industrial symbiosis initiatives [13].

However, the existence of a driver may not be sufficient to initiate the industrial symbiotic relationship. Often it is a set of factors that together provide the favourable conditions for synergy to develop. One of the examples, was what happened in Rotterdam in the Netherlands, where although the project of using waste heat from the refinery had been thought of for many years, it was only realized years later [14]. Although the pressure for the use of surplus heat was great, it was only when other actors came in, such as a company responsible for new developments and the industrial network was extended to other social actors, such as housing cooperation and energy suppliers, that synergy could be realized [14].

The dissemination of success cases, the existence of an entity that analyzes the potential of new symbiotic relationships and helps to create trust between companies to make them a reality, and the existence of companies that due to the amount of resources consumed and waste produced can act as an anchor for other companies, can also be decisive in the development of industrial symbiosis.

5 Conclusions, Limitations, and Future Work

The literature review presented in this article focused on oil refineries and aimed to analyze the main barriers and challenges that this type of industry faces when implementing industrial symbiosis. To this end, a survey was carried out of published case studies of industrial symbiosis, to study what wastes and the entities involved in the different cases, as well as the constraints present. From the geographical distribution analysis, the study revealed a higher number of published cases in Europe and Asia, followed by North America and Oceania. The study also revealed the enormous potential of refineries for establishing industrial symbiotic relationships given the huge diversity of economic activities involved and the nature of the flows. However, in most cases, in 63.2%, refineries were the source of these flows and not the sinks.

Despite this potential, refineries face many challenges, such as the complexity of their production system, the lack of expedient mechanisms in existing legislation, the constraints inherent in the use of some wastes, such as waste heat, unawareness of the practice, reluctance to disclose company data, and lack of trust.

Industrial symbiosis can make a very significant contribution to increasing sustainability in refineries and help mitigate the negative effects of their operation. However, more incentives and more studies are needed that can support the implementation and development of this practice in such an important sector of society.

This article presents some limitations, associated with the methodology used for the research. Only publications in English and peer-reviewed journals were considered. And although this last condition ensures the quality of the publications used, it has the limitation that other publications, such as industry reports, public documents, which could be an important source of information regarding case studies involving refineries, were not considered.

Thus, one of the future works would be a broader approach to various sources of information in order to collect more data on this industry and existing synergies. It would also be important in future research to deepen the study on ways to facilitate these industrial symbiotic relationships. Extending the study to other industries would also be very relevant. On the one hand, to study new possibilities of synergy and use of waste involving refineries and, on the other, to learn good practices with other similar industries that have well-established industrial symbiotic relationships and transpose them to refineries.

References

1. Eurostat, European Commission. Air emissions accounts by NACE Rev. 2 activity. https://appsso.eurostat.ec.europa.eu/nui/submitViewTableAction.do. Last accessed 2022/02/09.
2. Eurostat, European Commission. Generation of waste by waste category, hazardousness and NACE Rev. 2 activity[env_wasgen]. https://appsso.eurostat.ec.europa.eu/nui/show.do?dataset=env_wasgen&lang=en. Last accessed 2022/02/09.
3. Cervo, H., Ferrasse, J.-H., Descales, B., & Van Eetvelde, G. (2020). Blueprint: A methodology facilitating data exchanges to enhance the detection of industrial symbiosis opportunities – application to a refinery. *Chemical Engineering Science, 211*, 115254.
4. Henriques, J. D., Azevedo, J., Dias, R., Estrela, M., Ascenço, C., Vladimirova, D., & Miller, K. (2022). Implementing industrial symbiosis incentives: An applied assessment framework for risk mitigation. *Circular Economy and Sustainability, 2*, 669–692.
5. Neves, A., Godina, R., Azevedo, S. G., & Matias, J. C. O. (2020). A comprehensive review of industrial symbiosis. *Journal of Cleaner Production, 247*, 119113.
6. Wadström, C., Johansson, M., & Wallén, M. (2021). A framework for studying outcomes in industrial symbiosis. *Renewable and Sustainable Energy Reviews, 151*, 111526.
7. Herczeg, G., Akkerman, R., & Hauschild, M. Z. (2018). Supply chain collaboration in industrial symbiosis networks. *Journal of Cleaner Production, 171*, 1058–1067.
8. Mendez-Alva, F., Cervo, H., Krese, G., & Van Eetvelde, G. (2021). Industrial symbiosis profiles in energy-intensive industries: Sectoral insights from open databases. *Journal of Cleaner Production, 314*, 128031.
9. Liu, Z., Ashton, W. S., Adams, M., Wang, Q., Cote, R. P., Walker, T. R., Sun, L., & Lowitt, P. (2022). Diversity in financing and implementation pathways for industrial symbiosis across the globe. *Environment, Development and Sustainability*.
10. Golmohamadi, H. (2022). Demand-side management in industrial sector: A review of heavy industries. *Renewable and Sustainable Energy Reviews, 156*, 111963.
11. Ketabchi, E., Mechleri, E., & Arellano-Garcia, H. (2019). Increasing operational efficiency through the integration of an oil refinery and an ethylene production plant. *Chemical Engineering Research and Design, 152*, 85–94.
12. Park, H.-S., Rene, E. R., Choi, S.-M., & Chiu, A. S. F. (2008). Strategies for sustainable development of industrial park in Ulsan, South Korea—From spontaneous evolution to systematic expansion of industrial symbiosis. *Journal of Environmental Management, 87*, 1–13.
13. Baas, L. (2011). Planning and uncovering industrial symbiosis: Comparing the Rotterdam and Östergötland regions. *Business Strategy and the Environment, 20*, 428–440.
14. Baas, L. W., & Huisingh, D. (2008). The synergistic role of embeddedness and capabilities in industrial symbiosis: illustration based upon 12 years of experiences in the Rotterdam Harbour and Industry Complex. *Progress in Industrial Ecology, an International Journal, 5*, 399–421.

15. Chertow, M. R., Ashton, W. S., & Espinosa, J. C. (2008). Industrial symbiosis in Puerto Rico: Environmentally related agglomeration economies. *null, 42*, 1299–1312.
16. Chertow, M. R., & Lombardi, D. R. (2005). Quantifying economic and environmental benefits of co-located firms. *Environmental Science & Technology, 39*, 6535–6541.
17. Kuznetsova, E., Louhichi, R., Zio, E., & Farel, R. (2017). Input-output Inoperability Model for the risk analysis of eco-industrial parks. *Journal of Cleaner Production, 164*, 779–792.
18. Bansal, P., & McKnight, B. (2009). Looking forward, pushing back and peering sideways: Analyzing the sustainability of industrial symbiosis. *Journal of Supply Chain Management, 45*, 26–37.
19. Mirata, M. (2004). Experiences from early stages of a national industrial symbiosis programme in the UK: Determinants and coordination challenges. *Journal of Cleaner Production, 12*, 967–983.
20. Notarnicola, B., Tassielli, G., & Renzulli, P. A. (2016). Industrial symbiosis in the Taranto industrial district: Current level, constraints and potential new synergies. *Journal of Cleaner Production, 122*, 133–143.
21. Branson, R. (2016). Re-constructing Kalundborg: The reality of bilateral symbiosis and other insights. *Journal of Cleaner Production, 112*, 4344–4352.
22. Chopra, S. S., & Khanna, V. (2014). Understanding resilience in industrial symbiosis networks: Insights from network analysis. *Journal of Environmental Management, 141*, 86–94.
23. Domenech, T., & Davies, M. (2011). Structure and morphology of industrial symbiosis networks: The case of Kalundborg. *Procedia - Social and Behavioral Sciences, 10*, 79–89.
24. Ehrenfeld, J., & Gertler, N. (1997). Industrial ecology in practice: The evolution of interdependence at Kalundborg. *Journal of Industrial Ecology, 1*, 67–79.
25. Jacobsen, N. B. (2006). Industrial symbiosis in Kalundborg, Denmark: A quantitative assessment of economic and environmental aspects. *Journal of Industrial Ecology, 10*, 239–255.
26. Valentine, S. V. (2016). Kalundborg Symbiosis: fostering progressive innovation in environmental networks. *Journal of Cleaner Production, 118*, 65–77.
27. Zhang, X., & Chai, L. (2019). Structural features and evolutionary mechanisms of industrial symbiosis networks: Comparable analyses of two different cases. *Journal of Cleaner Production, 213*, 528–539.
28. Zhang, Y., Zheng, H., Chen, B., & Yang, N. (2013). Social network analysis and network connectedness analysis for industrial symbiotic systems: model development and case study. *Frontiers of Earth Science, 7*, 169–181.
29. Harris, S., & Pritchard, C. (2004). Industrial Ecology as a learning process in business strategy. *Progress in Industrial Ecology, an International Journal, 1*, 89–111.
30. Notarnicola, B., Tassielli, G., & Renzulli, P. A. (2014). Potential developments of industrial symbiosis in the Taranto Productive District. In R. Salomone & G. Saija (Eds.), *Pathways to environmental sustainability: Methodologies and experiences* (pp. 215–224). Springer International Publishing.
31. Nilsson, J., & Martin, M. (2022). Exploratory environmental assessment of large-scale cultivation of seaweed used to reduce enteric methane emissions. *Sustainable Production and Consumption, 30*, 413–423.
32. Chertow, M., & Miyata, Y. (2011). Assessing collective firm behavior: comparing industrial symbiosis with possible alternatives for individual companies in Oahu, HI. *Business Strategy and the Environment, 20*, 266–280.
33. Eckelman, M. J., & Chertow, M. R. (2013). Life cycle energy and environmental benefits of a US industrial symbiosis. *The International Journal of Life Cycle Assessment, 18*, 1524–1532.
34. Won, J., Kim, J., Lee, S., & Park, H. (2006). Industrial symbiosis as an integrated business/ environment management process: The case of ulsan industrial complex. In *2006 International forum on strategic technology* (pp. 423–428).
35. Chae, S. H., Kim, S. H., Yoon, S.-G., & Park, S. (2010). Optimization of a waste heat utilization network in an eco-industrial park. *Applied Energy, 87*, 1978–1988.

36. Cerceau, J., Mat, N., Junqua, G., Lin, L., Laforest, V., & Gonzalez, C. (2014). Implementing industrial ecology in port cities: International overview of case studies and cross-case analysis. *Journal of Cleaner Production, 74*, 1–16.
37. Yin, C.-Y., & Lee, L. Y. (2019). Teaching chemical engineering students industrial symbiosis using online resources: A Singapore case study. *Education for Chemical Engineers, 27*, 28–34.
38. van Beers, D., Bossilkov, A., Corder, G., & van Berkel, R. (2007). Industrial symbiosis in the Australian minerals industry: The cases of Kwinana and Gladstone. *Journal of Industrial Ecology, 11*, 55–72.
39. Harris, S. (2007). Industrial symbiosis in the Kwinana Industrial Area (Western Australia). *Measurement and Control, 40*, 239–244.
40. MacLachlan, I. (2013). Kwinana Industrial Area: Agglomeration economies and industrial symbiosis on Western Australia's Cockburn Sound. *null, 44*, 383–400.
41. Oughton, C., Anda, M., Kurup, B., & Ho, G. (2021). Water circular economy at the Kwinana Industrial Area, Western Australia—the dimensions and value of industrial symbiosis. *Circular Economy and Sustainability, 1*, 995–1018.

Towards a More Sustainable Use of the Portuguese Road Network: The A25 and IP5 Case Study

Hugo Ferreira, José Campos, Ângela Neves, and Francisco J. Lopes

Abstract The transport sector is an energy-intensive industry whose main purpose is to promote the mobility of people and goods. As such, a more rational use of the existing road infrastructures is unquestionably a way to reduce the environmental impact of this sector and to mitigate the depletion of energy resources. Some factors, however, may negatively affect the user's decision on which route to take, from an energy sustainability perspective. In this context, the introduction of tolls on modern highways can have the perverse effect of diverting some traffic to other routes that often do not favor the energy sustainability of the transport system. This work addresses this issue and is based on a Portuguese case study, aiming to compare the relative merits of two parallel routes, the A25 and the IP5, as to energy consumption, GHG emissions, and travel times. These roads cross the district of Viseu. The A25 is a modern 2 × 2 lane highway, with tolls, while the other is 2-lane toll-free road with severe slopes. These indicators were evaluated through a physics-based approach, considering detailed vehicle specifications and reported traffic volumes, supported by literature. The results show that for equal travel times the A25 is more environmentally friendly, for all vehicle classes, in terms of energy consumption and GHG emissions. However, the IP5 is more attractive from an economic point of view, except for heavy traffic.

Keywords Sustainability · Energy consumption · CO_2 emissions · Eco-routing · Traffic simulation

H. Ferreira · J. Campos · Â. Neves (✉) · F. J. Lopes
Department of Mechanical Engineering, Polytechnic Institute of Viseu, Viseu, Portugal
e-mail: hugomoreira@estgv.ipv.pt; aneves@estgv.ipv.pt

J. C. de Oliveira Matias et al. (eds.), *Quality Innovation and Sustainability*, Springer Proceedings in Business and Economics,
https://doi.org/10.1007/978-3-031-12914-8_22

1 Introduction

1.1 Study Motivation and Objectives

At the beginning of this century, the Portuguese public investment in road infrastructures was considerable, in order to provide the country with a modern road network. However, this effort worsened the republic's debt significantly. To mitigate the costs arising from the established private-public credit-funded partnerships, earlier in the second decade, the Portuguese government imposed tolls on all controlled access highways, with few exceptions. With a reference cost of 0.0667 €/km, excluding VAT, established by Decree-Law No. 111/2011, similar to the fuel costs of a passenger car, traffic volumes on these highways decreased considerably [1], being diverted to alternative toll-free roads, less efficient in terms of flow capacity, travel times, and fuel consumption. In fact, the vertical and horizontal alignments of these roads were designed for lower speeds and traffic volumes.

This political decision was defended as an application of the user-pays principle. However, the environmental and social impact of underutilizing a modern infrastructure in favor of an older network, while ignoring the suitability of the alternative routes, was not studied in light of any sustainability criteria. Additionally, some of these modern highways were built over preexisting roads, thus eliminating possible alternatives. This is the case of the A25, which overlaps what was the IP5 route along most of its length. This option had a negative impact on mobility and burdened toll-free local roads [2]. However, the comparison between the two routes is still possible between junctions the '14' and '19' of the A25 that runs parallel to the IP5, passing south of the city of Viseu, while the IP5 passes north of the city. The first is a 2 × 2-lane road, 32 km long, with a speed limit of 120 km/h and a maximum gradient of 5%, while the second is a 2-lane (undivided) toll-free road, 34 km long, with a speed limit of 90 km/h and a ruling gradient of 8%. Figure 1 shows the paths and the elevation profiles of these roads.

In this context, this case study has regional relevance and aims to quantify the user's gains and losses arising from diverting his route from the A25 to the IP5, while quantifying the environmental impact of such choice. The gains/losses computed are the direct costs with fuels and tolls. The indirect cost is the travel time. The environmental impacts considered are the energy resources consumed and greenhouse gas emissions (GHG) produced by the reported traffic for the year 2019 [3]. This is a study option, by reasons of representativeness, since the confinement measures enforced by the Portuguese government in 2020 and 2021, in response to the COVID-19 pandemic, have significantly altered traffic patterns.

In simple terms, the main research questions are:

1. Is it worth taking the alternative route IP5, and who will it serve best, light or heavy traffic?
2. What passive measures can be adopted in order to mitigate the environmental impact ensuing from an unregulated use of the available road network?

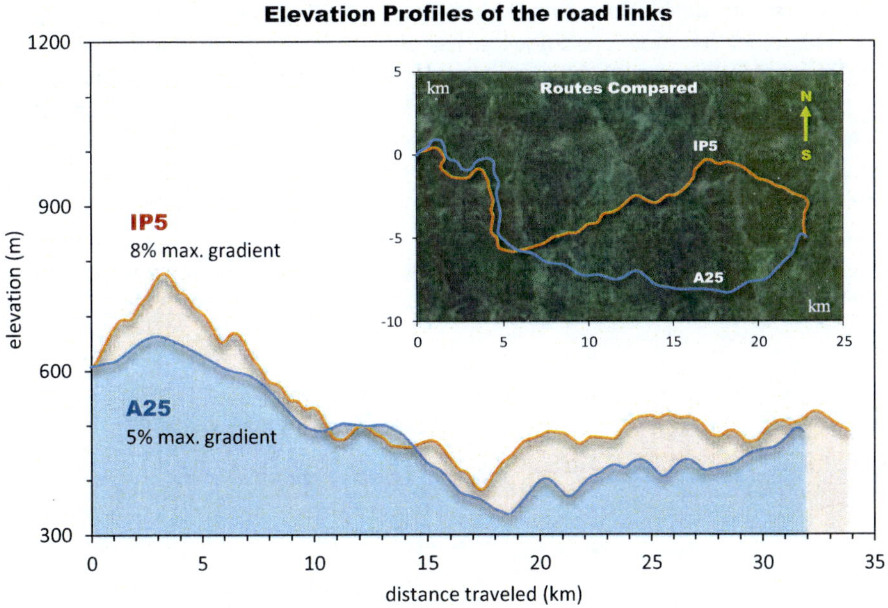

Fig. 1 Gradient profiles and horizontal alignments of the Portuguese roads A25 and IP5

In order to answer the first question, it is necessary to perform a cost analysis due to tolls charges and fuel consumption vs. travel time, since these factors impact on the user's decision on which route to take.

With regard to the latter, the measures mentioned fall within the scope of road policy decisions, such as the dynamic adjustment of speed limits and tolls, based on criteria of greater sustainability in the use of energy resources (thus mitigating environmental impacts), depending on the type of vehicle and fuel prices. These passive measures would not require the modification of existing infrastructure. The pertinence of this research topic is already established [4–9].

1.2 Literature Review

In order to answer the research questions above, it was necessary to consider representative vehicles, as to their type and technical specifications, and to estimate their fuel consumptions and GHG emissions for the routes under study, resorting to detailed gradient profiles, since they present long and severe slopes [10–14].

In this case study, constant speed duty cycles are assumed, as free-flow traffic conditions prevail throughout the day and year. In fact, the reported distance-weighted annual average daily traffic (AADT) for the A25 link under study is 9015 v/d (vehicles per day), for the year 2019 [3]. This is less than 100 v/d per lane, when free-flow traffic saturation occurs above 1000 v/d per lane [15]. Naturally, heavy trucks are unable to maintain the desired cruising speed on severe ramps. However, under free-flow conditions, it has already been shown that the constant speed simplification is valid, when equal to the actual average speed [16]. The modeling of the vehicle energy consumption, reflecting its motion resistance and the gradient profile of the road, was elaborated from the literature [16–21], considering the usual simplifications of $\sin\theta \cong \tan\theta$ and $\cos\theta \cong 1$ (Fig. 2). All equations presented were adapted in order to be coherent in the International System of Units (S.I.). The models used for the computation of the propulsive energy, fuel consumption, and GHG emissions are presented below in sequence, starting with the (positive) road work for free-flow constant speed operation over the road length L:

$$E_{\mathrm{P}} = \int_0^L F_{\mathrm{P}}^{+} dx \cong \Sigma\left(F_{\mathrm{R}}\Delta x + mg\Delta z\right)_{v=\mathrm{c.te}}^{+} \tag{1}$$

where the vehicle motion resistance, due to the drag force and rolling resistance, is

$$F_{\mathrm{R}} = C_{\mathrm{RR}} mg + \frac{1}{2}\rho_{\mathrm{air}} A C_{\mathrm{d}} v^2 \tag{2}$$

The rolling resistance coefficient C_{RR}, and the vehicle frontal area A, affected by the drag coefficient C_{d}, were retrieved from [19, 22]. Also, given the fact that the elevation profiles of the roads are considerably different, the air density ρ_{air} is computed as a function of altitude, in order to contemplate the local pressure. The correction factor in brackets reflects the barometric equation [23]

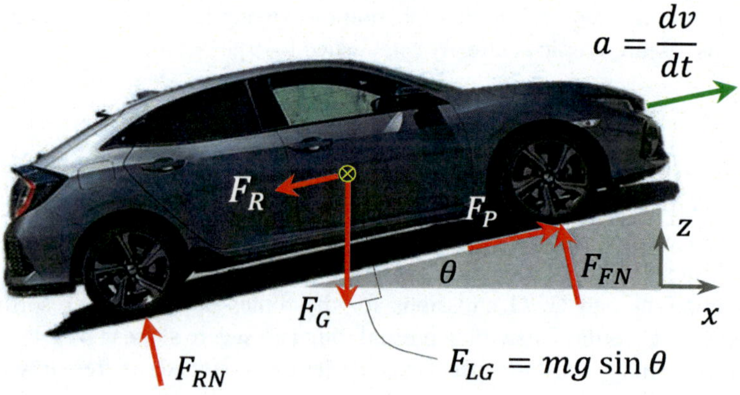

Fig. 2 Free-body diagram of a road vehicle over an incline plane

$$\rho_{\text{air}} = \rho_{\text{sea level}} \cdot \left(1 - \frac{z}{44330}\right)^{5.257} \tag{3}$$

Regarding the fuel consumption of an internal combustion engine, it depends almost entirely on the positive work developed, thus the integration in Eq. 2 considers only the action of the propulsive force when positive, F_{P}^{+}, where $X^{+} = \text{MAX}(0; X)$, being X any given quantity. The related fuel energy E_{f} is based on the thermodynamic concept of thermal efficiency, $\eta_{\text{th, w}}$ (Eq. 4), being the results accuracy-dependent only on the representativeness of the thermal efficiencies. In this study, the values proposed by [19, 22] are used, since they reflect thermal efficiencies at the wheels level obtained during field testing.

$$\eta_{\text{th,w}} = \frac{E_{\text{P}}}{E_{\text{fuel}}} = \frac{E_{\text{P}}}{m_{\text{f}} H_{\text{f}}} \tag{4}$$

where m_{f} is the fuel mass and H_{f} is its lower heat value (S.I. units). The values considered are 42.7 MJ/kg and 41.5 MJ/kg, respectively, for diesel and gasoline [19, 22].

The integration in Eq. 2 portrays an energy balance and was performed numerically (as depicted by the summation) with a spatial resolution Δx of one hectometer. The respective elevation changes Δz were obtained from *Google Earth* elevation data. This procedure was validated against the A25 design gradient profile provided by the road concessionaire ASCENDI [16]. This approach is already reflected in the literature [11, 24–26].

The main specifications of the most frequent vehicles encountered on these routes, according to [17], are summarized in Table 1, being retrieved from [19, 20, 22]. The rolling resistance coefficient of passenger cars is speed-dependent [19]

$$C_{\text{RR}} = 9 + 0.05v + 1.6 \times 10^{-6} v^{4} \tag{5}$$

where, v is the speed in "m/s" and C_{RR} is expressed in "N/kN."

Table 1 Specifications of the vehicles studied

Vehicle configuration	Fuel type	RVM (kg)	ACd (m²)	CRR (N/kN)	ASR (km/h)	TEW (%)
	D	13,098	5.39	5.50	68–85	35.6
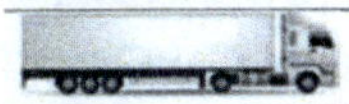	D	34,165	5.21	"	"	37.2
	D	1675	0.681	Eq. (5)	83–128	30.0
	P	1350	0.642	"	"	"

Notes: D diesel, P petrol, RVM running vehicle mass, ACd frontal area × drag coefficient, CRR coefficient of rolling resistance, ASR average speed range, TEW thermal efficiency at the wheels

Organization of the Paper This study is organized as follows: Sect. 2 establishes the methodology regarding traffic modeling and defines the simulated scenarios; in Sect. 3 the results of the case study are presented and discussed; and finally, in Sect. 4, the main conclusions are presented.

2 Methodology

2.1 Route Comparison Guidelines

To achieve the study objectives implicit in the first research question, the methodology followed considers, for equal travel times on both routes, the interests of the user with respect to fuel consumption, GHG emissions, and fuel and toll costs.

The propulsive energy and fuel consumption models were discussed in the context of the literature review (Eqs. 1 to 5). Regarding GHG emissions, they were computed from known emission factors of 3.04 kg_{CO2}/kg_{fuel} and 3.13 kg_{CO2}/kg_{fuel}, respectively, for petrol (gasoline) and diesel [20, 22].

Regarding the direct costs analysis, the known tolls charges per round trip were used, equal to 3.3, 5.6, and 8 euros, respectively, for passenger cars, 2-axel rigid trucks, and articulated 5-axle trucks. As for fuel costs, the average prices practiced in Portugal in 2019 were considered, these being 1.523 €/l for petrol and 1.390 €/l for diesel [27]. The respective fuel densities, needed to calculate the volume of fuel spent, are 750 kg/m^3 and 836 kg/m^3 [20, 22]. The years 2020 and 2021 were not considered due to representativeness issues, for the reasons previously discussed.

2.2 Traffic Characterization on the A25 and IP5 Routes

In order to quantify the overall economic and environmental impacts associated with the imposition of tolls, a methodology was developed in order to estimate the magnitude of the traffic that is being diverted from the A25 to the IP5. For this purpose, the traffic distribution was characterized in terms of average annual daily traffic volumes (AADT) per vehicle classes and route.

There are no published data regarding the current traffic volumes on the IP5, so they were inferred from the published AADT data along the A25 and from nodal flow balances. The main assumption is that the abrupt drop in traffic volumes observed on the A25 when it crosses the IP5 (at junctions '14' and '19') are being channeled to the alternative route. However, in order to mitigate possible sink and source effects, resulting from local traffic flows, the nodal balances were performed using the distance-weighted average traffic volumes observed west of node '14' and east of node '19', over no less than three road links, for each case. The mean AADT for these adjoining segments, 24 km long, is 14,108 v/d, the difference between the two groups being 3.4% of the mean value. However, the AADT for the A25 link

14–19 is 9015 v/d. These values were obtained from traffic reports published by the Institute of Mobility and Transport [3]. This represents a "fall" of 36%, presumably due to deterring effect of tolls. This finding is in line with the values reported for the metropolitan areas of Oporto and Lisbon [1]. As for the relative frequency distribution of traffic per vehicle class, this was obtained through traffic visualization at the ASCENDI control center [17].

The flowchart in Fig. 3 summarizes the described methodology, regarding the traffic modeling. In this context, the condition of equal travel times was eliminated, in order to increase the realism of the study. The AADT results, thus calculated, per route and vehicle class, are shown in Sect. 3, Table 2, in order to support and facilitate the critical analysis developed there.

However, to perform the required energy balances, it is necessary to know in advance the average speeds practiced on the two routes. In this regard, the average speeds of passenger cars were computed from the regulatory speed limits, being 116.4 km/h for the A25 and 90 km/h for the IP5. As for the heavy trucks, given the fact that they are unable to adhere to the road speed limits on the most severe climbs, a vehicle following strategy was adopted in order to obtain realistic speed values, considering only full length runs. The mean average speeds observed (in both directions) are 80.1 km/h and 73.9 km/h, with standard deviations (of the average speeds) of 3.6 km/h and 6.1 km/h, respectively, for the A25 and IP5 links under study. These

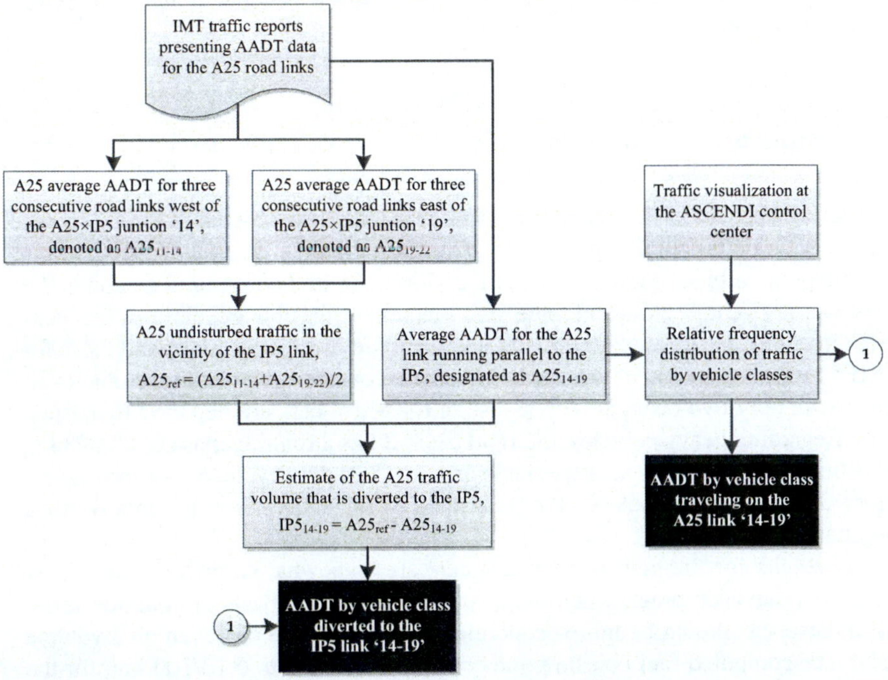

Fig. 3 Modeling of the traffic distribution by vehicle class and route

Table 2 Estimated traffic volumes by vehicle class and route for the year 2019

Vehicle class	A25 – AADT (v/d)	IP5 – AADT (v/d)	Total per Class (v/d)
Petrol car	3620	2155	5775
Diesel car	3620	2155	5775
Rigid truck	479	211	691
Tractor-trailer	1296	572	1868

Notes: The IP5 traffic volumes are those diverted from A25; AADT = Average Annual Daily Traffic (vehicles per day traveling in both directions); the AADT for petrol and diesel cars are presumed equal

values result from 37 observations, 20 eastbound and 17 in the westbound directions. Given the differences in the travel times, an indirect *time cost* was assumed, in order to adequately compare the two routes. This travel time cost was calculated as if it were paid work, considering a labor cost of 9.69 €/h. This value reflects the average gross salary earned by the Portuguese active population per hour of effective work in 2019 [28].

Traffic Scenarios Simulated In order to address the second research question, two traffic scenarios were considered: the current situation, and a hypothetical scenario assuming no traffic detour from the A25 to the IP5. This approach was implemented in order to quantify the impact that the elimination of tolls could have on the energy resources usage, and to assess which classes of vehicles would benefit most from this measure; and thus follow the path towards a more sustainable use of existing road infrastructure.

3 Results and Discussion

Comparing both routes in equal terms, with respect to travel times, it can be shown that the A25 is considerably more environmentally friendly than the IP5, regardless of the type of vehicle. Figure 4 shows the kilometric fuel consumptions and GHG emissions per vehicle class and route, as a function of feasible travel times (per running direction). The results for the A25 are identified by the blue lines, and those for the IP5 by the orange lines. The implicit scenario considers that the cars, vehicles A, travel with only two occupants. The results for petrol cars are depicted by dashed lines. Regarding heavy vehicles, the rigid truck B has a running mass of 13,098 kg, reflecting average conditions expectable on a regional duty cycle. As for the tractor-trailer combination, vehicle C, the total mass of 34,165 kg is representative of a long-haul duty cycle [22].

The results for the petrol and diesel cars are somewhat surprising, since they almost overlap each other, when it was expectable on the basis of common sense that a diesel car should be more economical than a gasoline car, even on a volume basis. The computed fuel consumptions on the A25 of 3.9 to 6.1 l/100 km, for the speed range of 83 to 119 km/h, are perfectly aligned with field results obtained in

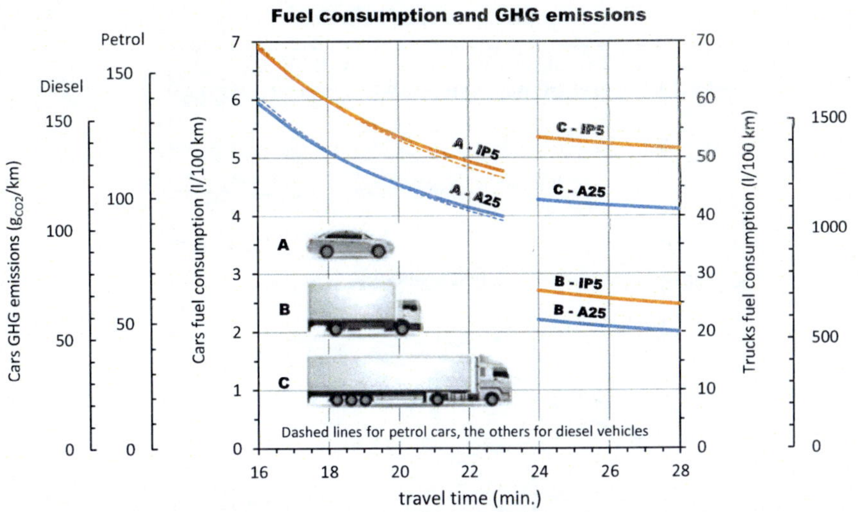

Fig. 4 Round trip performance of the vehicles studied for the two roads. The group A is subdivided according to the type of engine, diesel (continuous line) and petrol (dashed lines). The vertical scales in left are for cars, and right scales for trucks. The travel time is for one direction

free-flow traffic conditions with a gasoline car, only slightly lighter, on a similar link, although facing gradients of up to 6% [16]. This is a direct consequence of the difference in weights.

The results for the tractor-trailer combination are also significant, since they show that the energy consumption and GHG emissions of heavy vehicles depend little on speed on hilly highways, unlike those for lighter vehicles. However, the same cannot be said about the road gradients that have a noticeable negative impact on the energy efficiency and GHG emissions of heavier vehicles. This conclusion is well documented in the literature [5, 6, 11, 12, 16, 18]. However, the finding that reducing the speed of heavy trucks on hilly roads is of little value is not yet documented. It is thought that this is directly related to the gradients severity, since the aerodynamic drag of a heavy truck can be considerably lower than 1% of the vehicle weight, when loaded, thus much lesser than the gravitational resistance forces experienced on a hilly road.

Figure 5 highlights these findings. It shows the percentage increase in energy consumption and GHG emissions on IP5, when compared to A25, considering also the effect of the distance traveled, which is 6.8% higher on the IP5. The worsening, ranging from 23% to 34%, is such that when combined with an estimated 36% traffic detour from the A25 to the IP5, it becomes clear that the political decision to impose tolls on a modern road network does not favor energy sustainability.

The analysis above is based on the assumption of equal travel times. However, eliminating this condition, and assuming that users comply with the regulatory speed limits, the benefits of traveling on the A25 are considerably diminished. In fact, cars constitute the vast majority of the observed traffic and their energy

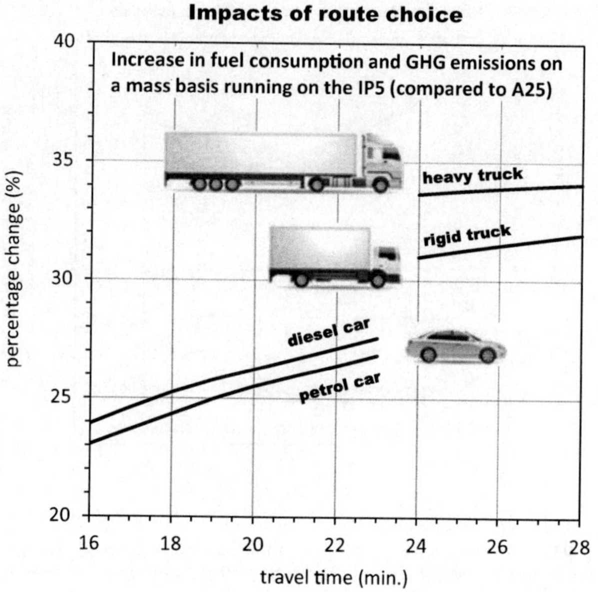

Fig. 5 Round trips percentage increase in energy consumption and GHG emissions on the IP5, when compared to A25, by vehicle class and travel time (between junctions)

consumption is highly dependent on speed, which is significantly higher on the A25. Figure 6 shows the direct travel costs and the CO_2 emissions produced according with the following traffic scenarios: the double color bars portray the current traffic distribution, characterized in Table 2, the A25 in blue and the IP5 in orange; and the light blue columns depict a hypothetical scenario that does not consider any traffic detour from the A25 to the IP5.

The results show that the concentration of heavy traffic on the A25 would be favorable to the environment. The opposite is true for light vehicles, due to a 29% increase in speed. As a result, the overall effect would be only marginally positive, with GHG emissions being reduced by 2.8%, from 51,468 mt/yr to 50,006 mt/yr (metric ton per year). Fuel costs would also be reduced by 2.5% (2019 prices). However, travel costs would increase by 8.3%, from 35.3 to 38.2 million euros per year, with the tolls paid accounting for 19.1% and 27.2% of these totals, respectively. On the other hand, the time spent on the road would be reduced by 182,400 hours in a year. This reduction of 10.7% is equivalent, in Portugal, to an average labor cost of 1.8 million euros. Nevertheless, the total equivalent costs would still be 2.3% higher for the A25 scenario, without traffic detour, than for the current traffic pattern, reflecting a traffic shift of 36% to the IP5. These values are significant for 30-km road links, especially in face of the low traffic volumes observed. Therefore, the deterring effect of the tolls still justifies the detour to alternative roads, despite the increase in travel time.

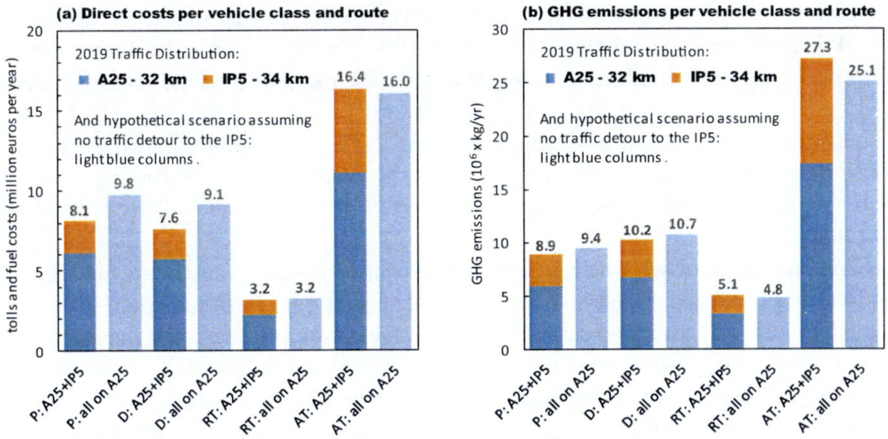

Fig. 6 (**a**) Direct travel costs and (**b**) CO_2 emissions for different traffic scenarios, routes, and vehicle classes: P petrol car, D diesel car, RT 2-axle rigid truck, AT 5-axle articulated truck

Conversely, for the tractor-trailer combination, it was found that the route IP5 is not favorable by any standard. In fact, the travel time increases by 15.8% without any economic advantage, since it is estimated that the increase in fuel costs exceeds tolls savings. The combined effect results in an increase in direct costs of 6.4%, and 7.7% if the cost of travel time is accounted for. Only empty trucks can benefit from such a rerouting strategy. The current fuel prices make this situation even worse. For long-haul services, it is common to refuel in Spain, where fuel prices are lower, but even so, this would not cancel the losses.

4 Conclusions

Route choice affects the energy consumption, GHG emissions, and travel costs. The present case study addresses this problem by analyzing the relative merits of two arterial routes crossing the Portuguese district of Viseu. These roads are the highway A25 and the IP5. The first has 2 × 2 lanes, a maximum gradient of 5% and a speed limit of 120 km/h. The other is an undivided road with only two lanes, has a maximum gradient of 8%, and a speed limit of 90 km/h. These roads run parallel between the junctions '14' and '19' of the A25. This link is 32 km long, against 34 km of the toll-free IP5, between the exact same nodes. Round-trip tolls on the A25 for cars, 2-axle heavy vehicles, and trucks with four or more axles are 3.3, 5.6, and 8.0 euros, respectively.

From the study results, it is possible to conclude that traveling on the IP5 is significantly more harmful to the environment than on the A25 for the same travel times. Energy consumption is also aggravated, with heavy vehicles being the most affected, the fuel bill increasing by 34%. Even for the average gasoline car, the aggravation ranges from 23% to 27%. However, given the magnitude of the tolls, the IP5 is still economically attractive to users, if time constraints are disregarded. This applies to all vehicle classes studied, with the exception of the tractor-trailer combination. In this case, for long-haul duty cycles, the benefits of the A25 are unquestionable, the total cost reduction being 7.7% (fuel, tolls, and travel time). Only empty movements will have lower costs. Another relevant finding is that reducing the speed of heavy trucks on hilly roads is of little value. This is not yet documented.

Considering the traffic as a whole, the impact of diverting traffic to the IP5 is small, since cars' speeds are considerably higher on the A25. If prevented, it would reduce the CO_2 emissions by only 2.8%, and fuel costs by 2.5%. Yet, taking tolls into account, total costs would increase by 8.3%. Hence, to take full advantage of the potential benefits offered by the A25, it will be necessary to reduce both the tolls and the speed limits of passenger cars. These passive measures can be effective if rationally introduced. For example, for a speed limit of 110 km/h, reducing light vehicle tolls by 57% would make both routes equally attractive to these users, while reducing the overall GHG emissions by 4.4%. This conclusion follows directly from the results presented in Fig. 5, showing fuel consumption and GHG emissions vs. travel time. Different compromises are possible, especially since heavy trucks have no visible advantage in diverting their route to the IP5. However, this is happening, certainly due to the psychological effect of tolls, i.e., the desire to avoid them, and to a misinformed route choice.

References

1. Concessões das Infraestruturas Rodoviárias, Portugal - 2019. (2021). Direção do Observatório do Ecossistema da Mobilidade e dos Transportes.
2. Barrosa, R. P. C. (2017). Impactos da cobrança de portagens na mobilidade rodoviária [MSc Thesis].
3. Relatório de Tráfego na Rede Nacional de Autoestradas. (2020). Instituto da Mobilidade e dos Transportes, I.P. (IMT, I.P.).
4. Boggio-Marzet, A., Monzon, A., Rodriguez-Alloza, A. M., & Wang, Y. (2021). Combined influence of traffic conditions, driving behavior, and type of road on fuel consumption. Real driving data from Madrid Area. *International Journal of Sustainable Transportation*, 1–13.
5. Booto, G. K., Bohne, R. A., Vignisdottir, H., Pitera, K., Marinelli, G., Brattebø, H., Wallbaum, H., & Ebrahimi, B. (2017). The effect of highway geometry on fuel consumption of heavy-duty vehicles operating in eco-driving mode, *8*.
6. Brunner, C., Giesen, R., Klapp, M. A., & Flórez-Calderón, L. (2021). Vehicle routing problem with steep roads. *Transportation Research Part A: Policy and Practice, 151*, 1–17.
7. Coiret, A., Deljanin, E., & Vandanjon, P.-O. (2020). Vehicle energy savings by optimizing road speed-sectioning. *European Transport Research Review, 12*, 41.
8. Croce, A. I., Musolino, G., Rindone, C., & Vitetta, A. (2020). Route and path choices of freight vehicles: A case study with floating car data. *Sustainability, 12*.
9. Romero, F., Gomez, J., Rangel, T., Jurado-Piña, R., & Vassallo, J. M. (2020). The influence of variable message signs on en-route diversion between a toll highway and a free competing alternative. *Transportation, 47*, 1665–1687.

10. Costagliola, M. A., Costabile, M., & Prati, M. V. (2018). Impact of road grade on real driving emissions from two Euro 5 diesel vehicles. *Applied Energy, 231*, 586–593.
11. Gallus, J., Kirchner, U., Vogt, R., & Benter, T. (2017). Impact of driving style and road grade on gaseous exhaust emissions of passenger vehicles measured by a Portable Emission Measurement System (PEMS). *Transportation Research Part D: Transport and Environment, 52*, 215–226.
12. Levin, M. W., Duell, M., & Waller, S. T. (2014). Effect of road grade on networkwide vehicle energy consumption and ecorouting. *Transportation Research Record, 2427*, 26–33.
13. Scora, G., Boriboonsomsin, K., & Barth, M. (2015). Value of eco-friendly route choice for heavy-duty trucks. *Research in Transportation Economics, 52*, 3–14.
14. Sentoff, K. M., Aultman-Hall, L., & Holmén, B. A. (2015). Implications of driving style and road grade for accurate vehicle activity data and emissions estimates. *Transportation Research Part D: Transport and Environment, 35*, 175–188.
15. PTV Vissim Manual. (2018). PTV AG, Karlsruhe, Germany.
16. Ferreira, H., Rodrigues, C. M., & Pinho, C. (2020). Impact of road geometry on vehicle energy consumption and CO2 emissions: An energy-efficiency rating methodology. *Energies, 13*, 119.
17. Ferreira, H. (2016). Simulação computacional do comportamento dinâmico de veículos pesados: classificação energética do traçado de autoestradas de montanha [PhD Thesis].
18. Loulizi, A., Rakha, H., & Bichiou, Y. (2018). Quantifying grade effects on vehicle fuel consumption for use in sustainable highway design. *International Journal of Sustainable Transportation, 12*, 441–451.
19. Rexeis, M., Hausberger, S., Kühlwein, J., & Luz, R. (2013). *Update of emission factors for EURO 5 and EURO 6 vehicles for the HBEFA version 3.2*. Institute of Thermodynamics and Sustainable Propulsion Systems, Graz University of Technology.
20. Rexeis, M., Quaritsch, M., Hausberger, S., Silberholz, G., Kies, A., Steven, H., Goschütz, M., & Vermeulen, R. (2017). *VECTO tool development: Completion of methodology to simulate Heavy Duty Vehicles' fuel consumption and CO2 emissions*. European Commission.
21. Rosero, F., Fonseca, N., López, J.-M., & Casanova, J. (2021). Effects of passenger load, road grade, and congestion level on real-world fuel consumption and emissions from compressed natural gas and diesel urban buses. *Applied Energy, 282*, 116195.
22. Rexeis, M., Röck, M., & Hausberger, S. (2018). *Comparison of fuel consumption and emissions for representative heavy-duty vehicles in Europe*. ICCT.
23. Wallace, J. M., & Hobbs, P. V. (2006). *Atmospheric science: An introductory survey*. Elsevier Academic Press.
24. Wang, Y., Du, X., Yu, H., Huang, J., & Li, Y. (2015). *Impacts of road grade on fuel consumption of light vehicles by use of Google earth DEM*. 2015 international conference on cyber-enabled distributed computing and knowledge discovery, pp. 360–363.
25. Wang, Y., Zou, Y., Henrickson, K., Wang, Y., Tang, J., & Park, B.-J. (2017). Google Earth elevation data extraction and accuracy assessment for transportation applications. *PLoS One, 12*, e0175756.
26. Wei, H., Luan, X., Li, H., Jia, J., Chen, Z., & Han, L. (2018). Elevation data fitting and precision analysis of Google Earth in road survey. *AIP Conf. Proc., 1967*, 020031.
27. Análise da evolução dos preços de combustíveis em Portugal. (2021). ENSE – Entidade Nacional para o Sector Energético, E.P.E.
28. Instituto Nacional de Estatística. https://www.ine.pt. Last accessed 2022/03/01.

Improvement of a Porcelain Baseboard Production Line Capacity Using Simulation: A Case Study

Inês Seabra Ferreira and Ana Luísa Ferreira Andrade Ramos

Abstract The current high and diversified demand from customers pushes companies to be more competitive, continuously looking for improvements and to enhance their productivity and efficiency. This paper presents a case study of a porcelain tile manufacturing industry, focusing on the processes involved in a porcelain baseboard production line. The main objective of the study was to identify and assess alternative scenarios of the system that could increase the production capacity. For this purpose, the current situation and possible alternatives were simulated using Arena Software. Simulation allowed to study different scenarios and compare potential improvements in order to pursue wise investment options, confirming the enormous utility of this decision-support tool for operations management. In the presented case study, the first alternative considered the addition of a parallel processing line and showed a 109% increase in the production capacity. The second presented scenario involved the addition of an operator to packing and showed an 83% increase in production. Lastly, the third alternative considered an adaptation of the system's layout and with only 46% capacity increase, proved to be better when considering the necessary investment. Lastly, the third alternative proposed layout changes to the system that increased the production capacity by 46%. Nevertheless, this last scenario proves to be better when considering the necessary investment.

Keywords Case study · Decision-making · Simulation · Scenario comparison · Ceramic baseboard production

---

I. S. Ferreira (✉)
DEGEIT – University of Aveiro, Campo de Santiago, Aveiro, Portugal
e-mail: ines.sferreira@ua.pt

A. L. F. A. Ramos
GOVCOPP/DEGEIT – University of Aveiro, Campo de Santiago, Aveiro, Portugal
e-mail: aramos@ua.pt

J. C. de Oliveira Matias et al. (eds.), *Quality Innovation and Sustainability*, Springer Proceedings in Business and Economics,
https://doi.org/10.1007/978-3-031-12914-8_23

1 Introduction

In today's modern era, companies seek to improve their efficiency in order to achieve a competitive advantage in a fast-changing market. To this purpose, Lean Thinking, Information Technologies (IT), Simulation, and other Industry 4.0 technologies have been applied to several sectors within companies as tools to improve their performance, eliminating non-added value activities and improving the process workflow [1].

Simulation is a key factor in Industry 4.0 [1], and when used in a company's context allows to evaluate the system's performance and can be of great help in making better and more supported decisions [2]. The elaboration of the current state model requires a thorough analysis and a comprehensive understanding of a company's processes, making it possible to identify wastes and improvement opportunities. Simulating different scenarios of new or improved processes is a cheap way of testing changes without implementing them in the real environment, leading to low investments [3].

Organizations have sought to use this method aimed to reach an agile, fast, and flexible production [4], reducing time and costs without compromising product quality. Furthermore, an investment in innovation, development, and improvement should be done to prevent eventual market fluctuation, deliver higher value to customers, and speed up the response time.

Considering all this, this paper aimed to simulate a porcelain baseboards production line, studying a possible change to the process that may improve the line capacity, in order to support the top management to make decisions without a great investment or interfering with the production.

This paper is structured in four sections, starting with an introduction followed by a theoretical background regarding simulation. Afterward, the case study is presented, including a brief contextualization, explanation of the methodology followed, and the exposure of results obtained with the simulation of the current state and proposed scenarios of the process. Lastly, it presents the conclusion that sums up the relevant aspects and limitations of the case study and proposes future work.

2 Literature Review

Simulation is a very powerful problem-solving methodology that mimics a real – world system in order to better comprehend its functioning, detect and analyze problems, test alternatives, and aid in decision making [5]. It can be applied in a variety of fields such as health systems, military systems, public services, and manufacturing and material handling applications [6]. In the manufacturing context, simulation is often used to assess the ideal number and type of machinery in a system, plan the production schedule, and perform a bottleneck analysis [7].

A successful simulation should have great concern with the formulation of the problem, the choice of the simulation software, the collection of data and logic of the system, and should consider reasonable randomness to the model as well as proper statistical interpretation of the output [8].

Some of the benefits of the application of simulation include the comparison of alternative system designs, to explore diverse possibilities and test "what if" scenarios [6, 9], the investigation of real – world systems that cannot be evaluated analytically by a mathematical model [10], and the study of systems in compressed or expanded time [6, 10]. Furthermore, simulation allows to estimate the performance of a system under specific conditions and identify problems and constraints within the system [6, 10]. Additionally, it tests and prepares changes without investing resources, making it possible not only to alter the plan without great expenses but also to investigate alternative procedures without disrupting the system's operations, thus leading to wise investments [6, 9]. Moreover, simulation makes clearer and understandable the reason behind certain phenomena, taking a closer look into the system [6], and the use of Operator Training Simulators (OTSs) provides reduced training costs for a new workforce [7].

However, simulation can have some limitations regarding the time consumed for collecting data [9], the special training needed to develop the model, and the potential difficulty to interpret the results [6]. Some of the pitfalls found by [10] are the misunderstanding of simulation by management, an inappropriate level of detail in the model, the collection of insufficient or inadequate data, the use of wrong performance metrics, a single replication run of the system to obtain results, and the misuse of animation.

Currently, in the manufacturing context, companies usually apply simulation software products instead of general – purpose programming language, taking concern mostly with the modeling flexibility and the usage easiness [7]. One of the most commonly used simulation software is Arena, developed by Rockwell Automation, which uses the discrete event method for most simulation efforts, and is the fastest simulation tool available [11].

One of the many reasons to resort to simulation tools is the pursuit for higher productivity and production capacity has been one of the many applications of simulation tools. This field has been widely explored in last years and is commonly sought by companies. Therefore, Table 1 presents a summary of the work developed in [12–15] and the results obtained.

In most cases, a simulation study follows the steps defined by [16], illustrated in Fig. 1.

The structure and summary of each step consist of:

1. Formulate the problem and plan the study: In this first stage, top management should define the problem to address and organize meetings with the project manager, simulation specialists, and Subject Matter Experts (SMEs) to plan the study. Topics such as the project's objectives, specific aspects, boundaries and domain of the study, system's performance indicators, resources necessary, and configuration of the system, should be defined [8, 10].

Table 1 Case studies summary

Case study	Summary
Gujarathi N., Ogale R., Gupta T. (2004)	This case study presents the application of a simulation study in a shock absorber assembly line. The proposed layout changes have increased the production capacity from 1,000 to 1,435 shock absorbers per day
Abed, S. (2008)	This case study applied the Arena Simulation tool to identify bottlenecks in a rusk production line and increase the production rate. Seven different scenarios were consecutively drawn to search for a good feasible solution. The changes resolved all bottlenecks, resulting in an increase of about 50% in production and a decrease of 11.4% in average total production time for a box of rusk
Talapatra, S., Tarannum, R., Shefa, J (2018)	This paper applied a simulation study and a statistical analysis with Excel to find the bottleneck in a bottle production line and rectify it to improve productivity. Several scenarios were formulated, but the addition of a semi-automated cap sealing machine in lieu of the manual operation in the workstation resulted in a reduction of 14.11% waiting time and an increase of 4% productivity
Nadir, U., Habib, T., Khattak, S., Noor, I. (2020)	A tile production facility was simulated using Simio software to study the current production process and analyze the cycle time, throughput, work in process, inventory, and production rate of tiles

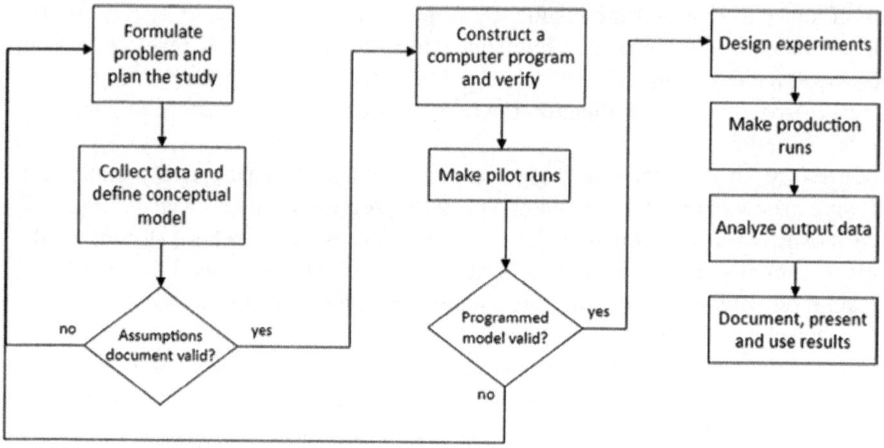

Fig. 1 Steps of a simulation study. (Adapted from Law et al. [16])

2. Collect data and define a conceptual model: Information related to the control logic of the system and operational procedures should be collected [8], and the parameters and probabilistic distribution of the process should be specified. The collected data should be grouped in a conceptual model with the assumptions, and mathematical and logical relations to produce a real system abstraction [10].
3. Is the assumptions document valid: Additionally, it's essential to measure the system's performance to be able to validate the conceptual model generated, making sure all the assumptions are correct, every intervenient is involved and,

if later, it won't be necessary to reprogram the model [10]. The model must be adjusted and redesigned until it proves to be valid [8].

4. Construct a computer program and verify: The valid conceptual model is transformed by a computer program into an operational model. To this purpose, a general – purpose programming language or simulation software is chosen, having a great impact on the simulation's success. Verification of the operational model (debug) must be included in this step [8, 10].
5. Make pilot runs: Execute pilot runs to determine if the model reflects correctly and accurately the analyzed system and evaluate the changes triggered by the abstractions and simplifications of the model [8].
6. Is the programmed model valid: The pilot runs serve to test if the programmed model is valid, leading to its redesign if proven not to be. The model results must be analyzed by the SMEs and simulation analysts and considered appropriate according to the simulation's objectives, for the model to be considered valid [8].
7. Design experiments: In this phase is established the length of each simulation run, the length of the warm-up period (if used), and the number of independent simulation runs. Different experiments are normally designed, reflecting different scenarios that vary in their logic or input data [8, 10].
8. Make production runs: The production runs are made with the parameters established for each experiment and will mimic the functioning of the system in each scenario designed. Each production run will use a different set of random values, resulting in unique output data that must be analyzed [8].
9. Analyze output data: The execution of the model originates output data that allows to evaluate the performance metrics defined for a system configuration and compare different alternatives between them. With the results obtained, the analyst can choose to make more production runs or additional scenarios [8, 10].
10. Document, present, and use results: The assumptions, software, and results of the study must be documented for future use, to facilitate its comprehension, communication with management, and aid in decision-making. For this purpose, animation is used to allow a better understanding of the model for people not familiarized with its details. The documentation will allow a review of the models and their parameters, provide credibility to the models developed, and facilitate and support the implementation of the chosen solutions [8, 10].

3 Case Study

The study focuses on the simulation of a ceramic baseboards production line at the transformation sector of a Portuguese ceramic company which produces porcelain tiles. The company produces a large variety of tiles, with different formats and sizes, currently having an overall installed capacity of 10,000,000 m^2 per year and employing more than 700 employees.

3.1 Problem Contextualization and Objectives

The study was carried out in a porcelain baseboards production line, which transforms porcelain tiles into baseboards through cutting and jiggling. The line is supplied with tiles produced within the company and is very flexible, being adaptable to transform tiles with different formats, lengths, and widths.

The transformation sector of the company has a production line that transforms ceramic tiles into baseboards, having a lower capacity than the market's demand. Therefore, the company showed interest to study alternatives to expand the line and increase production. Since the machinery to assemble an entirely new line represents a big investment, other alternatives to optimize the efficiency of the process with lower costs were also studied. Furthermore, the labor force associated with this process is poorly balanced, and the processing capacity is limited not by the velocity of the processing, but the whole system is adjusted to the capacity of the operator responsible for packing. On this note, relieving the operator would allow a faster machinery's velocity and an increase in the production capacity.

Thus, the objective of this study was to model the current process and appraise the possibility of expansion of the process, by adding a new operator, changing the system's layout, or adding a new parallel production line. Additionally, the study aimed to evaluate the results of such changes in comparison with the necessary investments, concluding on the impact of the presented alternatives.

3.2 Methodology

In order to achieve the study's objectives, the simulation conducted and presented in this paper followed the simulation study steps developed by [16], described previously. Information to prepare the simulation study was collected with the collaboration of the responsible for the transformation section, which is considered an SME. Afterward, data was collected from production documentation on the shop floor through direct observation and informal interviews with workers, revealing information about the layout of the system, the dislocations of operators, the velocity of conveyors, distances between machinery, and measuring processing and transportation times and input of material. This resulted in the construction of a conceptual model that proved to be valid, permitting the construction of the computerized simulation of the current state, using Arena 16.10 developed by Rockwell Software. After the operational model was verified and debugged, pilot runs of 30 min were done to evaluate if the model accurately represented the system. Because the results matched real production values and demonstrated a correct representation of reality, the model was considered valid, and experiments of the current state and three alternative scenarios were designed. These changes considered the addition of a parallel jiggling line, the addition of another operator to perform the choosing and packing operations, and the change of the layout. After running the experiments, results regarding production, resources occupation rate, and waiting times were obtained and analyzed. In addition, to better comprehend the systems, 3D animations were designed using Anylogic Software. To conclude, the scenarios were compared concerning the production capacity, line balancing, investments, and implications, and conclusions about the possible changes in the porcelain baseboards production line were drawn.

3.3 Results and Discussion

The variation of products transformed at the line requires an adjustment on the system's parameters such as the speed and width of the conveyors, the amount and speed of input supplied to the system, and the packaging method. These alterations are done by a tuner at the beginning of the process that settled the machinery speed to the established parameters, according to the type of production. Therefore, the study considered a warm-up period of 30 min to stabilize the production, and the parameters were settled for a continuous production phase. Because this line works shifts of 8 h, the study considered five replications with an 8 h length for each scenario studied.

The production line in study is almost fully automated, requiring only the assistance of operators in the supplying of the system and in the choosing and packing operations. The transformation process of tiles is conducted by machinery along the conveyor, not involving workforce. The production is limited not by the processing velocity of machinery and conveyors, but by the work capacity of the operators, being of interest to analyze changes to the system that can better balance the employees' operations.

For the development of the study, information was collected for the stabilized process in normal conditions during an afternoon. Additionally, both the operators and the tuner and the sector responsible were interviewed, because as SMEs, they have a deeper knowledge of the system. For the analysis, the performance indicators of interest defined were the resources occupation rate, cycle time, number of baseboards rejected, number of finished boxes produced, and waiting time in the choosing and packing operations.

3.3.1 Current Situation

Currently, in the baseboard production studied, porcelain tiles in the format 97 × 597 cm are supplied to the system by a supplier machine that has a capacity to hold up to 40 pieces. Operator 1 is responsible for moving the material from the pallet, unpacking the tiles, and placing them in the supplier machine, which constantly releases 1 tile every 18 s. The tile then enters a conveyor settled at 2.05 m per minute until it arrives at the cutting machine, where the material is cut breadthways in 3 baseboards by 2 blades while moving on the conveyor. The pieces are then separated, entering one at a time in a second conveyor settled at 6.15 m per minute, which leads to a first jiggle machine, responsible for smoothing one of the sides of the baseboard. The piece then moves to another jiggle machine that smooths the other side of the baseboard. These processes occur while the piece is being transported on the second conveyor, that ultimately leads to the choosing station, where a second operator is responsible to assess the quality of the tiles, put aside any damaged baseboard, pack sets of 20 pieces, and put them in a pallet, to be transported.

The production documentation showed that an average of 10% of the baseboards produced from this format have flaws, so only 90% of the pieces that arrive at the choosing station were effectively packed. The operator normally assessed five pieces at a time, taking on average 21.5 s. This reflected in a constant of 4.3 s per piece in the conceptual system due to the consistency of the task. The packaging process, because it depends on the packaging and pallet location, can vary, and was described for a triangular distribution: TRIA (15, 30, 60), based on the values registered in 30 measurements taken over an afternoon. Operator 1 normally adds to the supply machine 10 tiles at a time, taking on average 45 s to unpack and place it in the proper location, described by a constant in the model due to its minimal variation.

After developing the conceptual model, by measuring the operation time of the workers, gathering information on distances and velocities, and comprehending the system's dynamic on the shop floor, it was possible to design the logical model, using Arena. To better comprehend this model, Fig. 2 represents the 3D animation of the described system.

The results of this simulation for the performance indicators measured regarding resources occupancy, production, and time are presented in Tables 2, 3, and 4, respectively.

From the occupancy of the resources, it was concluded that Operator 2 can be considered the bottleneck of the system, since the velocity of the supply machine can be speeded up, as well as the remaining machinery. On the other hand, it was also possible to perceive that Operator 1 has high idle time, which shows an unbalanced work distribution. However, it is important to mention that Operator 1 is also responsible for transporting pallets with tiles and finished products in and out of the line, which may occupy at most 25% of his time.

The simulation also showed an average accepted production of 4036.00 ± 20.77 baseboards per 8 h-shift, and an average rejected production of 458.40 ± 13.30

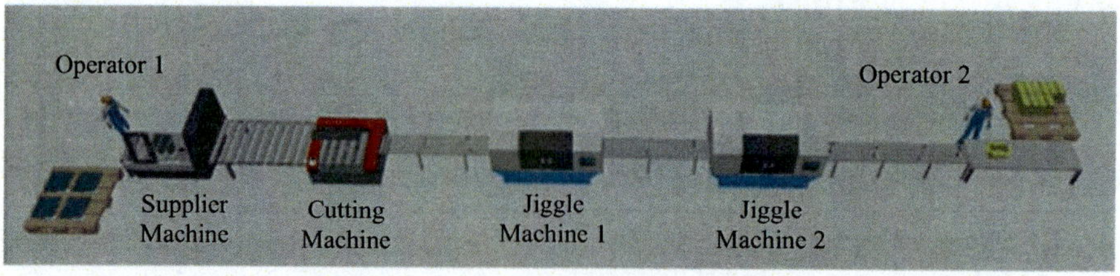

Fig. 2 3D animation of the AS-IS model of the system using Anylogic

Table 2 Current resources occupancy (%)

Resource	Occupancy
Supplier machine	100.00% ± 0.00%
Cutting machine	97.59% ± 0.00%
Jiggle machine 1	96.70% ± 0.00%
Jiggle machine 2	96.70% ± 0.00%
Operator 1	25.00% ± 0.00%
Operator 2	97.87% ± 0.00%

Table 3 Current production (number of pieces)

Average production	
N. baseboards OK	4036.00 ± 20.77
N. boxes produced	201.80 ± 1.04
N. baseboards rejected	458.40 ± 13.30
Total baseboard produced	4,494.40

Table 4 Current cycle time and OP2 waiting times (seconds)

Average time (sec.)	
Cycle time	6.006 ± 0.000
Waiting time choosing (OP2)	256.524 ± 39.000
Waiting time packing (OP2)	4.714 ± 1.800

baseboards. With a total of 4494.40 baseboards produced in 8 h, the average time between pieces manufactured was 6.006 ± 0.000 s.

3.3.2 New Scenario 1: Add Parallel Processing Line

In order to enhance the production capacity, the addition of a parallel processing line was studied, even though it was not the only possible solution of interest. This scenario considered the implementation of another pair of jiggle machines that would treat baseboards in another line, after leaving the cutting machine. Because the cutting process triples the number of pieces to address in the remaining processes, it needs to have a lower processing velocity than the following, however, by associating a new jiggle line, the cutting velocity could increase. This said, the alternative model considered that a supply machine delivers a tile every 9 s into a conveyor with a velocity of 4.1 m per minute, that conducts the material to the cutting machine. After cut, the baseboards are separated with a probability of 50% into two conveyors, that have a velocity of 6.15 m per minute. Each conveyor leads to one line of jiggle with 2 jiggle machines, and one choosing station in the end. The new choosing station included an additional operator associated, doing identical work to

Table 5 Alternative 1 resources occupancy (%)

Resource	Occupancy
Supplier machine	100.00% ± 0.00%
Cutting machine	97.59% ± 0.00%
Jiggle machine 1 – L1	95.87% ± 0.00%
Jiggle machine 2 – L1	95.87% ± 0.00%
Jiggle machine 3 – L2	97.48% ± 0.00%
Jiggle machine 4 – L2	97.48% ± 0.00%
Operator 1	50.01% ± 0.00%
Operator 2 (L1)	92.20% ± 0.00%
Operator 3 (L2)	93.70% ± 0.00%

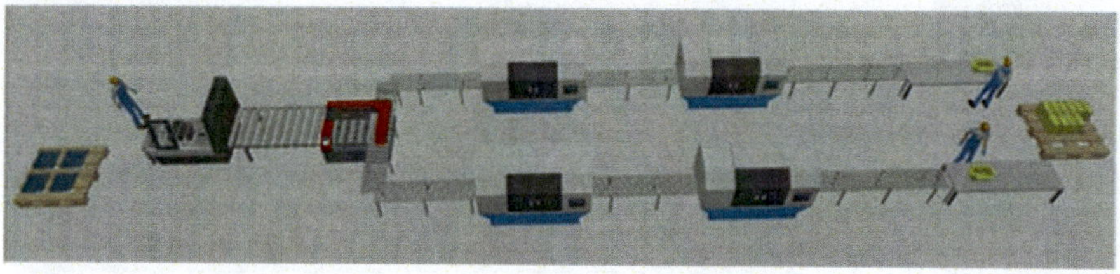

Fig. 3 3D animation of the TO-BE model of alternative 1 using Anylogic

Operator 2. To better understand the proposed model, Fig. 3 shows a 3D animation of the presented system.

The results obtained in this scenario are shown in Tables 5, 6, and 7, representing a 109% production increase when compared with the current scenario, with an average of 9387.8 baseboards produced per shift.

These results show an average accepted production of 8464.00 ± 32.37 baseboards, and an average rejected production of 923.80 ± 29.20 baseboards, in an 8 h simulation. Additionally, the average time between pieces manufactured was 3.003 ± 0.000 s, reducing 50% compared with the current system. Despite the increased production, this alternative required a very high investment that might not be viable for the company, being therefore relevant to study other alternatives for the problem, such as alternatives 2 and 3 presented below, before making an investment.

3.3.3 New Scenario 2: Add an Operator to Packing

Since the bottleneck of the system was the packing and choosing processes, performed by Operator 2, a possible way to increase production was simply the addition of another operator to perform the same tasks. Distribution of the checking and packing activities for 2 Operators would allow to speed the velocity of the conveyors and diminish the processing and transportation times, increasing production.

Table 6 Alternative 1 production (number of pieces)

Average production	
N. baseboards OK	8464.00 ± 32.37
N. boxes produced	423.20 ± 1.62
N. baseboards rejected	923.80 ± 29.20
Total baseboard produced	9387.80

Table 7 Alternative 1 cycle time and OP2 and OP3 waiting times (seconds)

Average time (sec.)	
Cycle time	3.003 ± 0.000
Waiting time choosing (OP2)	10.620 ± 0.600
Waiting time packing (OP2)	0.254 ± 0.000
Waiting time choosing (OP3)	10.908 ± 0.000
Waiting time packing (OP3)	0.277 ± 0.000

Table 8 Alternative 2 resources occupancy (%)

Resource	Occupancy
Supplier machine	100.00% ± 0.00%
Cutting machine	97.57% ± 0.00%
Jiggle machine 1	96.88% ± 0.00%
Jiggle machine 2	96.88% ± 0.00%
Operator 1	44.01% ± 0.00%
Operator 2	83.16% ± 2.00%
Operator 3	84.70% ± 1.00%

However, by speeding the velocity of the conveyors and the machines processing over 50%, the quality of the pieces would be affected. Additionally, the need to supply more tiles to the system diminished the idle time of Operator 1, making the system more profitable.

For the design of this alternative, the speed of the first conveyor, that carries the uncut tile to the cutting machine, was changed to 3.59 m per minute, and the speed of the other conveyors that transport the baseboards until the choosing station were changed to 10.76 m per minute, both increasing 75%. Also, the frequency of supply by the machine changed to 10.28 s between tiles. From the SME's perspective, it was estimated that these changes lead to a decrease of pieces' quality, being accepted only 82% of the pieces. In Tables 8, 9, and 10, the results obtained in the presented scenario regarding the resources occupancy, the waiting time for choosing and packing operations, the cycle time, and the output obtained from the system are given.

The simulation revealed an average accepted production of 6744.00 ± 32.37 baseboards per shift, and an average rejected production of 1480.00 ± 39.68 baseboards. This scenario led to a total of 4494.40 baseboards produced in 8 h, and cycle

Table 9 Alternative 2 production (number of pieces)

Average production	
N. baseboards OK	6744.00 ± 32.37
N. boxes produced	337.20 ± 1.62
N. baseboards rejected:	1480.00 ± 39.68
Total baseboard produced	8224.00

Table 10 Alternative 2 cycle time and OP2 and OP3 waiting times (seconds)

Average time (sec.)	
Cycle time	3.429 ± 0.000
Waiting time choosing (OP2)	13.770 ± 1.200
Waiting time packing (OP2)	4.051 ± 1.200
Waiting time choosing (OP3)	14.100 ± 0.600
Waiting time packing (OP3)	4.224 ± 0.600

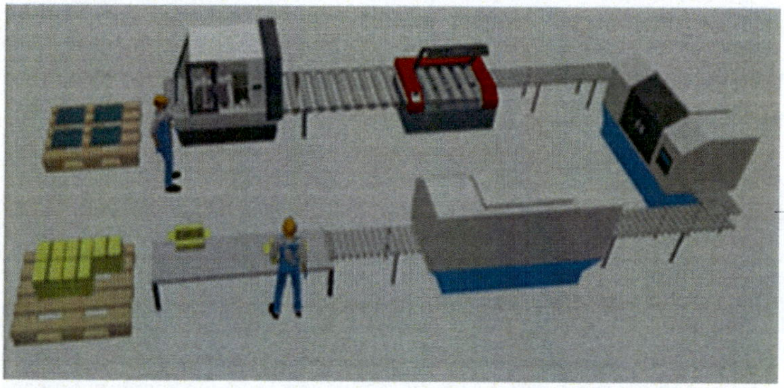

Fig. 4 3D animation of the TO-BE model of alternative 3 using Anylogic

time reduction of 42.90%, to 3.429 ± 0.000 s. This data shows us that this alternative resulted in increased 83% production when compared to the current state, involving, however, the investment of an additional operator.

3.3.4 New Scenario 3: Change the System's Layout

An alternative solution to increase production without an additional operator was to redesign the line, so that Operator 1, which had an average occupation rate of around 25%, could aid Operator 2 in the choosing and packing operations. For this purpose, the layout of the line should be rebuilt in a U shape, so that the supply point could be near the choosing station, for Operator 1 to work on both sides, as shown in Fig. 4.

This balance of resources allowed speeding up 40% the line velocity, making the supply machine deliver a tile every 12.86 s and speed up the conveyors' velocity to

Table 11 Alternative 3 resources occupancy (%)

Resource	Occupancy
Supplier machine	100.00% ± 0.00%
Cutting machine	97.53% ± 0.00%
Jiggle machine 1	96.55% ± 0.00%
Jiggle machine 2	96.,55% ± 0.,00%
Operator 1	82.64% ± 0.00%
Operator 2	82.62% ± 0.00%

Table 12 Alternative 3 production (number of pieces)

Average production	
N. baseboards OK	5912.00 ± 22.21
N. boxes produced	295.60 ± 1.11
N. baseboards rejected:	657.80 ± 18.94
Total baseboard produced	6569.80

Table 13 Alternative 3 cycle time and OP2 and OP1 waiting times (seconds)

Average time (sec.)	
Cycle time	4.286 ± 0.000
Waiting time choosing (OP2)	5.646 ± 1.800
Waiting time packing (OP2)	2.913 ± 0.600
Waiting time choosing (OP1)	4.395 ± 1.800
Waiting time packing (OP1)	5.416 ± 1.200
Waiting time manual supply	27.534 ± 0.600

2.87 m per minute, and 8.61 m per minute, for the first and second set, respectively. This alternative resulted in a higher baseboard production and operators' occupation rate, providing a lower cycle time and reduced waiting times, which is reflected in Tables 11, 12, and 13.

From these tables, it was possible to perceive a better balancing of the operators' work; however, the transportation work, which consumed up to 25% of Operator 1's time, would have to be supported by an external source or aided by Operator 2. Additionally, this alternative presented some implications concerning the space available in the transformation sector and the need for duct under the line that would need to be restructured in U shape. These alterations would require some costs related to the stoppage of the production during the implementation of the changes.

3.3.5 Scenarios Comparison

After analyzing the current situation of the production line and each of the 3 alternative scenarios presented, a simultaneous comparison of their summarized results was presented in Table 14, showing higher production capacity and lower cycle time

Table 14 Scenarios' results comparison

Scenario	Lead time (sec.)	Total production	N. box OK	Production/ operator	Box OK/ operator	Increase in production	Implications
Current system	6.006	4494.40	201.80	2247.20	101	–	–
Alternative 1	3.003	9387.80	423.20	3129.30	141	109%	+1 operator; +2 jiggle machines; + conveyors.
Alternative 2	3.429	8224.00	337.20	2741.30	112	83%	+ 1 operator; – 8% of quality pieces
Alternative 3	4.286	6569.80	295.60	3284.90	148	46%	Exchange of layout

Table 15 Resources occupancy rate comparison

	Resource usage (%)			
Scenario	OP1	OP2	(OP3)	Average
Current system	25.00%	97.87%	–	61.44%
Alternative 1	50.01%	92.20%	93.70%	78.64%
Alternative 2	44.01%	83.16%	84.70%	70.62%
Alternative 3	82.64%	82.62%	–	82.63%

in alternative 1. However, it was also necessary to consider the required investment, being specified not only the implications of each scenario, but also an analysis on the ratio of the overall production per operator and the number of packed boxes per operator. To compare the implications of the different alternatives, a rough estimate was calculated, considering an investment of 30,000 euros for the two additional jiggle machines and extra conveyors, a fixed cost of around 1,000 euros per month for an additional operator and a 1,500 euros investment in the labor and equipment necessary to adjust the line's layout. This allowed to perceive that, although alternative 3 presented a lower increase in production, it revealed to be the one with higher production rate per operator and the lowest costs.

Additionally, an analysis over the average occupancy of the operators and the waiting time for their tasks was done, in order to evaluate the line balancing in each scenario. From the results presented in Table 15, it was possible to perceive that alternative 3 makes better and more efficient use of resources, having a higher average of occupancy and a more balanced distribution of work between operators.

All in all, alternative 3 showed to be the best scenario presented, being a wise investment for the company, because, with minimal costs and no additional resources, the production increased 46%. However, if the company desires higher throughput, alternative 1 showed the best results, involving however the highest investment.

4 Conclusion

This paper illustrated the role that simulation can have in the evaluation of changes in a company's process and the aid it can be in decision-making. Simulation allows to study different scenarios and compare their results, to pursue wise investment and more efficient processes.

The presented case study demonstrated that better results do not always imply the best scenario, since limitations and consequences of the changes should also be considered. The three alternatives presented had the goal to increase a baseboard line production, showing different results and different levels of investment. Although scenario 1 showed a higher production, with a 109% increase, it also represented the highest investment. On the other hand, alternative 3, despite showing an increase of only 46%, demonstrated to be the most efficient, considering minimal investment required.

This analysis and comparison of alternatives will support the top management on the choice to expand the production and make it easier to comprehend the results and implications of each scenario. Due to the limitation of time in the study, it wasn't possible to explore the production of other formats and sizes in the production line in question to fully comprehend the improvements and limitations of the proposed changes. However, future work should include a broader study for all types of baseboards produced in the line and a detailed analysis of the prices and costs associated with the acquisition of additional machinery, the hiring of another employee, and the restructuring of the duct under the line.

References

1. de Paula Ferreira, W., Armellini, F., & De Santa-Eulalia, A. (2020). Simulation in industry 4.0: A state-of-the-art review. *Computers & Industrial Engineering, 149*, 106868.
2. Nordgren, W. B. (1995). Steps for proper simulation project management. In *Proceedings of the 1995 winter simulation conference* (pp. 68–73). IEEE.
3. Preston White, K., & Ingalls, G. (2019). The basics of simulation. In *Proceedings of the 2018 winter simulation conference* (pp. 147–161). IEEE.
4. Heilala, J., Montonen, J., Järvinen, P., Kivikunnas, S., Maantila, M., Sillanpää, J., & Jokinen, T. (2010). Developing simulation-based decision support systems for customer driven manufacturing operation planning. In *Proceedings of the 2010 winter simulation conference* (pp. 3363–3375). IEEE.
5. Banks, J. (1999). Introduction to simulation. In *Proceedings of the 1999 winter simulation conference* (pp. 7–13). IEEE.
6. Banks, J. (1998). Principles of simulation. In *Handbook of simulation* (1st ed.). Wiley.
7. Hosseinpour, F., & Hajihosseini, H. (2009). Importance of simulation in manufacturing. *International Journal of Economics and Management Engineering, 3*(3), 229–232.
8. Lyaw, A. M., & McComas, M. G. (1991). Secrets of successful simulation studies. In *Proceedings of the 1991 winter simulation conference* (pp. 21–27). IEEE.
9. Shannon, R. E. (1998). Introduction to the art and science of simulation. In *Proceedings of the 1998 winter simulation conference* (pp. 7–14). IEEE.

10. Law, A. M. (2015). *Simulation modeling and analysis* (5th ed.). McGraw-Hill Education.
11. Arena Simulation Homepage. https://www.arenasimulation.com. Last accessed 2022/02/01.
12. Gujarathi, N., Ogale, R., & Gupta, T. (2004). Production capacity analysis of a shock absorber assembly line using simulation. In *Proceedings of the 2004 winter simulation conference* (pp. 1213–1217). IEEE.
13. Abed, S. (2008). A simulation study to increase the capacity of a rusk production line. *WSEAS Transactions on Information Science & Applications, 9*(5), 1–10.
14. Talapatra, S., Tarannum, R., & Shefa, J. (2018). *Simulation modeling for productivity improvement of a production line: A case study*. Proceedings of the International Conference on Industrial Engineering and Operations Management.
15. Nadir, U., Habib, T., Khattak, S., & Noor, I. (2020). Performance improvement using simulation tool in a tiles production facility. *Mehran University Research Journal Of Engineering & Technology, 39*(1), 117–132.
16. Law, A. M., & Kelton, W. D. (1991). *Simulation modelling and analysis* (2nd ed.). McGraw-Hill.

Multifunctional Furniture for Tiny Houses–Design, Quality, Innovation and Sustainability in Advanced Materials

M. J. Félix, G. Santos, J. C. Sá, S. Dias, and M. Saraiva

Abstract The purpose of this paper is to demonstrate the innovative and most recent composite technology, developed by the R&D Wood Based Products Technology Group. The technology uses a combination of wild thistle (*Cynara cardunculus*) particles and polyurethane foam, proposed for the conception of lighter and mechanically resistant furniture, which competes with other existing materials on the market and is used in the construction of several types of furniture. Many studies point out that tiny houses are a sensible and innovative way of increasing urban density and improving affordability.

The main goal of this work is the contribution of knowledge based on the demonstration and application of more conscious choices by design professionals, on the selection of advanced materials, and on the design and development of a multifunctional furniture line that is sustainable and lightweight in the concept of circular economy. In the methodology, a theoretical argumentation was carried out, supported by bibliographical analysis, where the answer to the development of new innovative products was studied, which fits in with the trends and desires of our society, the research methodologies guide the project and use methods that clarify its application.

The result was the development of a line of multifunctional furniture based on an innovative material that offers greater lightness in the components, demonstrating it to be a viable and beneficial alternative for design projects, opening the way for future designers to be more aware of their choices regarding the materials used in furniture projects. In conclusion, it was intended to demonstrate the profound challenges of achieving a successful and sustainable consumer transition and it is hoped

M. J. Félix (✉)
Polytechnic Institute of Cávado and Ave, Design School, Barcelos, Portugal

CIAUD - Research Centre for Architecture, Urbanism and Design, Faculty of Architecture, University of Lisbon, Lisbon, Portugal
e-mail: mfelix@ipca.pt

G. Santos · S. Dias · M. Saraiva
Polytechnic Institute of Cávado and Ave, Design School, Barcelos, Portugal

J. C. Sá
Enginering School, Polytechnic of Porto, Porto, Portugal

J. C. de Oliveira Matias et al. (eds.), *Quality Innovation and Sustainability*, Springer Proceedings in Business and Economics,
https://doi.org/10.1007/978-3-031-12914-8_24

that the project will culminate in the design and development strategies of sustainable and lighter multifunctional furniture products carried out in partnership with industry and in line with the expectations of all those involved in this process.

Keywords Multifunctional Furniture · Advanced Materials · Tiny Houses · Sustainability · Significant Experiences

1 Introduction

In the last few years, tiny houses have been promoted as a new eco-friendly solution to face the current wastefulness of the housing industry [1]; thus, tiny home life can correct the excesses that have occurred and contribute to an increase in conscious environmental thinking, encouraging practices that link practicality and contributing to what has been found to be a positive correlation in more conscious consumption.

Living well with less, is the purpose of the architectural and social movement Tiny Houses, based on the issue of sustainability of the planet. It is a way of being able to live in a small space, but without giving up comfort, and having the opportunity to contribute to a healthier life. The Tiny Houses are built on top of a trailer, allowing them to be mobile and also to be placed in places fenced off from building construction.

The big difference compared with a motorhome is that if you enter a motorhome you feel like you are inside a car, whereas when you enter a Tiny House you feel like you are entering a house, because of the way it is decorated and the well-defined spaces, making it a real house in a very reduced space. The only similarity is that both forms of housing are on wheels. Another motivation is the increasing number of people who have difficulty in obtaining housing; therefore, it seems necessary to create a wide range of affordability measures.

The socioeconomic situation has brought with it a number of challenges, such as mobility, adaptability and the functionality of structures and systems as we know them. People who opt for these Tiny Houses highlight the interest in different attributes as essential factors of choice, including the type of unit and energy label, with young, single-person households in urban areas being most likely to occupy these dwellings [2].

Commercial prices and increasing urbanisation force people to occupy less space, but human needs remain the same. Within this context, we can highlight furniture, which is a necessity and not a luxury. Furniture occupies almost 50% of a normal area [3] in a house; therefore, living in a small space requires smart and space-saving furniture solutions to maintain human well-being. An approach to flexible and multifunctional furniture design [4, 5] is important and necessary when we talk about small houses. The design of multifunctional furniture responds effectively to the circumstances addressed, and is intended to be easy to manage and maintain, universal and sustainable.

The present project arises from the challenge of developing multifunctional furniture for the new concept of the Tiny House that is lighter than the solutions that are currently on the market. The proposal starts with the innovative and recent composite technology, developed by the R&D Wood Based Products Technology Group of the Polytechnic Institute of Viseu [6], proposed for the design of lighter and mechanically resistant furniture, which competes with other existing advanced materials in the market, and is used in the construction of various types of furniture. This composite is a low-density, naturally based agglomerate (300 kg/m^3) that can be produced by moulding each piece and manufactured by cold compression of the mixture of the thistle and foam particles, fitted into a previously manufactured mould [7].

Although multifunctional furniture is usually designed with light advanced materials, the intention of this project was to develop quality products that were still light, easy to use and easy to transform, avoiding excessive effort when performing tasks, with extendable or foldable components.

The main contribution of this paper was the proposed design and development of a multifunctional furniture line that was sustainable and lightweight in the concept of circular economy and adaptable to the residents' way of life. The aim of the research will also be the contribution of knowledge based on the demonstration and application of more conscious choices by design professionals in the selection of advanced [8, 9], innovative and environmentally responsible materials for the design of furniture for interior spaces.

The result is present in the design and development of a multifunctional furniture line based on wild thistle particles and a foam composite, which offer greater lightness in the components [10], demonstrating it to be a viable and beneficial alternative for design projects, paving the way for future designers to be more aware of their choices regarding the materials used in furniture projects [11]. This work arises from the need to unify in a single project a more sustainable and conscious product and contribute to the materialisation of physical, comfortable, harmonious and healthy spaces.

Promoting meaning and value, which are fundamental requirements, to activate a symbolic relationship with the objects [12], compete with the banalisation brought by the familiarity of the object and a useful strategy in a consumer society, stimulated by the designer's ability to propose lasting solutions that can continuously feed the user's interest.

In conclusion, it was intended to demonstrate the profound challenges of achieving a successful sustainable consumption transition [13] and it is hoped that the project will culminate in the design and development of strategies for more sustainable and lighter multifunctional furniture products, which are carried out in partnership with the industry [14] and in line with the expectations of all those involved in this process.

2 Literature Review

The Tiny House is the result of a social and architectural movement that advocates a reflection on how to build and live. This movement, which is increasingly common in the USA, draws attention to the reduction of housing areas, and has, over the years, gained supporters from all over the world. They can be built like most houses on a permanent base or on wheels, or on a trailer to offer greater mobility. This architectural style can vary in type, size and configuration [15].

The Tiny House has comfort features like those of a conventional house [16] and is built on the principles of sustainability [17], accessibility and social inclusion [18]. The movement has grown based on people's need to reduce costs [19], simplify their way of life and reduce their environmental impact. It is based on the philosophy of living with sustainable and ecological characteristics, advocating the quality and simplicity of living with the same comfort as in a traditional house, referring to an adaptation of nomadic citizens and presenting new architectural proposals inspired by the idea of micro-housing that has emerged in recent years.

As a strong trend for the future in single-family residential architecture [20], this housing concept has an important social role to play. Prefabricated houses became a trend owing to the confinement forced by the pandemic. It does not seem to be a temporary phenomenon, as questions started to be asked about whether the houses meet real needs and a desire emerged to adapt them to the longer time spent inside, to having a garden, for example.

From Ikea [21] to the Liten project of the president of Mercadona [22], a 30-m^2 house made of wood, customisable and using an industrialised system that provides good insulation, everyone wanted to enter the new boom in the housing market. And there is already a brand with Portuguese DNA, turtle houses [23].

Ikea's, 'Tiny Home Project' seeks to promote sustainability through inclusion and innovation, thinking about meeting the needs of current generations without compromising the needs of future generations. The designers of this tiny home on wheels believe that minimalist living in small spaces will take hold in the post-pandemic period, with thousands of people leaving cities and sharing flats to find their own safe space. Their models are equipped with solar panels and toilets that use composting to enable a self-contained existence.

According to Mutter [18], Tiny Houses are proven to have a reduced impact on the environment compared with ordinary dwellings, and a vision is shown of these dwellings with a focus on the environmental considerations of the movement, referring to health concerns related to the choice of materials and finishes used in furniture and decorative elements in interior spaces. The study conducted by Rashdan and Ashour [24] is aimed at proposing a set of criteria for the selection of sustainable design solutions for interior environments. The suggested set of proposals presents supportive sustainable selection criteria with the intention of better management for designers in the processes of research, selection and specification of interior finishes applied to fixed and movable structures, in furniture and/or equipment [24]. One of the pioneers of design with environmental concerns was Victor Papanek, who saw design as being one of the professions with the greatest potential to affect both the natural and human environments [25].

The relationship between sustainability and design [26] is an ongoing process of discovery that identifies involvement and recognition by designers, researchers and students. Today, many designers already have knowledge, regarding the specific properties of the advanced materials [11] they use in their projects.

Furniture plays an important role in the efficiency of a space as a main element in interior design. Made of various materials with surfaces in different textures and colours, the main sustainable criteria are the materials used in the production process and the long-term use of the furniture.

Wood is one of the most widely used materials for furniture production [27]. Sustainable furniture is any type of furniture manufactured with production processes that have little or no impact on the environment. The concept of sustainability covers not only production factors but also other factors such as the extraction of raw materials, the way furniture is transported and its usefulness. The sustainability of homes is reflected in meeting energy and efficiency standards or providing energy from renewable sources as well as through the materials used in furniture [28].

Flexible furniture in comparison with conventional furniture has less mechanical resistance owing to the number of times it is manipulated daily when changing environments, which leads to more wear and less durability of the accessories and systems connecting components. Thus, products that require greater resistance emphasise the functional aspect more than the aesthetic aspect. Furniture for the flexible use of space is that which has multiple functions related to space management achieved through strategies. These strategies allow the compactness of the form and through solutions secondary functions for the same product to be achieved.

In recent years, there has been a progressive trend in the housing market with increasingly compact and adaptable spaces. The growing development of this new reality captures the attention of professionals, increasing the challenge for architects and designers to develop [29] compact and multifunctional solutions in design that make living and working spaces compatible.

Based on the type of project it is important to select the material according to the particularities of the function because each function has specific needs [27], as well as the promotion of materials easily available in the region, which contributes to the reduction of the overall price of projects.

Owing to the large quantities of forest residues and residues from the wood processing industries, the interest of different types of industries in polymeric composite materials using this plant biomass as reinforcement in polymeric matrices has been increasing [30] and the furniture industry is an example of this interest.

Wood composites result in products with more dominant properties than other materials. They incorporate lightness, flexibility and durability, acquiring increasingly improved mechanical behaviour, good resistance to performance, low cost, they are easy to recycle and have a great aesthetic aspect and texture. These are materials that allow a greater exploration of shapes, curves and details, as they are easily adaptable to production processes [31] and also allow the possibility of obtaining tailor-made products.

Researchers from the R&D Wood Based Products Technology Group, an applied research group in innovative technology for processing wood-based materials that has been developing solutions for the wood composites industry for over 20 years and is based at the Polytechnic Institute of Viseu, through the Lightwood project, aimed to leverage the R&D results obtained in the LightFillers (QREN), project, which was aimed at producing low-density particles and incorporating them into low-density wood composite panels for application in the automotive and furniture industries. This project allowed the development of a lightweight wood-based composite panel based on the combination of wild thistle (*Cynara cardunculus*) particles and polyurethane (PU) foam [32].

The raw material (wild thistle) tested in the context of Lightwood consists of an unused agro-forestry residue, which is found in the region of "Serra da Estrela" in the central part of Portugal. Only the fluorescent flower part of the plant is used in the region for cheese production; its stem is not reused in other production areas.

The bush has beneficial characteristics: it is light and easy to work with. This new wood composite technology, is nationally patented [32]. (PPP 115374, Low density polyurethane wood composites and their manufacturing method). The inventors of this composite have observed that the use of PU foam represents a significant portion of the composite of synthetic origin and propose its subsequent replacement by a binder system entirely of biological origin.

Inserted into the concept of circular economy, this new composite is a natural-based agglomerate of low density (300 kg/m^3). It is possible to be produce it by moulding, it can be shaped, and each piece can be manufactured by the process of cold pressing of the mixture of cardboard particles with the foam, adjusted to a previously manufactured mould. This new technology of wood composites demonstrates technical viability of the application in furniture, to the producers of wood derivatives and manufacturers of furniture, in order to expand the competitiveness of the companies [7]. This will facilitate the combination of other materials, such as other woods, wood by-products and accessory materials (metals, glass, components, etc.)

The Tiny House achieves the maximum use of practical and functional interior space and manages to be fully customisable according to the inhabitant, generically consisting of a single room, with or without attic waters (loft), which combines living room, bedroom and kitchen to maximise space. In the interior design multifunctional furniture, such as convertible, folding furniture and smart storage solutions, is incorporated [19].

However, it is our intention in this project to look at other fundamental aspects for the harmony and sensibility of the users, in this case the meaning and value as stimuli for the design of meaningful experiences [12]. These are fundamental requirements for activating a symbolic relationship with the objects, qualities that come both from the level of commitment assumed by the person during the interactive communication process and from the durability of the solutions proposed by the designer, competing with the banalisation brought about by the familiarity of the object, a useful strategy in a consumer society.

In this context, it is up to the designer to propose lasting solutions that can continuously feed the user's interest. We live in a big cage that is an open space where everything is the same, the houses, the times of life, the entertainment, the activities, and the differences are only apparent.

We are individual beings; we need to live in consonance with the way we see the world, seeking meaningful experiences adjusted to our options.

Considering the current context as a society that is increasingly global and diverse, but that is also paradoxically proving to be lonelier and emptier, the way in which our society has been structured, based on an offer of products for a public eager to participate in their consumption, has led to increased expectations of finding new emotional levels of often immediate experiences [12]. The sense of familiarity is a very useful factor for the current economic model, it is in its interest that a certain trend quickly becomes tedious to awaken a new desire in the consumer.

3 Methodology

In order to carry out the present work, which is based on an Polytechnic Institute of Cávado and Ave master's degree study, it was necessary to collect information from the applied research group in Innovative Wood Derivative Processing Technology of the Wood Engineering Department (DEMad) of the School of Technology and Management, Polytechnic Institute of Viseu, and a theoretical basis and bibliographic analysis, which was based on bibliographic collection and analysis as a reliable documental support, through which the tiny house, the design and development of multifunctional and composite wood and polyurethane furniture, sustainability and meaningful experiences were studied and systematised. A content analysis was carried out, based on primary information-gathering instruments, which consisted of information gathering and clarifications in the research laboratory for the development of the new material, direct observation of phenomena and articles, among others.

As secondary sources, books, scientific articles, news articles, theses and dissertations were consulted. All these instruments resulted in an analysis and treatment of the vast literature review and the information collected. This was followed by the definition of a research gap that led us to the investigation that gave rise to this work and the deduction of the hypothesis pointed out, which bridges the gap between theoretical and empirical constituents. Based on the interventionist methodology, an active investigation of furniture products was transversally developed with the intervention of the researcher using an information source that allowed the evolution of research on furniture made of wood composite materials to be documented.

This analytical model was supported; the whole development process was decisive for the consolidation of the questions and hypotheses related to the construction of multifunctional furniture lighter than that already on the market. Therefore, the non-interventionist methodology is continued using different documentation sources (written and non-written), outlining methods of data collection from documentary research, with brief expositions of the main authors' reflections approached reaching the study object.

Some conclusions were reached through the synthesis and analysis of the information obtained that originated strategic reflections that could be useful in the development of research in this area.

4 Results

A multifunctional furniture project was developed, to make it possible to obtain furniture models for data comparison, with quantifiable results, through a practice of weighing prototypes previously selected, built in two different raw materials, classified as low-density materials, with the purpose of obtaining valid and reliable results for analysis and scientific knowledge.

With the application of this method in the data measurement, what was intended was to confirm a greater lightness in multifunctional furniture pieces and to ascertain the validation of the new wood composite material, as a viable and beneficial alternative for this kind of project against the existing materials used on the market in interior design projects for the Tiny House.

For the implementation and testing phase a piece of furniture was chosen to prototype with the real materials chosen for the project. The test requires the choice of another material for comparison of results. It is important to stress that the materials being compared have the same characteristics with regard to density for the reliability of the results. They fall into the category of lightweight materials.

In order to carry out the weighing test and obtain credible and reliable results, a second material was chosen for comparison of density between the specimens (cube module).

The material selected for the comparison of results with the lightweight composite chip-board panel, based on the combination of wild thistle particles and polyurethane (PU) foam, was the poplar plywood. This material is a wood derivative obtained by gluing adjacent layers with crossed wires, at right angles, with faces in uncoiled wood, entirely made of poplar, which is characterised by its lightness, dimensional stability, quality and ease of machining. The choice of plywood is based on its use in the construction of multifunctional furniture owing to its physical and mechanical characteristics. The purpose of the weighing test is to develop an experiment to prove that the cube module constructed in Lightwood can weigh less than the cube module constructed in poplar plywood. The piece of furniture selected for the test was the cube module. This cube module in the project is a multifunctional, removable and transportable or movable (on wheels) piece of furniture (Figs. 1 and 2). The reason why this piece of furniture in particular was chosen for the prototype is the dimensional limitation (maximum length–2 m) in the process of manually manufacturing panels in Lightwood.

The results show that Cube 2, which was constructed entirely of Lightwood, is lighter than Cube 1. Thus Cube 2 weighs 1257.7 g less than Cube 1.

Fig. 1 Cube weighing cube 1 made of poplar wood plywood

Fig. 2 Cube weighing cube 2 made of Lightwood

From these results we can observe that changing the material in one component of the cube module significantly decreases the weight of the piece of furniture. In this study we observed through the weighing values a greater lightness in the Lightwood material and less lightness in the poplar plywood. The thistle Lightwood will attribute more lightness to the multifunctional furniture than most materials on the market. In this way it proves to be valid for the materialisation of the project. These results derive from the physical properties that compose the materials chosen for its construction.

From this study other (secondary) conclusions were drawn: the Lightwood material is shown to be valid for the materialisation of the project, but it cannot be used structurally owing to its weak resistance to compression; it does not offer great adherence to the application of fittings; but it does offer aesthetic advantages. Throughout the process of construction, assembly and use/handling of the two modules, mainly the cube 2 module, this study became important in reformulating the application of the materials in the project.

The furniture line consists of a set of multifunctional pieces of furniture, sustainable in the concept of circular economy and adaptable to the resident's way of life. The furniture was designed to be integrated into the interior of the Tiny House, but it can also be adapted to other types of compact environments, and makes the most of the space available, without having to sacrifice functionality and aesthetics. This new furniture line provides a layout solution for sleeping, living, cooking and office areas. The collection is composed of an eight-piece (Fig. 3) cube module, drawer block, kitchen furniture, laundry furniture, console furniture, bookcase, bed furniture and sofa. The cube module is versatile because it can incorporate several types of furniture; it can be used as a stool, a suspended shelf and a bookcase with wheels (mobile) or without wheels (transportable on your lap). The block of drawers is a piece that fits into different environments; it can be an auxiliary piece of furniture in the kitchen, a support in the dining area or a nightstand in the bedroom.

The kitchen units with sliding doors contribute to practicality and saving space, giving the small area a greater sense of space. The top unit can be a versatile shelf or a closed cupboard when combined with a cube unit and the corner unit has a double function: to store a removable waste bin or storage unit. The auxiliary

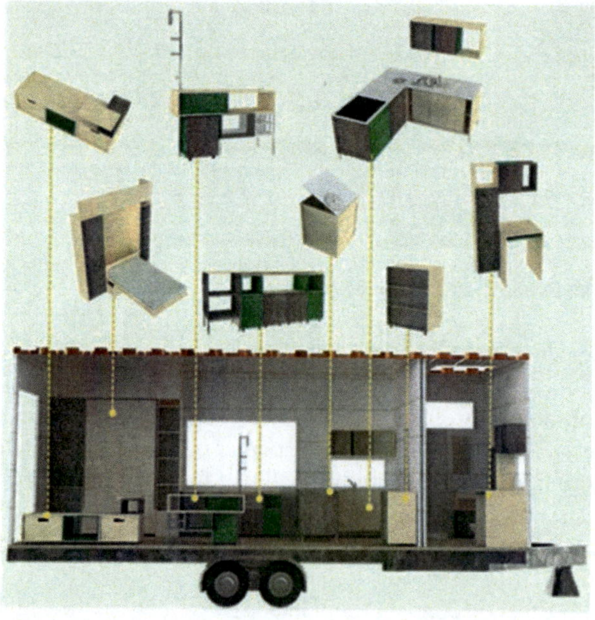

Fig. 3 Poster presenting the multifunctional furniture line, developed by the authors

laundry unit can be used as a pantry unit, ironing board or side table and as storage in high cupboards to save space. The multi-purpose console unit simultaneously groups together three units: a wine cellar, a glass case and a coat rack.

The bookcase unit with a built-in table contributes to saving space, making it quicker and more convenient to change environments for work, meals and social activities. The retractable bed, hidden between cupboards, incorporates wardrobe and shoe rack functions. It allows the environment to be changed without the need to move furniture and frees up more space in the room. The multifaceted movable sofa encases three double function stools, can transfigure a coffee table or stool, or couples a swivel coffee table with an open compartment arm.

5 Conclusion

The results obtained in this research can be translated into the development of a viable project to be produced in an industrial context.

After the literature review, it was possible to develop a proposal of multifunctional furniture with the application of the new material Lightwood thistle, which is expected to be part of a contribution to the innovation of the wood and furniture industries, as an innovative solution and an alternative to what is available on the market.

We can conclude that the furniture pieces entirely made of the Lightwood composite are lighter than other pieces made of light wood composites, in this case, the poplar plywood.

In this way, it is possible to affirm that the Lightwood material provides more lightness to multifunctional furniture than most currently used materials on the market, Lightwood being the one that presents the lowest density of all the materials presented in the study.

The construction of the prototypes allowed the observation of phenomena in order to gather conditions to make decisions regarding the application of the material.

The present project intended to make the best use of the Lightwood material, of the aesthetic advantages of the material, to apply the material in components free of hardware application and not to apply it in components with large dimensions subject to compression forces. It is important to mention that the future of this project is to continue to investigate and deepen it, through the realisation of more prototypes that will prove the project viability.

References

1. Ford, J., & Lanier, L. (2017). Are tiny homes here to stay? A review of literature on the tiny house movement. *Family and Consumer Sciences Research Journal, 45*, 394–405. https://doi.org/10.1111/fcsr.12205
2. Glumac, B. (2021). Tiny portable home: Measuring the rental preferences. *Cities, 116*, 103279. https://doi.org/10.1016/j.cities.2021.103279
3. Thøgersen, K. (2017). Small spaces need smart solutions. Designing furniture for small spaces, in connection with human wellbeing. In *Norwegian University of Science and Technology* (pp. 1–12).
4. Siu, K. W. M., & Wong, K. S. L. (2015). Flexible design principles street furniture design for transforming environments, diverse users, changing needs and dynamic interactions. *Facilities, 33*, 588–621. https://doi.org/10.1108/F-02-2014-0021
5. Gómez-Carmona, O., Casado-Mansilla, D., & López-de-Ipiña, D. (2018). Multifunctional interactive furniture for smart cities. In *12th international conference on ubiquitous computing and ambient intelligence* (pp. 1–12).
6. Carvalho, L., Martins, J., Coelho, C., Magalhães, F., Ribeiro, E., Ferreira, N., & Almeida, M. LightWood. https://www.arcp.pt/pt/projetos/prestacao-de-servico/lightwood. Last accessed 27 Feb 2022.
7. Departamento de Engenharia de Madeiras: Linha de mobiliário criada com nova tecnologia que combina partículas de madeira e cardo. https://site.ipv.pt/saladeimprensa/comunicados/InformaçãoIPV_74-19_ProjetoLightwood.pdf. Last accessed 4 Mar 2022.
8. Ferreira, P., Félix, M. J., Simoes, R., Silva, O., & Santos, G. (2022). Material selection guidelines for the product designer. In J. Machado, F. Soares, J. Trojanowska, & E. Ottaviano (Eds.), *Innovations in mechanical engineering* (pp. 182–190). Springer.
9. Karana, E., Hekkert, P., & Kandachar, P. (2008). Material considerations in product design: A survey on crucial material aspects used by product designers. *Materials and Design, 29*, 1081–1089. https://doi.org/10.1016/j.matdes.2007.06.002
10. Wood, M. *Lightweight wood enables ultimate architectural freedom*. https://www.metsawood.com/global/news-media/articles/Pages/Architectural-lightness.aspx. Last accessed 8 Mar 2022.
11. Delibaş, H., Uzay, C., & Geren, N. (2017). Advanced material selection technique for high strength and lightweight spur gear design. *European Mechanical Science, 1*, 133–140. https://doi.org/10.2139/ssrn.3425651
12. Dias, S., & Baptista, A. (2020). Play as a trigger for designing significant experiences. In *International conference on design and digital communication* (pp. 300–309). Springer.
13. Witt, U. (2011). The dynamics of consumer behavior and the transition to sustainable consumption patterns. *Environmental Innovation and Societal Transitions, 1*, 109–114. https://doi.org/10.1016/j.eist.2011.03.001
14. Félix, M. J., Gonçalves, S., Jimenez, G., & Santos, G. (2019). The contribution of design to the development of products and manufacturing processes in the Portuguese industry. *Procedia Manufacturing, 41*, 1055–1062. https://doi.org/10.1016/j.promfg.2019.10.033
15. Coleman, R. (2018). *Are tiny houses useful and feasible to help address homelessness in alameda county? How could tiny houses be used, and under what conditions?* https://ternercenter.berkeley.edu/wp-content/uploads/2020/11/Tiny_Houses_Study_FINAL_Terner.pdf
16. Koutalianos, A., & Gambling, S. (2018). *Designing tiny*. Vancouver.
17. Teixeira, P., Sá, J. C., Silva, F. J. G., Ferreira, L. P., Santos, G., & Fontoura, P. (2021). Connecting lean and green with sustainability towards a conceptual model. *Journal of Cleaner Production, 322*, 129047. https://doi.org/10.1016/j.jclepro.2021.129047
18. Mutter, A. (2013). *Growing tiny houses- motivations and opportunities for expansion through niche markets*. https://lup.lub.lu.se/luur/download?func=downloadFile&recordOId=4196241&fileOId=4196242

19. Kilman, C. (2016). *Small house, big impact: The effect of tiny houses on community and environment*. https://d31kydh6n6r5j5.cloudfront.net/uploads/sites/111/2019/07/charlie_kilman_tinyhouses__4_.pdf
20. Alexander, L. (2019). Community in property: Lessons from tiny homes villages. *Minnesota Law Review, 104*, 385–463.
21. IKEA. Tiny house. https://www.ikea.com/ch/en/ideas/profile-tiny-house-pub0e4c2c16. Last accessed 4 Mar 2022.
22. Ônarc. Liten. https://onarc.es/liten. Last accessed 4 Mar 2022.
23. Houses, T. *Turtle houses*. https://turtlehouses.pt/main/. Last accessed 4 Mar 2022.
24. Rashdan, W., & Ashour, A. (2017). Criteria for sustainable interior design solutions. *WIT Transactions on Ecology and the Environment, 223*, 311–322. https://doi.org/10.2495/SC170271
25. Papanek, V. (1985). *Design for the real world: Human ecology and social change*. Academy Chicago.
26. Moxon, S. (2012). *Sustainability in interior design*. Hachette UK.
27. Ayalp, N. (2013). Environmental sustainability in interior design elements. In N. Altawell, K. Volkov, C. Matos, P. F. de Arroyabe (Eds.), *Recent researches in environmental and geological sciences environmental* (pp. 163–167). WSEAS Press. https://www.wseas.org/multimedia/books/2012/Kos/WEGECM.pdf
28. Bianco, I., Thiébat, F., Carbonaro, C., Pagliolico, S., Blengini, G. A., & Comino, E. (2021). Life Cycle Assessment (LCA)-based tools for the eco-design of wooden furniture. *Journal of Cleaner Production, 324*, 129249. https://doi.org/10.1016/j.jclepro.2021.129249
29. Félix, M. J., Santos, G., Simoes, R., & Surrador, C. (2022). Product design and development strategies: Construction of regional innovation systems, for new achievements brought by the academy for industry. In J. Machado, F. Soares, J. Trojanowska, & V. Ivanov (Eds.), *Mechanical engineering* (pp. 174–184). Springer.
30. Jimenez, G., Santos, G., Félix, M. J., Hernández, H., & Rondón, C. (2019). Good practices and trends in reverse logistics in the plastic products manufacturing industry. *Procedia Manufacturing, 41*, 367–374. https://doi.org/10.1016/j.promfg.2019.09.021
31. Doiro, M., Fernández, F. J., Félix, M. J., & Santos, G. (2019). Machining operations for components in kitchen furniture: A comparison between two management systems. *Procedia Manufacturing, 41*, 10–17. https://doi.org/10.1016/j.promfg.2019.07.023
32. Martins, J., Coelho, C., Ferreira, N., Oliveira, M., Ribeiro, E., Barracosa, P., Santos, R., Flávio, S., Magalhães, F., Barreiro, M., & Carvalho, L. (2019). Lightwood: Innovative wood and polyurethane based composites. In R. Fangueiro (Ed.), *Proceedings of the 4th International Conference on Natural Fibers – Smart Sustainable Solutions. Tecminho–Associação universidade-empresa para o desenvolvimento*.

Angolan Cement Industry: Marketing Channel and Distribution Channel Strategies

Pedro Manuel Campos, Carina Maria Oliveira Pimentel, and José Dias Lopes

Abstract The Angolan Cement Industry (ACI) operates in a market economy that requires it to formulate appropriate strategies to better overcome market challenges. The problem of distribution has assumed great importance, both as a decisive element of the context and as a factor for cost reduction and profitability recovery. Thus, the objective of this paper is to identify marketing channels and distribution channels strategies that are appropriate for the Angolan market context, to make the cement supply process in the Angolan market more efficient and effective. This research focused on a qualitative approach, based on a literature review and application of the Delphi technique (applied in 2021). To this end, a survey was developed and applied to identify respondents' perceptions of which distribution strategies and marketing channels suit the Angolan Cement Market context. As a result, market expansion, diversification, and the installation of distribution centers, as well as the direct channel and the multiple channel marketing were identified as the strategies and channels that best suit the context of the Angolan Cement Market.

Keywords Marketing channels · Distribution channel strategies · Angolan Cement Industry

---

P. M. Campos (✉)
Department of Economics, Management, Industrial Engineering and Tourism (DEGEIT), University of Aveiro, Aveiro, Portugal
e-mail: pmcampos@ua.pt

C. M. O. Pimentel
GOVCOPP Research Unit, Department of Economics, Management, Industrial Engineering and Tourism (DEGEIT), University of Aveiro, Aveiro, Portugal
e-mail: carina.pimentel@ua.pt

J. D. Lopes
Lisbon School of Economics & Management, Lisbon University & ADVANCE, Lisboa, Portugal
e-mail: diaslopes@iseg.ulisboa.pt

J. C. de Oliveira Matias et al. (eds.), *Quality Innovation and Sustainability*, Springer Proceedings in Business and Economics,
https://doi.org/10.1007/978-3-031-12914-8_25

1 Introduction

Knowledge about marketing channels and distribution channels strategies is an indispensable tool to overcome product distribution problems, because the decision-making regarding the allocation of goods and services requires knowing the market in which companies operate or intend to operate. ACI's installed production capacity is 8.6 million tons/year and, in 2015, the consumption was close to 6 million tons/year [1]. In 2016, the problem of cement distribution in the Angolan market resurfaced, worsening until today causing a drastic production fall. Currently, the production of ACI is 30% of its installed production capacity and consumption has decreased to almost two million tons/year, while in 2015, its production was 70% of ts total capacity [2].

One of the main decisions related to distribution is the choice of the distribution channel [3]. By analyzing the integration of distribution channels [4], it was noted that today's market competition has become aggressive and the distribution channels are key for companies to remain competitive. In the same vein [5], sought to understand how distribution channels influence the position and performance of companies in markets in a context of global change, raising the following starting question: how should companies develop their distribution channels strategies in national and international markets to meet customer expectations and respond appropriately to global competition?

There are several articles on marketing channels and distribution channel strategies, but none of them address their implementation in the cement industry (CI) supply chain (SC). Thus, the existence of sparse literature signals a deficit that somewhat hinders the understanding of the problem. Nevertheless, this kind of gap has led [6] to acknowledge that there are still contributions that could be made to CI management. The current state of ACI is worrisome. Given its socioeconomic character, the following research question that guides this research was raised: which marketing channels and distribution channel strategies best suit the context of the Angolan cement market?

This article is organized into five sections. Section 1 presents the nature of the problem, the research question, and the justification of the need for the research to be carried out. Section 2 presents the literature review. Section 3 presents the research methodology. Section 4 presents the results of the first round of the Delphi study, and finally, Sect. 5 presents the final considerations.

2 Literature Review

2.1 *Marketing Channels*

A distribution channel consists of producers, distributors, retailers, resellers, and consumers linked by the exchange process. The greater the number of intermediaries in the process/the longer the channel, the higher the price of the product for the final consumer [7].

2.1.1 Direct and Indirect Channel

The direct channel represents the type of channel where producers sell/distribute their products directly to consumers, i.e., they do not use intermediaries to get their products to the consumer. On the other hand, the indirect channel incorporates intermediaries to get the final product to the final consumer. According to [8–11], there are greater benefits in the direct channel than in the indirect one. In the direct channel, there is greater assurance that the product will be delivered to the customer smoothly. The indirect channel is long and requires less investment (money and time) by the supplying company. However, the authors conclude that the literature does not provide conclusive information about which is the best distribution channel.

The development of e-commerce has pushed retailers and consumers to use multiple media, involving websites, physical shops, catalogs, the Internet, kiosks, mobile applications, and call centers [12]. Online distribution channels have changed retail perspectives [13]. The choice of a distribution channel depends on the economic objectives and constraints of each market [7].

2.1.2 Multichannel

Multiple channels are derived from the combination of online and offline channels. Multiple channels allow companies to market their products simultaneously in different distribution channels and make a higher profit. The emergence of multiple channels is not enough to distribute products in different market segments [9, 14–17].

One of the advantages of using the Internet is the reduction of entry barriers into new market segments from a distance and direct communication with customers. With the Internet, companies connect in real-time, share information and knowledge – the Internet and information technologies have boosted e-commerce [18, 19]. Among other advantages are sales growth and cost reduction. Companies that use multiple channels have higher performance and profit compared to companies that use only a single channel [14, 15]. According to [3], the use of multiple channels is increasing. But one of its disadvantages is consumer dissatisfaction with the difference in prices in the different distribution channels and the conflicts that arise from these channels.

2.2 *Distribution Channel Strategies*

Distribution channel strategies are as important as marketing channels for the allocating of products to consumers. However, identifying which ones best suit the Angolan cement market context and how to select them requires research. According to [4], currently, companies can produce products of high quality, but if they do not adopt an adequate distribution strategy, their market share may decrease.

2.2.1 Geographic Marketing Expansion Strategy

In general, market expansion is a strategy that aims to increase the growth of a given company by identifying and penetrating other local or international markets. A study was conducted by [20] with the following starting questions: (1) what is the market? what is marketing expansion? What factors affect market expansion? The authors define market expansion as a strategy that increases primary demand for a product category and converts non-customers into customers of an industry. For the authors, market antecedents are factors that drive market expansion. The authors identified the following three antecedents: (1) unmet needs and desires of potential customers; (2) their purchasing powers; and (3) access to desired products and services.

The recognition of consumers' consumption needs is a key point for the implementation of this strategy because it requires understanding what a market is and how to position oneself in it [21].

Access to products and services desired by consumers is an important antecedent. The availability of products in the market depends on the firm's willingness to reach unexplored and/or underserved market segments. To implement the market expansion strategy, the firm must define its market scope (customers and competitors) [20, 22], as decisions on market expansion are made based on the analysis of the market environment and the firm's available resources. The main strategic decisions include: (1) goal setting; (2) decisions about the strategic focus to be followed (such as market expansion, market penetration, or productivity improvement), generic strategies (cost leadership versus differentiation); and (3) the desired competitive position (the combination of price, quality, service, innovation, customization, speed, responsiveness) [23].

2.2.2 Diversification Strategy

Diversification is driven by the availability of raw material, accumulation of experience, and the sharing of know-how. According to [24], know-how as well as other intangible factors can be used to produce, promote, and distribute different products. Companies make more profit when they diversify their production. According to [25], a diversification strategy drives organizational conduct and performance.

Companies can diversify their related or unrelated products and can diversify by seeking international markets [26]. This strategy requires targeted distributions to a market segment [27–29]. As a rule, companies resort to diversification as it is a way to leverage their business horizons and capabilities [30]. The classification into horizontal, vertical, concentric, and conglomerate diversification is much addressed on the impact of related versus unrelated diversification and the mode of entry (organic growth, acquisitions, or alliances) [31]. External factors (industry characteristics, laws, culture, and transaction costs) also influence diversification.

2.2.3 Generic Strategies

Generic strategies (cost leadership, differentiation, and focus) are related to competitiveness. They focus on specific market segments [32]. Firms can achieve competitive advantage through differentiation, cost leadership, or focus and differentiate themselves from each other by their ability to compete [33]. In exploring the relationship between firms' strategic positioning, sustainability, and performance [34], concluded that differentiation generates better financial performance compared to cost leadership. Differentiation requires developing products and services with unique qualities desirable to customers.

Differentiation requires agility in service and products with high quality [33, 35, 36], as cost, delivery time, and product quality influence purchasing decisions [37]. Cost leadership requires investing heavily in technologies, staff training, benchmarking, and expense control. This requires the company to have high financial health [34] to stay in the low-cost system and continue to produce [35]. Focus strategy, on the other hand, is focused on a market segment to then implement differentiation or cost leadership [38]. The focus strategy targets specific markets [32].

2.2.4 Lean, Agile, and Leagile Strategy

In conducting their study [39], posed the following starting question: when should a company adopt lean or agile management? The authors noted that the complexity of lean and agile strategy arises during its implementation. Agile requires a flexible workforce, a trained human capital capable of performing the tasks. In general, make-to-order-based SC are shot, with little or no intermediary.

Lean and Agile SC aims to make logistics more flexible given market competition. Lean aims to reduce waste and minimize losses in SC. Agile SC, on the other hand, refers to the capacity to react and adapt to changes in the market environment. In general, the Agile paradigm is applied for specific products with a relatively short lifespan [37, 40, 41]. Combining the lean and agile extremes, it gave rise to the Leagile [hybrid] strategy [42]. Leagile SC uses the advantages of both lean and agile paradigms. This concept is in general applied in the case of innovative and unique end products with a more customized design [37].

3 Methodology

Given the nature of this study, was used the Delphi technique. The choice of this technique was based on the statement of objective, assuming the expansion of knowledge that might result in a significant contribution to the problem under study. Different authors who used the Delphi technique point to it as an efficient and effective technique for fieldwork in different sectors. For instance [43], point out that the Delphi technique is an appropriate methodological tool for researchers whose complex issues benefit from the insights and consensus of a group of experts. However, to be able to implement the Delphi technique in this research, the following starting questions were posed:

1. What is the Delphi technique and what makes it a choice over other methods?
2. How to select the panelists and why?
3. How to design the questions in the questionnaire?

The Delphi technique assumes that group consensus is more reliable than an individual decision. Furthermore, the Delphi technique has implications in various sectors such as public health, public transport, education, logistics chains, public policy, strategic decision-making in companies, among others [44–46]. This technique assumes that several heads think better than one [47]. The Delphi technique is a method of obtaining consensus on a given topic through sequences of questionnaires that are sent to specialists in each area [46, 48, 49]. Interactions, anonymous responses, sequences of questioning, and feedback to the group in each sequence are features that distinguish this technique [50].

Each characteristic provides the Delphi method with a unique approach to gaining, consensus on a topic. The samples in a Delphi study vary according to the purpose of the study, its complexity and available resources. A smaller, homogeneous panel reaches consensus quickly and, on the other hand, a heterogeneous panel takes a long time and may not reach consensus [49]. The number of respondents in the Delphi technique is an individual criterion [51]. Representativeness is assessed based on the quality and diversity of the panel, not its quantity [52].

This technique can produce data that would otherwise be difficult to obtain [53]. The Delphi technique begins with the formulation of the problem and the choice of experts, followed by an initial questionnaire that acts as a strategy to obtain the expert's appreciation of the subject under study [52]. Afterward, the responses from the first round are used to construct the second-round questionnaire [54].

Since this is an exploratory research, aimed at obtaining a consensus among a panel of experts, regarding the most suited marketing channels and distribution channel strategies for the Angolan cement Supply Chain, the Delphi technique was the chosen instrument. Thus, using the Delphi technique, 22 experts belonging to entities linked to the SC of the ACI were surveyed. All the selected entities have extensive work experience in the areas in which they operate and can contribute directly or indirectly to achieving the goal.

The questionnaire was composed of eight questions aimed to identify marketing channels and distribution channel strategies that are most appropriate to the Angolan cement market context. The questionnaires were divided into two sets, on the first one the goal was to classify distribution channel strategies and marketing channels, on other hand, the second one was created for the respondents to justify the classification given, on the first set. The questionnaire before being sent to the respondents was designed and previously validated.

In the first set of questions, respondents were asked to choose and justify their choices for the following questions, using a 5-point Likert scale:

1. How would you rate each of the marketing channels listed below, considering their suitability in the context of the Angolan market to enable the short-, medium-, and long-term efficient and effective supply of cement to the Angolan market? (On a scale of 1 [Unsuitable], 2 [Little suitable], 3 [Reasonably suitable], 4 [Suitable], and 5 [Very suitable]). Justify your answer.
2. How would you rate the installation of distribution centers as a mechanism to enable an efficient and effective supply of cement to the Angolan market in the short, medium, and long term? (On a scale of 1 [Unsuitable], 2 [Little suitable], 3 [Reasonably suitable], 4 [Suitable], and 5 [Very suitable]). Justify your answer.
3. What rating would you give to each of the distribution channel strategies listed below, considering their suitability in the context of the Angolan market to enable the short-, medium-, and long-term efficient and effective supply of cement to the Angolan market? (On a scale of 1 [Unsuitable], 2 [Little suitable], 3 [Reasonably suitable], 4 [Suitable], and 5 [Very suitable]). Justify your answer.
4. Regarding the following aspects, how would you classify the Angolan cement industry regarding its capacity to allocate cement to communities in order to generate added value, especially for customers located in the interior of Angola? (On a scale of 1 [Unsuitable], 2 [Little suitable], 3 [Reasonable suitable], 4 [Suitable], and 5 [Very suitable]). Justify your answer.

The results of the first sequence of questions were assessed as follows:

- The results "Unsuitable" have been multiplied by a factor (−2).
- The results "Little suitable" have been multiplied by a factor (−1).
- The results "Reasonably suitable" were multiplied by a factor (0).
- The "Suitable" results were multiplied by a factor (1).
- The "Very suitable" results were multiplied by a factor (2).

The 5-point Likert Type Scale was used to obtain the results described in the Tables (Sect. 4). The values were multiplied and then summed, giving rise to the results in the last column of the Tables. According to [55], despite decades of research, the optimal number of response categories for the Likert rating scale is still undetermined. According to [56], the 5-point scale is preferred because it provides fewer model mismatches than the 4-point format.

In fact, the multiplication factor used increases the numerical value of the results, in terms of significance, it influences the result, clearly showing the respondents' ideas. At the same time, it cancels out the respondents' answers in the center of the scale when they are multiplied by zero.

4 Results and Discussion

In the first sequence of questions using the Delphi technique, respondents were invited to indicate and justify their choices as to the marketing channels and distribution strategies that they considered to best suit to the Angolan cement market. Of the 22 selected entities, 11 responded to the questionnaire, which corresponds to 50% of the total sample. Concerning the question of how they would classify the marketing channels in terms of their suitability in the context of the Angolan market, the results are shown in Table 1.

The respondents' choices reveal that overall, the direct and multiple channels are considered the ones that best suit the Angolan market context, compared to the indirect channel, although the results are distributed along the scale, showing that there is no consensus among respondents.

Given the nature of the problem of cement distribution in the Angolan market and the importance of distribution centers, the following question was introduced in the script: (1) how would you rate the establishment of distribution centers in the Angolan market? The results are shown in Fig. 1. In general, a higher percentage of respondents (55%) consider that distribution centers are very suitable or suitable for the ACI and 27% consider that they are not at all suitable or not very suitable. In this case, there is also some dispersion in the respondents' answers, although less expressive than in marketing channel strategies.

The results of the question of what rank you would give to the appropriateness of distribution channel strategies in the context of the Angolan cement market are shown in Table 2.

The answers reveal that in general, respondents consider that the diversification strategy, geographic expansion of the market, and focus suit better the Angolan market context, unlike the differentiation and lean strategies.

Concerning the question, how would you rate ACI's ability to allocate cement to communities to generate added value, especially for the customer located in the interior of Angola, the respondents' answers are shown in Table 3.

Table 1 Marketing channels results

	Unsuitable	Little suitable	Reasonably suitable	Suitable	Very suitable	Results
Direct channel	3	1	1	3	3	2
Indirect channel	3	3	4	1	0	-8
Multiple Channel	2	1	3	3	2	2

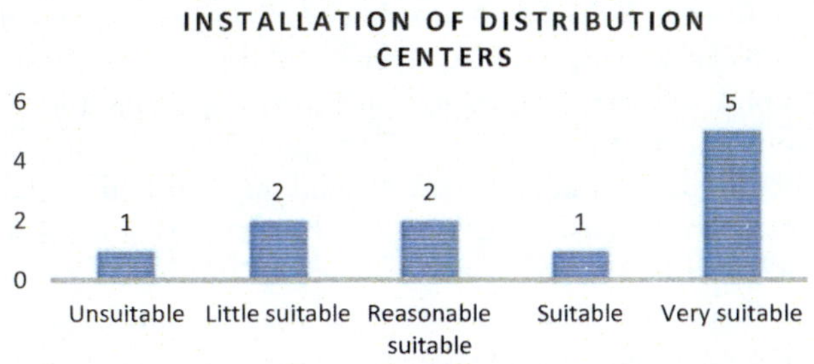

Fig. 1 Installation of distribution centers

Table 2 Distribution channel strategies

	Unsuitable	Little suitable	Reasonably suitable	Suitable	Very suitable	Results
Costs leadership	2	2	1	3	3	3
Differentiation	2	3	1	2	3	1
Focus	1	1	4	1	4	6
Diversification	2	–	1	3	5	9
Lean	2	2	2	3	2	1
Agile	1	3	1	2	4	5
Leagile	1	3	1	3	3	4
Market expansion	1	1	2	2	5	9

Table 3 Voting results

	Unsuitable	Little suitable	Reasonably suitable	Suitable	Very suitable	Results
Capacity	4	2	1	1	3	−3
Performance	3	3	2	2	1	−5
Effectiveness	5	2	1	2	1	−8
Efficiency	5	1	2	2	1	−7

The results obtained reveal that more than 50% of the respondents consider that for each factor under evaluation, the ACI is not at all empowered or is poorly empowered. The responses reveal that respondents generally consider ACI to have a weak capacity, low performance, low efficiency, and low effectiveness in allocating cement to communities to generate added value, particularly for the customers located in the interior of the country. Three respondents point out that the weak performance in the Angolan cement market is due to the poor quality of transport infrastructures that makes the distribution of cement impossible. On the other hand, some respondents pointed out that the weak capacity to allocate cement to communities in the interior of Angola, the low performance, poor efficiency and effectiveness in cement distribution in the Angolan market is partly due to weak implementation of distribution channel strategies.

Based on the answers derived from the first sequence, we believe that the work is progressing well. Thus, to achieve the objective of this research, it is our intent to apply soon a second sequence of questions aimed at improving the consistency of the answers obtained in the first sequence. The main goal of the study (sequences of surveys) is to identify the marketing channels and the distribution channels strategies that the respondents believe to be the most appropriate for the Angolan market context. In this way, the first sequence has provided an elucidation on which marketing channels and distribution channels strategies the respondents judge to be most appropriate for the Angolan cement market context.

However, to complete the study, at least a second sequence must be applied, which aims to question the respondents to evaluate if, after knowing the global results, they still maintain the same opinion about their previous answers. According to the Delphi technique, the answers of the first sequence are used to construct the following sequences of questions. Therefore, some points will be used derived from the first sequence to compose the questionnaire of the second sequence. Basically, the questions will be formulated to determine initial positions on the questions in which all the experts who participated in the study agree and the unimportant ones to be discarded.

5 Conclusion

The distribution strategies and marketing channels identified in this study aim to enable the short-, medium – and long-term distribution of cement in the Angolan territory efficiently and effectively, which will help improve the lives of the population and the financial health of companies. Considering the results achieved so far, there is not a single channel that is considered the most efficient and effective. However, market expansion, diversification, and the installation of distribution centers are identified as strategies that best suit the context of Angolan cement distribution. Further, the respondents consider that the distribution process is not adequately developed regarding its capacity, performance, effectiveness, and efficiency. The golden rule used in choosing the best marketing channel and the best distribution strategy for a given cement company seems to be to know the market environment (customers, competitors, products, win loyal intermediaries, and empower human capital).

The results of the first set of questions provide valuable insights into the distribution of cement in the Angolan market. We believe that this study will add value to the decisionmakers regarding the distribution of cement in the Angolan market, as well as in developing countries that present a similar context. As for its value, although it needs further generalization, this work is one of the few to identify distribution strategies and marketing channels appropriate to the cement market context.

References

1. Wolf, C. (2017). Industrialization in times of China: domestic-market formation in Angola. *Publ. by Oxford Univ. Press behalf R.* African Soc. All rights Reserv, *4*, 435–461.
2. de Angola, P. J.. Lack of market lowers cement sales capacity across the country. https://www.jornaldeangola.ao/ao/noticias/falta-de-mercado-baixa-capacidade-de-venda-de-cimento-em-todo-o-pais/. Last accessed 6 Mar 2022.
3. Chen, M. S., & Lai, G. C. (2010). Distribution systems, loyalty and performance. *International Journal of Retail & Distribution Management, 38*(9), 698–718. https://doi.org/10.1108/09590551011062448
4. Wang, Y.-H., & Liao, i. C. (2011). Design of SOA integration for 3C distribution channel. *WSEAS Transactions on Information Science and Applications, 8*(1), 39–53.
5. Jurše, M., & Jager, J. (2017). Marketing channel strategy management in international markets. *International Journal of Innovating and Learning, 21*(2), 127–148.
6. Agudelo, I. (2009), *Supply chain management in the cement industry. Tesis submitted to the engineering systems division in partial fulfillment of the requirements for the degree of Master of Engineering in Logistics at* the Massachusetts Institute of Technology.
7. Imam, Y. B. (2014). Channels of distribution of cowpea in Borno state of Nigeria. *Developing Country Studies, 4*(17), 55–63.
8. Chen, M. S., & Chang, P. L. (2010). Distribution channel strategy and efficiency performance of the life insurance industry in Taiwan. *Journal of Financial Services Marketing, 15*(1), 62–75.
9. Segetlija, Z., Mesarić, J. & Dujak, D. (2012). Importance of distribution channels-marketing channels–for national economy, *Univ. J.J.Strossmayer, Fac. Econ. Osijek, Croat.*, pp. 785–809.
10. Szopa, P., & Pękała, W. (2012). Distribution channels and their roles. *Polish Journal of Management Studies, 6*(1), 143–150.
11. Casali, G. L., Perano, M., Presenza, A., & Abbate, T. (2017). Does innovation propensity influence wineries' distribution channel decisions? *International Journal of Wine Business Research, 30*(4), 446–462. https://doi.org/10.1108/IJWBR-06-2017-0037.
12. Zhang, G., Dai, G., Sun, H., Zhang, G., & Yang, Z. (2020). Equilibrium in supply chain network with competition and service level between channels considering consumers' channel preferences. *Journal of Retailing and Consumer Services, 57*(4), 102–199.
13. Katsikea, E. S., Theodosiou, M., Morgan, R. E., & Papavassiliou, N. (2005). Export market expansion strategies of direct-selling small and medium-sized firms: Implications for export sales management activities. *Journal of International Marketing, 13*(2), 57–92.
14. Thornton, J., & White, L. (2001). Customer orientations and usage of financial distribution channels. *Journal of Services Marketing, 15*(3), 168–185.
15. Wright, A. (2002). Technology as an enabler of the global branding of retail financal services. *Journal of International Marketing, 10*(2), 83–98.
16. King, R. C., Sen, R., & Xia, M. (2004). Impact of web-based e-commerce on channel strategy in retailing. *International Journal of Electronic Commerce, 4415*(3), 103–130.
17. Adesoga, A., & James, A. (2019). Cannel strategy and marketing performance of selected consumer goods firms in Lagos State, Nigeria. *Academy of Marketing Studies Journal, 23*(1), 1–18.
18. Aldin, N., & Stahre, F. (2003). Electronic commerce, marketing channels and logistics platforms–a wholesaler perspective. *European Journal of Operational Research, 144*, 270–279.
19. Yu, D. Z., Cheong, T., & Sun, D. (2017). Impact of supply chain power and drop-shipping on a manufacturer's optimal distribution channel strategy. *European Journal of Operational Research, 259*(2), 554–563. https://doi.org/10.1016/j.ejor.2016.11.025
20. Bang, V. V., & Joshi, S. L. (2008). Conceptualization of market expansion strategies in developing economies. *Academy of Marketing Science Review, 12*(4).
21. Blankson, C., & Crawford, J. C. (2012). Impact of positioning strategies on service firm performance. *Journal of Business Research, 65*(3), 311–316.
22. Bang, V. V., & Joshi, S. L. (2010). Market expansion strategy – Performance relationship. *Journal of Strategic Marketing, 18*(1), 57–75. https://doi.org/10.1080/09652540903511316

23. Hooley, G., Greenley, G., Fahy, J., Cadogan, J., & Greenley, G. (2001). Market-focused resources, competitive positioning and firm performance. *Journal of Marketing Management, 17*(5–6), 503–520. https://doi.org/10.1362/026725701323366908
24. Jovanovic, B., & Gilbert, R. J. (1993). The diversification of production. *Brookings Papers on Economic Activity. Microeconomics, 1993*(1), 197–247.
25. Antoncic, B. (2006). Impacts of diversification and corporate entrepreneurship strategy making on growth and profitability: A normative model. *Journal of Enterprising Culture, 14*(1), 49–63.
26. Narasimhan, R., & Kim, S. W. (2002). Effect of supply chain integration on the relationship between diversification and performance: Evidence from Japanese and Korean firms. *Journal of Operations Management, 20*(2002), 303–323.
27. Hitt, M., Ireland, D., Hoskisson, R., & Miller, T. (1994). A mid-range theory of the interactive effects of international and product diversification on innovation and per. *Journal of Management, 20*(2), 297–326. https://doi.org/10.1016/0149-2063(94)90018-3
28. Hitt, M. A., & Hoskisson, R. E. (1997). International diversification: Effects on innoation and firm performance in product-diversified firms. *Academy of Management Journal, 40*(4), 767–798.
29. Tustin, D. H., & Strydom, J. W. The potential impact of formal retail chains' expansion strategies on retail township development in South Africa. *Southern African Business Review, 10*(3).
30. Matsusaka, J. G. (2001). Corporate diversification, value maximization, and organizational capabilities. *Journal of Business, 74*(3), 409–431.
31. Dhir, S., & Dhir, S. (2015). Diversification: Literature review and issues. *Publ. online Wiley Online Library, 24*(11), 569–588. https://doi.org/10.1002/jsc.2042
32. Kim, B. (2013). Competitive priorities and supply chain strategy in the fashion industry. *Qualitative Market Research: An International Journal, 16*(2), 214–242. https://doi.org/10.1108/13522751311317602
33. Qi, Y., Zhao, X., & Sheu, C. (2011). The impact of competitive strategy and supply chain strategy on business performance: The role of environmental incertainty. *Decision Sciences Journal, 42*(2), 371–389.
34. Banker, F. D., Mashruwala, R., & Tripathy, A. (2014). Does a differentiation strategy lead to more sustainable financial performance than a cost leadership strategy? *Management Decision, 52*(5), 872–896. https://doi.org/10.1108/MD-05-2013-0282
35. Kim, E., Nam, D., & Stimpert, J. L. (2004). The applicability of Porter's generic strategies in the digital age: Assumptions, conjectures, and suggestions. *Journal of Management, 30*(5), 569–589. https://doi.org/10.1016/j.jm.2003.12.001
36. Kim, S. W., & Park, S. (2008). Development of a three-echelon SC model to optimize coordination costs. *European Journal of Operational Research, 184*(2008), 1044–1061.
37. Kovács, G., & Kot, S. (2017). Economic and social effects of ovel supplyy chain concepts and virtual enterprise. *Journal of International Students, 10*(1), 237–254.
38. Van Donk, D. P., & Van Der Vaart, T. (2006). Responsiveness through buyer-focused cells: Exploring a new supply strategy. *International Journal of Operations & Production Management, 27*(12), 1362–1379.
39. Goldsby, T. J., Griffis, S. E., & Roath, A. S. (2006). Modeling lean, agile and leagile supply chain strategies. *Journal of Business Logistics, 27*(1), 57–80.
40. Christopher, M., Peck, H., & Towill, D. (2006). A taxonomy for selecting global supply chain strategies. *International Journal of Logistics Management, 17*(2), 277–287.
41. Kovács, G., & Kot, S. (2016). New logistics and production treds as the effect of global economy changes. *Polish Journal of Management Studies, 14*(2).
42. Ben Naylor, J., Naim, M. M., & Berry, D. (1999). Leagility: Integrating the lean and agile manufacturing paradigms in the total supply chain. *International Journal of Production Economics, 62*, 107–118.
43. Meskell, P., Murphy, K., & Casey, D. (2014). Insights into the use and complexities of the Policy Delphi technique. *Nurse Researcher, 21*(3), 32–39.

44. Giannarou, L., & Zervas, E. (2014). Using Delphi technique to build consensus in practice. *International Journal of Business Science & Applied Management (IJBSAM), 9*(2).
45. Makkonen, M., Hujala, T., & Uusivuori, J. (2016). Policy experts' propensity to change their opinion along Delphi rounds. *Technological Forecasting and Social Change, 109*, 61–68.
46. Hsu, C., & Sandford, B. A. (2007). The Delphi Technique: making sense of Consensus. *Practical Assessment, Research & Evaluation, 12*.
47. Rowe, G., & Wright, G. (2001). *Expert option in forecasting: The role of the Delphi technique*, Inst. Food Res. Norwich Res. Park. UK Georg.
48. Keeney, S., Hasson, F., & Mckenna, H. (2006). Consulting the oracle: Ten lessons from using the Delphi technique in nursing research. *Journal of Advanced Nursing, 53*(2), 205–212.
49. Chalmers, J., & Armour, M. (2019). *The Delphi Technique* (pp. 715–735). Springer.
50. Murry, J. W., & Hammons, J. O. (1995). Delphi: A versatile methodology for conducting qualitative research. *The Review of Higher Education, 18*(4), 423–436.
51. Duffield, C. (1993). The Delphi technique: A comparison of results obtained using two expert panels. *International Journal of Nursing Studies, 30*(3), 227–237.
52. Paré, G., Cameron, A.-F., Poba-Nzaou, P., & Templier, M. (2013). A systematic assessment of rigor in information systems ranking-type Delphi studies. *Information Management, 50*, 207–217. https://doi.org/10.1016/j.im.2013.03.003
53. Beech, B. (1999). Go the extra mile–use the Delphi technique. *Journal of Nursing Management, 7*, 281–288.
54. Beretta, R. (1996). A critical review of the Delphi technique. *Nurse Researcher, 3*(4), 79–89.
55. Preston, C. C., & Colman, A. M. (2000). Optimal number of response categories in rating scales: Reliability, validity, discriminating power, and respondent preferences. *Acta Psychologica, 104*, 1–15.
56. Adelson, J. L., & Mccoach, D. B. (2010). Measuring the mathematical attitudes of elementary students: The effects of a 4-point or 5-point Likert-type scale. *Educational and Psychological Measurement, 70*(5), 796–807. https://doi.org/10.1177/0013164410366694

How Can Customer Experience Improve Retail Operations Sustainability?

Rita Silva e Sousa, Tiago Pinho, and David Simões

Abstract Consumer needs drive supply chains, so they are arguably the main actors in the process. Nonetheless, consumers are unaware of their ability to contribute to stock management and the sustainability of retail operations, from the reduction of stockouts and waste to the minimization of energy-environmental impacts, through the centralization of stock in distribution and consolidation of last-mile delivery in pooling systems. For this to happen, companies must provide channels that allow consumers to actively participate in the process and negotiate delivery times and prices through sustainable purchase options, through a crowdsourcing strategy in phygital stores. This paper explores two alternative strategies, maintaining or changing the current physical retail business model, based on the increase in online commerce and the use of mobile devices and applications in the purchase process. The first is applied in physical stores in a gaming context through a consumer-facing augmented reality mobile application that rewards users for identifying stockouts and informing the need to replace products in the shelves. The second involves the transformation of physical stores into a showroom format, where desired products are read through a QR code or using artificial intelligence through a mobile application, in which virtual shopping carts are created and deliveries can be fulfilled via home distribution centers, collection points, or drive-in.

Keywords Customer experience · Sustainability · Inventory management · Retail operations · Logistics and supply chain · Innovation

R. Silva e Sousa
Instituto Politécnico de Setúbal, Campus do IPS, Setúbal, Portugal
e-mail: 200350002@estudantes.ips.pt

T. Pinho (✉) · D. Simões
Instituto Politécnico de Setúbal, Campus do IPS, Setúbal, Portugal

CICE–Centro de Investigação em Ciências Empresariais, Campus do IPS, Setúbal, Portugal
e-mail: tiago.pinho@esce.ips.pt; david.simoes@esce.ips.pt

J. C. de Oliveira Matias et al. (eds.), *Quality Innovation and Sustainability*, Springer Proceedings in Business and Economics,
https://doi.org/10.1007/978-3-031-12914-8_26

1 Introduction

Despite the growth of e-commerce, according to TimeTrade [1], consumers continue to prefer to buy products in physical stores, as they value product experimentation and immediate purchase. Digital technologies have changed the nature of business-to-consumer relationships and the role of the consumer has increased in terms of participation and interactivity [2], as mobile devices and connectivity became part of marketing strategies and in-store experiences [3]. Currently, more than anticipated needs or simple experiences of consuming products or services, consumers value the interaction with the product, a more active participation [4] with new and unique experiences [5].

Considering that the result of these interactions improves customer shopping experience, which increases the probability of having a positive impact on sales, retail must adapt its strategy and business model in response to consumer behavior, which implies the hybridization of the physical spaces of the stores through the integration of intelligent technologies in the physical channel.

Bearing in mind that stockouts result in a bad shopping experience and have a negative impact on the value of retailer sales, the causes have been studied. Literature refers not only to the difficulty in predicting real needs and converting them into physical flows on-time but also to the inefficiency of operations, both in terms of inventory control and management and in-store operational management. For this reason, this paper explores the extent to which new experiences can allow direct involvement between the consumer and the store, providing customers with the ability to play an active role in the management of stocks and transport, in order to minimize disruptions and the underlying environmental footprint, ensuring the sustainability of retail operations.

2 Literature Review

Stock and Boyer [6] define supply chain management as the management of a network of relationships within the company and between interdependent organizations and business units consisting of material suppliers, purchasing processes, production sites, logistics, marketing, and related systems that facilitate the forward and backward flows of materials, services, financials, and information from the original producer to the final consumer with the benefits of adding value, maximizing profits through efficiencies and achieving customer satisfaction.

According to Rousseau [7], we can define retail as the activity of selling goods and services to final consumers, and retailer as an economic agent that sells products and services to final consumers. Retailers purchase goods and services using different sources of supply and deliver them to consumers in small quantities appropriate to their purchasing power and to meet the needs of the final consumer. To ensure the efficiency of operations, supply chains need to configure the intended strategy in the long term and plan in aggregate form in the medium term to guarantee the success of the activities necessary for the day-to-day of retail stores. Figure 1 presents a framework that aligns supply chain planning with demand and the respective physical and informational flows. According to Grewal et al. [3], to achieve success, retailers must allow their business models to evolve and create experiences and offers through all channels.

According to Hoogveld and Koster [9]; the evolution of omnichannel retail is a major challenge as retailers need to change their business model to integrate their key resources across all channels, through an analysis of customer behaviors both digitally and at physical touch points, studying their decision-making process. According to Hübner et al. [10] and Marchet et al. [11], backend fulfillment, last mile distribution, and returns processing management are some of the key processes of omnichannel logistics. According to Lorente et al. [12], online orders are prepared for shipping in fulfillment centers such as: distribution centers/warehouses which encompass physical store supply processes simultaneously with orders from online; fulfillment centers/dark stores that act as warehouses allocated to respond exclusively to online orders; and physical stores, which are additionally integrated with the online channel and have the capacity to respond to requests from this channel.

In retail, the consumer experience indicator can be considered as Customer Shopping Experience (CSE), defined according to Choi et al. [13] as a summary of the interaction between the customer and the retail company that begins before the customer enters the store and ends long after leaving the store. According to Longo et al. [14], the improvement of CSE can be a benefit and a decisive factor with the objective of regaining market share in the physical spaces of retail stores. In line with this perspective, Kale and Ulusoy [15] state that retail professionals have adopted a vision oriented towards building experiences where retail environments are transformed into theaters of interactions with the consumer to provide entertainment experiences. According to the Forrester report [16]; marketing and technology play a driving role in CSE growth, with 72% of companies saying that investing in

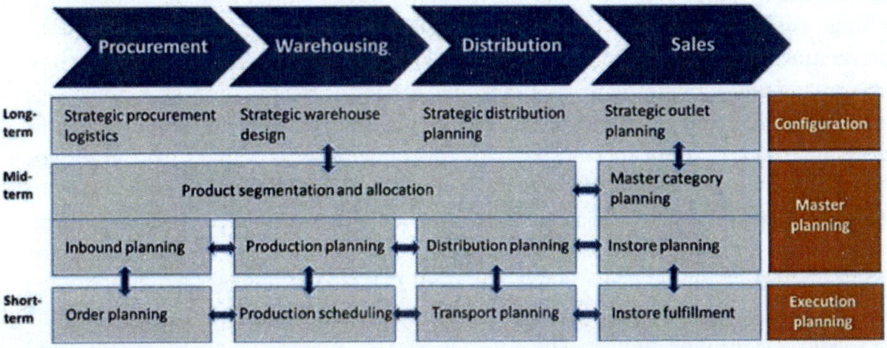

Fig. 1 Retail demand and supply chain planning framework. (Source: Hübner et al. [8])

the customer experience is their priority and 63% of business professional marketers valuing investments in technology to achieve this same goal.

A report by Conversant [17] highlights that 58% of consumers start their shopping journey online and complete their purchase offline (webrooming), while 46% start their journey in-store and end online (showrooming). According to Marktest E-Commerce Barometer [18], in Portugal, the number of respondents who admit to searching online and then buying in a physical store dropped from 83.7% to 81.4%, while the number of respondents who say viewing products in physical stores before buying online increased from 56.1% to 65.7%.

In this context, mobile commerce (m-commerce) practices, showrooming, phygital, and smart stores are mentioned in the literature as the main trends in the scope of omnichannel retail, with a special focus on the physical space of the store, with the objective of differentiating the offer and the consumer experience.

The phygital experience is defined, according to Belghiti et al. [19] and Verhoef et al. [20], by the hybridization process of the physical (point of sale, products) and the digital (touch screens, connected mirrors, NFC, QR codes) at the same time and in the same place. From a marketing point of view, according to Moravcikova & Kliestikova [21], the phygital concept – agglutination of physical and digital – brings together conventional sales promotion tactics with digital brand activation strategies in an integrated solution, in order to respond effectively to the consumer's alternation between the offline (store) and online worlds (search for information about products, prices, and respective reviews), offering consumers the convenience of online channels together with the authentic experience from the physical store. According to Goethals [22], there are four different types of phygital experiences: informational, transactional, entertainment, and support. Informational experiences serve to provide the client with additional information, such as object recognition and voice assistants; transactional experiences stimulate the customer throughout the purchase journey and is the most used in retail, examples can be self-service at checkout and coupons through beacons; entertainment experiences aim to entertain and engage the customer in the store and are mainly used during campaigns, usually in the form of interactive games using various types of technology; and finally, support, in which the objective is to add value to the customer, helping them during their purchase journey in the store, for example, through augmented reality – wayfinding. Different types of phygital experiences can be combined, for example, entertainment (fun) and informational (educational) through gamification systems. Zichermann and Cunningham [23] consider the term gamification as a process of "game-thinking" and game mechanics that aim to involve users and solve problems. Subsequently, according to Huotari [24], this concept was defined from a service perspective as a process of strengthening a service with a view to enhancing the "gameful experiences" to add value to the service user experience. Today, consumers are already committed to phygital experiences. According to research conducted by International Council of Shopping Centers [25], mobile devices are already used in stores during the purchase journey. Customers aged between 20 and 37 are the ones who use mobile devices the most in stores. According to Emarketer [26] regarding US digital consumers, the main action of consumers when using in-store retailer mobile applications is to look for promotions and personalized offers.

Technological development related to the Internet of Things (IoT) can create dynamic environments – intelligent spaces – capable of responding to human behavior and needs, in which objects and people interact. According to Kotler et al. [4], sensor technologies within the IoT launch new opportunities for marketers and retail companies that can take advantage of their potential to monitor consumer behavior, communicate with consumers through highly targeted and personalized offers based on collected data, and promote a convenient, interactive, and technological consumer experience in the physical space of the stores, through machine-to-machine communication. According to Sahni et al. [27], different software and applications are being developed within the scope of IoT and computer science not only to capture data related to consumer behavior within the store but also to respond contextually to consumer interactions. According to Moravcikova and Kliestikova [21], smart stores can be seen as an evolution of the phygital store, in the sense that the central objective is the same, that is, a hybrid, more dynamic and faster shopping experience. Hwangbo et al. [28] consider this type of store as one of the most recent evolutions of retail stores that deploy new technologies that create immersive and authentic experiences for the consumer.

3 Supply Chain vs. Retail Operations

In retail, distribution centers manage stocks and, in accordance with replenishment policies, send the products to the stores. Stores, in turn, receive, store, and stock shelves with products for sale. This business model can lead to excess stock and high inventory costs if the expected sales in the store do not occur, or stockouts, if shelf replenishment is not carried out or the product does not exist in the store due to delays in deliveries or inaccuracies of existing quantities. To supply stores according to expected needs, based on historical sales data, distribution centers, in addition to trying to minimize inventory costs through centralization and obtaining economies of scale in the purchasing processes, try to minimize transport costs in shipping of products to stores by maximizing transport capacity and optimizing routes. For this to happen successfully, retail companies need to integrate as much information as possible so that it is possible to meet consumer needs with the desired level of service across all channels, supported by constant trade-offs between inventory costs and transport costs to ensure the sustainability and efficiency of operations. With data analysis tools, it is now possible to integrate and relate large volumes of information from multiple sources to support the best decisions for the business based on outlined assumptions and considering existing constraints. However, on the other hand, despite technological advances also in terms of product traceability, such as the Radio Frequency IDentification (RFID), retailers do not separate, in their systems, the existing stock in the store's warehouse from the existing stock on the shelf, which causes the products to be dispersed throughout the store in more than one location. This fact brings disadvantages as, in addition to compromising the accuracy of the existing stock and consequently the needs for replenishment, it forces an increase in the time spent internally in handling the products and in controlling stock ruptures on the shelf. Last year, according to NielsenIQ [29], supermarket shelves registered an availability rate of 92.6% and the disruptions cost US retailers 82 billion dollars. When the customer perceives that the desired product is not available, a fifth of individuals postpone the purchase, while 10% go to another retailer and 16% go to online. According to Gruen et al. [30], having a direct impact on financial performance, stockouts are estimated to represent as much as 4% of sales, which is equivalent to the almost 5% average that retailers spend on logistics operations in accordance with PwC [31]. Most of these costs are caused by store operations, represented in 38% by product handling and 7% by having items in stock in the store [32], with the remaining 28% due to handling in the distribution center, 5% to inventory, and 22% to shipping [33].

4 Strategies to Reduce the Impact of Stockouts

Through a crowdsourcing strategy in phygital stores, pursuing common goals such as minimizing environmental impacts, improving the shopping experience, and avoiding stockouts and waste, consumers can and should have an active participation. This strategy, according to Nakatsu et al. [34], involves the outsourcing of problems and tasks, which have already been performed by the employees of a company, to an undefined and anonymous group of people who come together to solve these problems and tasks. Customers can participate by helping to forecast real demand, both in the current business model, through a mobile application in a gaming format that rewards interaction by identifying stockouts and triggering the need for replacement, as in the store business in showroom format, in which customers scan product QR codes (Fig. 2 – Amazon GO), or use NFC (Fig. 3 – Pingo Doce GO) and create virtual carts where online purchase value and the delivery periods vary according to what social responsible customers are willing to contribute to the sustainability of operations.

In this scenario, stock is shipped directly from the distribution center to the agreed delivery point, avoiding inventory costs and transport costs regarding the continuous shipment of products to the stores, as it allows transport partnerships to be created in the last mile that minimize environmental impacts through flexibility in the delivery time agreed by consumers. In this format, multiple purchase options can be activated in the application, with varying purchase price quotes depending on the selected delivery period, based on the consolidation of loads and shipment to the customer, and calculations subject to the expected demand and respective gaseous emissions.

Fig. 2 QR code. (Source: Flexible Packaging [35])

Fig. 3 NFC. (Source: NIT [36])

5 Main Conclusions and Future Work

Given the growth of e-commerce driven by the pandemic, the growing use of mobile devices in retail stores and the use of technology as a source of information and interaction for personalized offers, companies were forced to strategically reconfigure their logistics and marketing operations to secure profit margins, including the transformation of physical stores. In this context, considering the issue of disruptions and the impact it generates on customers and retail companies, this paper addressed the major retail trends in terms of customer interaction and satisfaction in physical stores and the extent to which these can help to minimize stockouts at the point of sale and ensure the sustainability of operations through personalized offers and purchase options based on environmental impacts. Noting that QR codes and NFC technology are already part of the retail industry and that the objective is to improve customer shopping experience, this paper suggests the use of mobile applications featuring virtual carts that could be operated by pointing the cameras in customers' mobile devices or by reading product QR codes in a showroom format. In a gaming scenario, an application could ask customers to carry out inventory-related "missions" inside the store, rewarding them with loyalty points and/or product offers, while allowing them to contribute to the sustainability of operations. In this way, customers could improve their shopping experience and that of the next customer, as they could help to avoid stockouts by intervening in the process or by practicing showrooming, improving the efficiency of shelf replenishment and ongoing supply operations to stores. Given the importance of improving customer shopping experience, future work toward the development of the alternatives under study includes analyzing the environmental impact of existing operations and studying the financial and environmental impact of incorporating the new business model in showroom format with direct deliveries from distribution centers.

References

1. TimeTrade. (2017). *The state of retail report 2017*. www.timetrade.com/wpcontent/uploads/2017/02/TimeTrade_State_of_Retail_2017_Report.pdf
2. Vrontis, D., Thrassou, A., & Amirkhanpour, M. (2017). B2C smart retailing: A consumer focused value-based analysis of interactions and synergies. *Technological Forecasting and Social Change, 124*, 271–282. https://doi.org/10.1016/j.techfore.2016.10.064
3. Grewal, D., Roggeveen, A. L., Runyan, R. C., Nordfält, J., & Vazquez Lira, M. E. (2017). Retailing in today's world: Multiple channels and other strategic decisions affecting firm performance. *Journal of Retailing and Consumer Services, 34*(1), 261–263. https://doi.org/10.1016/j.jretconser.2016.01.007
4. Kotler, P., Kartajaya, H., & Setiawan, I. (2017). *Marketing 4.0–Mudança do tradicional para o digital. Actual Editora.*
5. Papagiannidis, S., & Bourlakis, M. (2010). Staging the new retail drama: At a metaverse near you! *Journal For Virtual Worlds Research, 2*(5). https://doi.org/10.4101/jvwr.v2i5.808
6. Stock, J. R., & Boyer, S. L. (2009). Developing a consensus definition of supply chain management: A qualitative study. *International Journal of Physical Distribution & Logistics Management, 39*(8), 690–711.
7. Rousseau, J. (2020). *Manual de Distribuição*. 3ª Ed. Princípia.
8. Hübner, A. H., Kuhn, H., & Sternbeck, M. G. (2013). Demand and supply chain planning in grocery retail: An operations planning framework. *International Journal of Retail & Distribution Management, 41*(7), 512–530. https://doi.org/10.1108/IJRDM-05-2013-0104
9. Hoogveld, M., & Koster, J. (2016). Implementing omnichannel strategies. The success factor of agile processes. *Advances in Management & Applied Economics, 6*(2), 25–38. https://doi.org/10.1016/j.jretconser.2019.02.024
10. Hübner, A., Kuhn, H., & Wollenburg, J. (2016). Last mile fulfilment and distribution in omnichannel grocery retailing: A strategic planning framework. *International Journal of Retail & Distribution Management, 44*(3), 228–247. https://doi.org/10.1108/IJRDM-11-2014-015
11. Marchet, G., Melacini, M., Perotti, S., Rasini, M., & Tappia, E. (2018). Business logistics models in omni-channel: A classification framework and empirical analysis. *International Journal of Physical Distribution & Logistics Management, 48*(4), 439–464. https://doi.org/10.1108/IJPDLM-09-2016-0273

12. Lorente, J., Gabor, A., & Cueto, E. (2020). Omnichannel logistics network design with integrated customer preference for deliveries and returns. *Computers & Industrial Engineering*, 144. https://doi.org/10.1016/j.cie.2020.106433
13. Choi, S. H., Yang, Y. X., Yang, B., & Cheung, H. H. (2015). Item-level RFID for enhancement of customer shopping experience in apparel retail. *Computers in Industry, 71*(2015), 10–23. https://doi.org/10.1016/j.compind.2015.03.003
14. Longo, S., Kovacs, E., Franke, J., & Martin, M. (2013). *Enriching shopping experiences with pervasive displays and smart things [Conference]*. UbiComp '13 Adjunct: Proceedings of the 2013 ACM conference on Pervasive and ubiquitous computing adjunct publication. Zurich. www.ubicomp.org/ubicomp2013/adjunct/adjunct/p991.pdf~
15. Kale, G., & Ulusoy, E. (2014). Experience marketing at retail environments. In O. Cengel & Y. Oztek (Eds.), *Handbook of research on the impact of culture and society on the entertainment industry* (pp. 314–331). IGI Global.
16. Forrester. (2016). *72% of businesses name improving customer experience their top priority*. https://bit.ly/2lP3e9
17. Frasquet, M., & Romero, M. (2021). Competitive (versus loyal) showrooming: An application of the push-pull-mooring framework. *Journal of Retailing and Consumer Services, 60*. https://doi.org/10.1016/j.jretconser.2021.102639
18. Marktest. (2022). www.marktest.com/wap/a/grp/p~116.aspx
19. Belghiti, S., Ochs, A., Lemoine, J., & Badot, O. (2017). The phygital shopping experience: An attempt at conceptualization and empirical investigation. In *Academy of marketing science world marketing congress* (pp. 61–74). Springer. https://doi.org/10.1007/978-3-319-68750-6_18
20. Verhoef, P., Kannan, P., & Inman, J. (2015). From multi-channel retailing to omni-channel retailing: Introduction to the special issue on multi-channel retailing. *Journal of Retailing, 91*(2), 174–181. https://doi.org/10.1016/j.jretai.2015.02.005
21. Moravcikova, D., & Kliestikova, J. (2017). Brand building with using phygital marketing communication. *Journal of Economics, Business and Management, 5*(3), 148–153. https://doi.org/10.18178/joebm.2017.5.3.503
22. Van Tichelen, B. (2019). *The role and opportunities of Phygital in the digital omni-channel strategy*. [Dissertação de Mestrado, Louvain School of Management, Université catholique de Louvain]. Coeurderoy, Régis. http://hdl.handle.net/2078.1/thesis:21074
23. Zichermann, G., & Cunningham, C. (2011). *Gamification by design: Implementing game mechanics in web and mobile apps*. O'Reilly Media.
24. Huotari, K., & Hamari, J. (2012). *Defining gamification: A service marketing perspective [Conference]*. MindTrek '12 proceeding of the 16th international academic mind trek conference, Tampere. www.academia.edu/27414242/Defining_gamification
25. ICSC (2018). *Industry insights: Gen Z shopping behaviors support retail real estate*. https://www.icsc.com/uploads/t07-subpage/Gen_Z_2018_Consumer_Series.pdf
26. eMarketer. (2019). *Two-thirds of shoppers check phones in-store for product information, Skipping Store Associates*. www.emarketer.com/content/two-thirds-of-internet-users-check-phones-in-store-for-product-information-skipping-store-associates
27. Sahni, Y., Cao, J., & Shen, J. (2018). Challenges and opportunities in designing smart spaces. In *Internet of things* (pp. 131–152). Springer. https://doi.org/10.1007/978-981-10-5861-5_6
28. Hwangbo, H., Kim, Y. S., & Cha, K. J. (2017). Use of the smart store for persuasive marketing and immersive customer experiences: A case study of korean apparel enterprise. *Mobile Information Systems*, 1–17. https://doi.org/10.1155/2017/4738340
29. NielsenIQ. (2022). *NielsenIQ*. https://nielseniq.com/global/en/insights/
30. Gruen, T., Corsten, D., & Bharadwaj, S. (2002). *Retail out of stocks: A worldwide examination of extent, causes and consumer responses*. https://docplayer.net/168570-Retail-out-of-stocksa-worldwide-examination-of-extent-causes-and-consumer-responses.htm
31. PricewaterhouseCoopers. (2010). *Strategic issues for retail CEO–perspectives on operating in India's retail sector*. https://www.pwc.in/assets/pdfs/rcpublications/final_strategic_issues_for_retail_ceos.pdf

32. Ehrenthal, J. (2012). *A service-dominant logic view of retail on-shelf availability* [Doctoral dissertation, University of St.Gallen]. University of St.Gallen. https://www.alexandria.unisg.ch/216724/1/dis4008.pdf
33. van Zelst, S., van Donselaar, K. H., Van Woensel, T., Broekmeulen, R. A. C. M., & Fransoo, J. C. (2009). A model for shelf stacking in grocery retail stores: Potential for efficiency improvement. *International Journal Production Economics, 121*(2), 620–632. https://doi.org/10.1016/j.ijpe.2006.06.010
34. Nakatsu, R. T., Grossman, E. B., & Iacovou, C. L. (2014). A taxonomy of crowdsourcing based on task complexity. *Journal of Information Science, 40*(6), 823–834. https://doi.org/10.1177/0165551514550140
35. Flexible Packaging. (2022). QR CODE TRENDS: Sistema de descoberta para FCCN (ebscohost.com)
36. NIT. (2019). https://www.nit.pt/comida/gourmet-e-vinhos/ja-fomos-ao-novo-pingo-doce-sem-filas-ficamos-impressionados

Optimization of Last Mile Logistics Process Combining Passenger and Freight Flows

Bruno Machado, Amaro de Sousa, and Carina Pimentel

Abstract This research addresses the operational planning of an urban logistics problem where an existing bus passenger transportation network is integrated with the urban freight process, and with a last mile delivery service, to send freight to city centers. The aim is to reduce the number of fossil combustion powered commercial vehicles traveling within city boundaries, solely for goods transportation, thus contributing to reduce negative effects of urban logistics activities, namely, pollution, noise, and traffic congestion. An integer linear programming model is proposed to support the planning of the distribution process, aiming to minimize the delivery time of the last mile operator. Results considering instances based on a real bus network of Porto city, Portugal, show the efficiency of the proposed model.

Keywords City logistics · Urban logistics · Integrated urban logistics · Mathematical model

1 Introduction

The continuous increase of population and e-commerce within urban areas enhances the importance of adopting solutions that reduce the impact of associated logistic activities, contributing to an improved and healthier quality of life within city

B. Machado (✉) · C. Pimentel
Governance, Competitiveness and Public Policies Research Unit (GOVCOPP), Department of Economics, Management, Industrial Engineering and Tourism (DEGEIT), University of Aveiro, Aveiro, Portugal
e-mail: brunocruzmachado@ua.pt

A. de Sousa
Instituto de Telecomunicações (IT), Department of Electronics, Telecommunications and Informatics (DETI), University of Aveiro, Aveiro, Portugal

J. C. de Oliveira Matias et al. (eds.), *Quality Innovation and Sustainability*, Springer Proceedings in Business and Economics,
https://doi.org/10.1007/978-3-031-12914-8_27

centers. The increasing transportation of freight directly to consumers' houses instead of conventional retail stores will result in a significant increase in logistics vehicles operating within urban areas and, consequently, an increase in the traffic congestion, noise and environmental pollution, road accidents, and the emissions of greenhouse gases [1].

Currently, the transportation of goods in city centers is carried by private companies using dedicated vehicles. The high number of small orders to be delivered into the city centers intensifies the need of dedicated vehicles to transport them, since these small orders are, typically, distributed to many consumers which are dispersed in different parts of the urban area. On the other hand, most cities have a dedicated network to support the passenger flow, such as, taxis, trams, metro, or buses. During a day, these networks face high variations in the number of passengers using the network, but the network remains operational uninterruptedly leading to a lower system's capacity utilization.

Urban logistics (UL) is being introduced as a popular term whose solutions aim to reduce the aforementioned negative impacts and to offer better and faster deliveries [2]. However, even though UL has been investigated for several years, there is no universal definition of it [2]. According to Savelsbergh and Woensel [3], all existing definitions have in common that UL is about finding efficient and effective ways to transport goods in urban areas while considering the negative effects on congestion, safety, and environment. With this perspective, the integration of passenger and freight flows is gaining attention as it ensures that goods distribution is more efficient and environmental friendly, since the same transportation network is shared. According to Neghabadi et al. [4], the integration of networks and infrastructures to combine freight and passenger transportation remains a big challenge.

This paper addresses an UL service which integrates the bus passenger and urban freight transportation flows. The main motivation for this approach is to use the spare capacity of the bus network during the periods when the number of passengers using the network is lower. The expected customers for this service are residents from e-commerce and small business with their daily needs of transportation. The main contributions of this paper are twofold: (1) the presentation of a new optimization model to support the operations planning problem under study and (2) the validation and efficiency demonstration of the proposed model through a real bus network from the city of Porto, Portugal.

The remainder of this paper is organized as follows. In Sect. 2, a brief literature review is presented. In Sect. 3, the problem description and assumptions are given. Section 4 presents the optimization model. Section 5 details the generation of instances and presents the results and the performance analysis of the model for the generated instances. Finally, the paper concludes with Sect. 6, where the main conclusions of the research are presented.

2 Literature Review

The integration of passenger and freight flows has been investigated from different perspectives and objectives. Such investigations can have an impact not only in terms of socioeconomic objectives of cities but also in terms of the impact on city living conditions, in particular on the congestion, emission, and pollution levels [5].

Li et al. [6] formulate the train service design problem on a single urban rail line with passenger and freight flows. Freight can be transported by adding dedicated trains for freight transportation or using the spare capacity of the passenger trains. The number of trains, train stops, and timetables are determined in the train service design. A mixed-integer nonlinear programming model is formulated to maximize the total profit. Linearization techniques are applied to transform the model from nonlinear to linear, enabling the model to be handled with commercial linear solvers. Additionally, two iterative heuristic algorithms are proposed to solve the problem.

Liu and Dessouky [7] consider the daily problem of scheduling and routing both passenger and freight trains jointly for complex railway network, where the objective is to minimize the tardiness of passenger trains at station stops and the delay of freight trains. The authors present a mathematical model formulation highlighting that it is hard to be solved by exact methods. Thus, the authors propose a vertical decomposition of the original problem into subproblems and then present optimization or heuristic techniques to be applied on each of the subproblems.

Li et al. [8] have studied the share-a-ride problem (SARP) considering a taxi network, where, for a given set of passengers and parcel requests, the best taxi route is determined. The authors present Mixed Integer Linear Programming (MILP) formulations for the problem and solved it exactly using the GUROBI solver. Later, the same authors have addressed the same problem, including heuristics approaches with stochastic transportation time [9, 10].

There are a few studies addressing the integration of bus passenger networks and freight flows within the cities, and therefore, for small distances. Ghilas et al. [11] suggested that part of the requests' journey can be performed by a passenger transportation service, synchronizing it with dedicated pick-up and delivery vehicles (PD). MILP formulations with the objective to minimize the total operating costs are presented and solved, using CPLEX solver. Later, Ghilas et al. [12] and [13] have addressed the same problem including delivery time windows for the PD.

The work of Masson et al. [14] is one of the most cited articles in this field. Their work considers a two-tiered problem, where the freight is transported from one hub by a single bus line and, afterwards, collected at bus stops and delivered at the customer destination using a fleet of small vehicles. The authors solved their problem using a mathematical model and a meta-heuristic named Adaptative Large Neighborhood search, focusing on the number of vehicles assigned to the last mile deliveries and the routes performed by them.

3 Problem Description

In this paper, we extend the work of Pimentel and Alvelos [15] presenting a different optimization model and applying it to a real bus network dataset of the city of Porto. The problem considers the use of a bus transportation infrastructure to transport requests into the city center. From there, the requests are delivered to the final customer through a fleet of small environmentally friendly vehicles using a last mile operator.

Consider a set of bus hubs positioned on the outskirts of the city, and a set of passenger bus services departing from each hub, performing a predetermined route through the city center, and stopping at respective bus stops of their route to offload either passengers or freight requests. Also, consider a set of requests of freight to be delivered to the city center. These requests release is triggered by different clients (typically logistics companies), indicating the demand, the final customer destination addresses of the request and the time window for delivery at the destination. Lastly, the clients also indicate which bus hubs can be used to deliver their request, so it can continue the transportation on the bus network. Since this problem is solved every day, we consider the timespan of one day. The requests characteristics vary from one day to another. Also, consider a last mile operator that performs the last mile delivery of the requests from the offloading bus stop to the destination address of the final customer. Below each of the relevant entities are characterized.

3.1 Clients

Clients are the entity that starts the process, conducting a request release to transport their freight into the city center. During a request release, the client indicates the request demand, the request destination address within the city center, and the time window to deliver the request at the destination. Also, each client must indicate the possible bus hubs he/she is available to drop his/her request, so it can continue the transportation into the city center.

3.2 Bus Hub Operations

Bus hubs are the place where the bus services start/end and where the requests are dropped off by the clients and loaded to bus services. Thus, at the bus hub, three operations occur: (1) receive requests that are dropped by the clients, (2) sort and package of requests on standard packaging containers and (3) load of standard containers to bus services, fulfilling the bus capacity of containers. We assume that the demands and the drop-off times at the bus hubs for each request are known in advance. Also, for the logistics operations, we assume a service time to package a request that can vary from one bus hub to another. Furthermore, no transshipments of requests between different bus hubs are allowed.

3.3 Bus Service

Bus services represent the transportation service performed by the buses. The schedule and route of each bus service is known in advance. The capacity of each bus to transport requests is limited and can vary from one bus to another. Thus, each bus service can transport more than one request, stop in several bus stops of its route, and unload more than one request at each bus stop, fulfilling the unloading capacity of the bus stop.

3.4 Bus Stop Operations

The packaged requests are transported by bus services from the bus hub to a bus stop. At each bus stop, a certain number of requests can be offloaded. The offloading process is limited by the bus stop offloading capacity that is determined to not significantly deteriorate the passenger's service. We assume that offloading operations take a constant time. To maintain a fast and smooth process, the reader is referred to the work of Machado et al. [16] that considers an automated process for the offloading process from the bus. The offloading operations are conducted by the last mile operator that must be at the bus stop when the bus arrives. This involves synchronizing buses and last mile operators. So, we assume that there is no storage capacity at the bus stops (the last mile operator offloads the requests directly from the bus). Additionally, no transshipments of requests between different bus services are allowed.

3.5 Last Mile Delivery

The last mile delivery is performed by the last mile operator, which uses an environmentally friendly fleet. This operator is responsible to collect the requests at the bus stops and deliver them to the destination address within the delivery time window provided by the client during the request release. We assume that the last mile operator partitions the city center in different zones and defines a maximum delivery time to deliver any request from each bus stop of the city to each zone. This time represents the maximum time needed to deliver any request from the unloading bus stop to its destination address and varies according to which zone the request's destination address is located. The last mile operator routes are managed by itself, and so are not considered in the current research. This provides flexibility to the last mile operator to integrate this operation with its daily operations.

3.6 Final Customer

Final customer represents one who receives the request at the destination address within the agreed time window.

4 Integer Linear Programming (ILP) Model

Consider a bus network with a set of hubs T (where requests can be dropped by clients), a set of bus stops S (where the requests can be offloaded by the last mile operator) and a set of bus services P (where requests can be transported from hubs to stops). Each hub $t \in T$ is characterized by a maximum logistic service time F_t (the maximum time interval required from the moment a request is dropped by a client until the request is loaded in a bus service) and has an associated set of bus services $P(t) \subset P$. Each bus service $p \in P(t)$ has an associated load capacity U_{tp}, a departing time H_{tp} from hub t and a set of bus stops $S(p) \subset S$. Finally, each bus stop $s \in S(p)$ of bus service $p \in P(t)$ has an associated arrival time H_{tps} (according to the route of the bus service) and an offload capacity U_{tps}.

Consider a set of requests K. Each request $k \in K$ has an associated demand D_k, a destination address B_k and a delivery time window $[E_k, L_k]$ defining the earliest E_k and the latest L_k delivery time instant of the request at its destination address. Moreover, the hubs at which the client of request $k \in K$ can drop it are modelled by the binary parameters A_{kt} that are equal to 1 if request $k \in K$ can be dropped in hub $t \in T$ or equal to 0, otherwise. Consider a last mile operator whose service is characterized by the maximum time T_{ks} to deliver request $k \in K$ from each bus stop $s \in S$ to the request destination address B_k. All sets and parameters are summarized in Tables 1 and 2.

The optimization problem under study considers that each request must be assigned with one bus service and one of its bus stops, among the bus services starting on one of the possible hubs of the request. Moreover, each assignment must meet the delivery time window of the request and the whole set of assignments must meet the capacity of the bus services and stops. Concerning the objective function, we focus on the last mile delivery phase of the requests aiming at minimizing the

Table 1 Sets of the ILP model

Set	Description
T	Set of bus hubs t
S	Set of bus stops s
P	Set of all bus services p
$P(t) \subset P$	Set of bus services departing from hub $t \in T$
$S(p) \subset S$	Set of bus stops s of bus service $p \in P$
K	Set of requests k

Table 2 Parameters of the ILP model

Parameter	Description
F_t	Maximum time on hub t to load any incoming request into any bus service
H_{tp}	Departing time of bus service $p \in P(t)$
H_{tps}	Arrival time of bus service $p \in P(t)$ to bus stop $s \in S(p)$
U_{tp}	Load capacity of bus service $p \in P(t)$
U_{tps}	Capacity of bus service $p \in P(t)$ to offload requests in bus stop $s \in S(p)$
D_k	Demand of request k
B_k	Destination address of request k
E_k	Earliest delivery time of request k at its destination address
L_k	Latest delivery time of request k at its destination address
A_{kt}	Binary parameter indicating if request k can be dropped in hub t
T_{ks}	Maximum delivery time of request k from bus stop s to the address of k

delivery times of the last mile operator from the bus stops to the destination address of the requests. However, as defined before, the last mile service is characterized by a maximum delivery time (from each bus stop to the request destination address) and, therefore, we do not know the expected delivery time. Using the reasonable assumption that the maximum delivery time is strongly correlated with the expected delivery time, we consider the objective function as the average of the maximum delivery times among all requests and the aim is to minimize this function.

To define the ILP model of the optimization problem under study, we first define the following additional parameters: the binary parameter h_{ktps} is equal to 1 if, for request $k \in K$, hub $t \in T$ is one of the possible hubs for the request (i.e., A_{kt} is equal to 1) and it is possible to meet with delivery time window $[E_k, L_k]$ of the request when it is dropped at hub $t \in T$, loaded in bus service $p \in P(t)$, and offloaded in stop $s \in S(p)$:

$$h_{ktps} = \begin{cases} 1, & A_{kt} = 1 \wedge E_k \leq H_{tps} + T_{ks} \leq L_k \\ 0, & \text{otherwise} \end{cases}$$

Then, we consider the following decision variables:

z_{ktps} – binary variable that is equal to 1 if request $k \in K$ is dropped in hub $t \in T$, loaded in bus service $p \in P(t)$ and offloaded in bus stop $s \in S(p)$, and equal to 0, otherwise.

Thus, a solution such that variable z_{ktps} is equal to 1 defines that request $k \in K$ must be dropped in hub $t \in T$ no later than time instant $H_{tp} - F_t$ to be loaded in bus service $p \in P(t)$ whose departing time is H_{tp} and offloaded (by the last mile operator) in bus stop $s \in S(p)$ at time instant H_{tps}. The ILP model is as follows:

$$Minimize \frac{1}{|K|} \sum_{k \in K} \sum_{t \in T} \sum_{p \in P(t)} \sum_{s \in S(p)} T_{ks} z_{\text{ktps}} \quad (1)$$

Subject to:

$$\sum_{t \in T} \sum_{p \in P(t)} \sum_{s \in S(p)} h_{\text{ktps}} z_{\text{ktps}} = 1, \quad \forall k \in K \quad (2)$$

$$\sum_{k \in K} \sum_{s \in S(p)} D_k z_{\text{ktps}} \le U_{tp}, \quad \forall t \in T, \forall \ p \in P(t) \quad (3)$$

$$\sum_{k \in K} D_k z_{\text{ktps}} \le U_{tps}, \quad \forall t \in T, \forall p \in P(t), \forall s \in S(p) \quad (4)$$

$$z_{\text{ktps}} \in \{0,1\}, \quad \forall k \in K, \forall t \in T, \forall p \in P(t), \forall \ s \in S(p) \quad (5)$$

The objective function (1) minimizes the average of the maximum delivery times of the last mile operator. Constraints (2) guarantee that each request $k \in K$ is assigned with one bus service $p \in P(t)$ in one hub $t \in T$ to be offload in one bus stop $s \in S(p)$. Recall that, due to parameters h_{ktps} in constraints (2), these constraints guarantee that the assignment is in one of the possible hubs for each request and meets its delivery time window. Constraints (3) guarantee that the requests loaded on each bus service $p \in P(t)$ of each hub $t \in T$ are within the bus capacity U_{tp}. Constraints (4) guarantee that the requests offloaded at each bus stop $s \in S(p)$ of each bus service $p \in P(t)$ of each hub $t \in T$ are within the bus stop capacity U_{tps}. Finally, constraints (5) are the variable domain constraints.

5 Computational Results

All problem instances used to test the proposed ILP model are based in a real bus network dataset in the city of Porto, Portugal, defined and provided in SOLFI project. As part of the project, a pilot is planned to run in Porto and this dataset is the current plan for the pilot, defined by the main project contractor company and based on the bus network of the bus operator in Porto. The bus network to be used in the demonstration is based on four hubs (with IDs 34, 42, 107, and 305) strategically selected among the ones with high connectivity from outside the city and seven bus stops (labelled from 1 to 7) strategically selected nearby the areas with higher concentration of potential final customers, see Fig. 1. Within a weekday, a total of 220 bus services can be used: 66 services departing from hub 34, 54 services departing from hub 42, 12 services departing from hub 107, and 86 services departing from hub 305.

Concerning the last mile operator, the city is partitioned in three destination zones (highlighted in Fig. 1 with different colors). Then, the maximum delivery

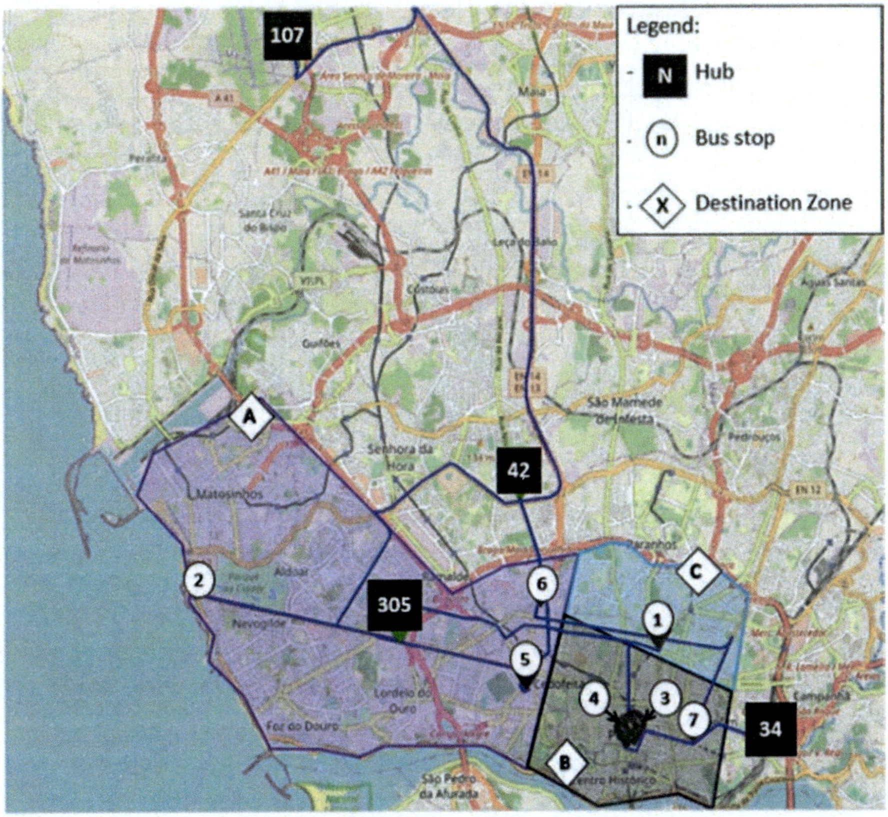

Fig. 1 Bus network and destination zones of Porto city

time of each request is computed considering the zone of the request destination address and the zone of each bus stop location using the values shown in Table 3 (for example, if one request is offloaded at bus stop 1, the last mile operator guarantees a maximum delivery of 60, 45, or 30 min if the request destination address is located at zone A, B, or C, respectively).

On this dataset, we have defined two sets of six problem instances each. The first set is composed by instances 1–6 and consider (1) all requests with a demand value of 1; (2) all bus services with a load capacity of 7; and (3) all bus stops (of all bus services) with an offload capacity of 5. The second set is composed by instances 7 to 12 and consider (1) the request demand values randomly generated between 1, 2, and 3 with equal probability; (2) all bus services with a load capacity of 14; and (3) all bus stops (of all bus services) with an offload capacity of 10. In both sets, all requests can be dropped in any hub (the worst case in terms of optimization complexity) and their destination zone is randomly generated considering all zones with equal probability.

Table 3 Maximum delivery time (in minutes) from each bus stop to each destination zone

	Destination zone		
Bus stop	A	B	C
1	60	45	30
2	30	45	60
3	60	30	45
4	60	30	45
5	30	45	60
6	30	45	60
7	60	30	45

Table 4 Computational results of all problem instances

Inst.	$\|K\|$	W	Opt.	Time (sec.)	DZ	Inst.	$\|K\|$	W	Opt.	Time (sec.)	DZ
1	100	4 hours	30	0.28	0	7	100	4 hours	30	0.30	0
2	100	2 hours	30	0.20	0	8	100	2 hours	30	0.25	0
3	200	4 hours	30	0.58	0	9	200	4 hours	30	0.63	0
4	200	2 hours	30.60	0.44	6	10	200	2 hours	30.525	0.61	6
5	300	4 hours	30.20	0.89	4	11	300	4 hours	30.15	1.09	3
6	300	2 hours	31.95	0.77	26	12	300	2 hours	31.70	0.92	24

Concerning the number of requests, we have considered three values on each set of instances: 100, 200, and 300 requests. These values represent the number of requests for a single day, which is much higher than the expected values in the SOLFI project pilot. Finally, for each of the three values, we have considered two problem instances: one instance where the delivery time windows are with a duration of 4 hours around a central time instant (randomly generated with a uniform distribution) and another instance where the previous delivery time windows are shortened to a duration of 2 hours by delaying in 1 hour the earliest delivery time and anticipating in 1 hour the latest delivering time. All problem instances were generated ensuring that the optimization problem is feasible.

The ILP model was implemented in IBM ILOG CPLEX Studio IDE 12.10 and the optimization of the 12 problem instances was run in an ASUS VivoBook, Intel Core i7 processor 1.80 GHz and 16 GB of RAM. Table 4 summarizes the obtained optimization results. In Table 4, 'Inst' indicates the instance ID, '|K|' indicates the number of requests, 'W' indicates the delivery time window duration, 'Opt.' indicates the value of the optimal solution, 'Time' indicates the CPLEX runtime (in seconds) to obtain the optimal solution, and 'DZ' indicates the number of requests in the optimal solution that are offloaded at a bus stop located in a zone different from the request destination zone.

Concerning the computational times, the main conclusion is that the problem can be efficiently solved using the proposed ILP model for all problem instances (with the largest runtime value around one second for the hardest instance). As expected, the instances with a higher number of requests are harder to be solved as the resulting ILP model has a larger number of variables. There are no significant differences in runtimes between the instances where all requests have the same demand (instances from 1 to 6) and the equivalent ones (i.e., with the same number of requests and the same delivery time window duration) where the requests have different demand values (instances from 7 to 12). Finally, instances with the delivery time window of 2 hours are (slightly) easier to solve than the equivalent ones with the delivery time window of 4 hours. This is because the former ones have a smaller number of parameters h_{ktps} equal to 1 and the variables z_{ktps} whose parameters are equal to 0 are eliminated by the solver in its pre-processing phase.

Concerning the optimal solutions, recall from Table 3 that 30 min is the smallest maximum delivery time (i.e., when the offload bus stop and the request destination address are in the same zone). All instances with 100 requests have an optimal value of 30 min, which means that the bus network capacity is enough to offload all requests in a bus stop located in the requests' destination zones (see column 'DZ' in Table 4). On the other hand, for larger number of requests, this is not the case as bus service (and bus stop) capacities become a constraint. In the instances with 200 requests, the delivery time window of 4 hours still allows all requests to be offloaded in a bus stop located in the request destination zone, but with the delivery time window of 2 hours, 6 out of 200 requests (i.e., 3% of the requests) are offloaded in zones different from the request destination zone (in both instances 4 and 10). Moreover, in instances with 300 requests, there are always some requests offloaded in zones different from the request destination zones and such number of requests is always higher for the instances with the shortest delivery time window duration.

6 Conclusions

This paper addressed an alternative urban logistics service for freight transportation within the city center, combining the passenger transportation flow, that already takes place in the city, with freight transportation flow. It focused on the problem of assigning parcel requests to bus hubs, bus routes, and bus stops to perform their transportation, while balancing the freight demands with the system capacities. An ILP model was proposed with the objective of minimizing the delivery times of the last mile operator. The fact that the bus network used by the model was based in a real dataset allowed us to get accurate results about the impact of adopting this combined logistic solution. Results have shown that adopting this combined logistic solution could be handled by the infrastructure of the city, even though it requires some adjustments to the bus fleet to guarantee the cargo safety. The results showed that the ILP model is efficient since it can deal with large instances, solving them in short computational times.

Nevertheless, we identify as limitations of the study the usage of artificial data related to the requests in our instances and, therefore, a pilot test is needed to test realistic requests datasets. A practical implication of this study is the improvement of the operational decision-making process applied to the context of urban logistics with integrated flows. Also, as theoretical implications of this study, besides the presentation of the new ILP model, we use real data related to the bus network services that can be used in future studies to compare approaches on this topic.

Acknowledgements This work was co-financed by the European Regional Development Fund (FEDER) through COMPETE 2020 (Operational Program for Competitiveness and Internationalization) through the project SOLFI-Urban logistics optimization system with integrated freight and passenger flows (POCI-01-0247-FEDER-039870). The work was also supported by the research units Governance, Competitiveness and Public Policy (UIDB/04058/2020) funded by national funds through FCT.

COMPETE 2020 PORTUGAL 2020 UNIÃO EUROPEIA Fundo Europeu de Desenvolvimento Regional

References

1. Demir, E., Huang, Y., Scholts, S., & Van Woensel, T. (2015). A selected review on the negative externalities of the freight transportation: Modeling and pricing. *Transportation Research Part E: Logistics and Transportation Review, 77*, 95–114. https://doi.org/10.1016/j.tre.2015.02.020
2. Lagorio, A., Pinto, R., & Golini, R. (2016). Research in urban logistics: A systematic literature review. *International Journal of Physical Distribution and Logistics Management, 46*(10), 908–931. https://doi.org/10.1108/IJPDLM-01-2016-0008. Emerald Group Publishing Ltd.
3. Savelsbergh, M., & Van Woensel, T. (2016). City logistics: Challenges and opportunities. *Transportation Science, 50*(2), 579–590. https://doi.org/10.1287/trsc.2016.0675
4. Dolati Neghabadi, P., Evrard Samuel, K., & Espinouse, M. L. (2019). Systematic literature review on city logistics: Overview, classification and analysis. *International Journal of Production Research, 57*(3), 865–887. https://doi.org/10.1080/00207543.2018.1489153. Taylor and Francis Ltd.
5. Behiri, W., Belmokhtar-Berraf, S., & Chu, C. (2018). Urban freight transport using passenger rail network: Scientific issues and quantitative analysis. *Transportation Research Part E: Logistics and Transportation Review, 115*, 227–245. https://doi.org/10.1016/j.tre.2018.05.002
6. Li, Z., Shalaby, A., Roorda, M. J., & Mao, B. (2021). Urban rail service design for collaborative passenger and freight transport. *Transportation Research Part E: Logistics and Transportation Review, 147*, 102205. https://doi.org/10.1016/j.tre.2020.102205
7. Liu, L., & Dessouky, M. (2017). A decomposition based hybrid heuristic algorithm for the joint passenger and freight train scheduling problem. *Computers and Operations Research, 87*, 165–182. https://doi.org/10.1016/j.cor.2017.06.009
8. Li, B., Krushinsky, D., Reijers, H. A., & Van Woensel, T. (2014). The Share-A-Ride Problem: People and parcels sharing taxis. *European Journal of Operational Research, 238*(1), 31–40. https://doi.org/10.1016/j.ejor.2014.03.003
9. Li, B., Krushinsky, D., Van Woensel, T., & Reijers, H. A. (2016). An adaptive large neighborhood search heuristic for the share-a-ride problem. *Computers and Operations Research, 66*, 170–180. https://doi.org/10.1016/j.cor.2015.08.008
10. Li, B., Krushinsky, D., Van Woensel, T., & Reijers, H. A. (2016). The share-a-ride problem with stochastic travel times and stochastic delivery locations. *Transportation Research Part C: Emerging Technologies, 67*, 95–108. https://doi.org/10.1016/j.trc.2016.01.014

11. Ghilas, V., Demir, E., Van Woensel, T., & Van Woensel, T. (2013). *Integrating passenger and freight transportation: Model formulation and insights*. (BETA Publ. Work. Pap. Vol. 441). Technische Universiteit Eindhoven.
12. Ghilas, V., Demir, E., & Van Woensel, T. (2016). An adaptive large neighborhood search heuristic for the pickup and delivery problem with time windows and scheduled lines. *Computers and Operations Research, 72*, 12–30. https://doi.org/10.1016/j.cor.2016.01.018
13. Ghilas, V., Cordeau, J. F., Demir, E., & Van Woensel, T. (2018). Branch-and-price for the pickup and delivery problem with time windows and scheduled lines. *Transportation Science, 52*(5), 1191–1210. https://doi.org/10.1287/trsc.2017.0798
14. Masson, R., Trentini, A., Lehuédé, F., Malhéné, N., Péton, O., & Tlahig, H. (2017). Optimization of a city logistics transportation system with mixed passengers and goods. *EURO Journal on Transportation and Logistics, 6*(1), 81–109. https://doi.org/10.1007/s13676-015-0085-5
15. Pimentel, C., & Alvelos, F. (2018). Integrated urban freight logistics combining passenger and freight flows-mathematical model proposal. *Transportation Research Procedia, 30*, 80–89. https://doi.org/10.1016/j.trpro.2018.09.010
16. Machado, B., Teixeira, L., Ramos, A. L., & Pimentel, C. (2021). Conceptual design of an integrated solution for urban logistics using industry 4.0 principles. *Procedia Computer Science, 180*, 807–815. https://doi.org/10.1016/j.procs.2021.01.330

Analysis of Pharmacy and Laboratory Reagents Stock Management in a Portuguese Public Hospital: A Case Study

José Sequeiros, Maria Teresa Pereira, Marisa Oliveira, and Fernanda A. Ferreira

Abstract The current situation of healthcare units is characterised by the increasing cost of providing the respective care, the consequent deterioration of the financial situation, and the complicated and time-consuming processes. Together with rising demand, they may become factors contributing to a decrease in service demand. Due to this situation, more efficient and effective logistics and supply chain management are widely recognised as one of the main areas for improvement. To provide insight on which areas to improve, several objectives were looked at in this work, including the analysis of the methods and criteria for the selection of medicines in hospital pharmacies, the definition of obstacles to the rational management of stocks, and the analysis of historical data to forecast future demand for a Portuguese public hospital. The study revealed that some of the 1346 products present on the pharmacy's ERP do not have sufficient historical data to create an accurate forecast. In this context, and with a service level of 95%, 41% of products have a stock higher than what should be the maximum stock, amounting to approxi-

J. Sequeiros
School of Engineering of Porto (ISEP), Polytechnic of Porto, Porto, Portugal

M. T. Pereira (✉)
School of Engineering of Porto (ISEP), Polytechnic of Porto, Porto, Portugal

Instituto de Ciência e Inovação em Engenharia Mecânica e Engenharia Industrial (INEGI), Porto, Portugal
e-mail: mtp@isep.ipp.pt

M. Oliveira
Instituto de Ciência e Inovação em Engenharia Mecânica e Engenharia Industrial (INEGI), Porto, Portugal

Department of Mathematics, School of Engineering of Porto (ISEP), Polytechnic of Porto, Porto, Portugal

F. A. Ferreira
School of Hospitality and Tourism of Polytechnic Institute of Porto, Applied Management Research Unit (UNIAG), Vila do Conde, Portugal

J. C. de Oliveira Matias et al. (eds.), *Quality Innovation and Sustainability*, Springer Proceedings in Business and Economics,
https://doi.org/10.1007/978-3-031-12914-8_28

mately € 147.908.87 in fixed assets, and 11% of products were at risk of being out of stock at the time. The importance of the evolution of the core information system of hospitals was at stake, ensuring the technological sustainability of the ongoing digital transformation, alignment with ICT rationalisation measures, improvement of customer service, and improvement of the quality of information available to the user.

Keywords Supply chain · Hospital pharmacy · Forecasting methods · Decision support systems · COVID-19

1 Introduction

In the healthcare sector, the value can be largely based on the ability to efficiently coordinate the activities of the pharmacy supply chain with that of the hospital. Awareness of logistics is becoming more widespread, with many initiatives and studies on the integration of the supply chain, such as the provision of outsourcing strategies [1]. However, the internal supply chain, unlike the external one, remains the weakness of many organizations. Representing the costs of materials and services as the second – largest group of costs in a hospital, it is recognized that supply chain management is one of the main areas of improvement, where we achieve important impacts on the organization's performance [2]. The adequate acquisition and management of medicines and other pharmaceutical products are closely related to the ability of a given healthcare institution to provide care inherent to its mission and functions. Thus, medicines and other pharmaceutical products are considered essential goods for the proper functioning of healthcare institutions [3]. Pharmaceutical stock management emerges, which encompasses the set of activities that aim to guarantee the stock of medicines and other pharmaceutical products, to meet the therapeutic needs of patients, but also the interests of the health institution [4]. This complexity of processes and data analysis coupled with the current activity of the pharmacy may be an obstacle to more efficient resource management [5]. Thus, the management of hospital material is identified as a central point in containing the constant increase in health costs in industrialized countries [3]. This work intends to analyze the consumption of a hospital unit, both at the pharmaceutical and laboratory levels, projecting the consumption of the following month (using Linear Regression and Holt – Winters) to conjecture about the levels of stocks and inefficiency in the management of same.

2 Literature Review

As Carvalho puts it, the central dimensions of Logistics are time, cost, and quality of service. This means that logistics management is done through a management instrument that includes these dimensions and that promotes reasoning and decisions, essentially through balances and exchanges, among them [6]. A greater contribution of the logistics area can be offered to the healthcare sector, through a wide range of activities. The hospital supply chain does not differ when compared to the supply chain of any other organization [7]. Considering any healthcare provider, the stock management must be balanced, since the excess of material or the lack of it can cause problems to the organization and, most importantly, the patient [8]. Excess stock increases ownership costs and might outdate or obsolete products, but, on the other hand, the lack of it can delay treatments, worsening health status or death [4]. So, the management of materials that includes the acquisition, reception, and internal distribution is fundamental [9]. As Carvalho puts it, the central dimensions of Logistics are time, cost, and quality of service. This means that logistics management is done through a management instrument that includes these dimensions and that promotes reasoning and decisions, essentially through balances and exchanges, among them [6]. Additionally, one of the most important steps in the purchasing process is the selection of suppliers, which integrates and influences these central dimensions [10]. Generally, a buyer selects suppliers that meet requirements for delivery times, quality as specified in the requested project, and satisfactory payment terms. In the hospital context, according to Paterno, procurement is the most important function of logistics and can be defined as a service that aims to provide the materials needed by the hospital, plan the correct quantities and satisfy them at the right time, in the best quality and at the lowest cost [11]. Regarding the optimization of the use of resources, the difficulties in transposing the industry's best practices to the hospital environment are evident. One of the biggest causes of inefficiency is due to the existence of hidden stocks to avoid stock – outs, which are based on procedures that are more focused on policies and user experience than on the institution's data analysis. As previously mentioned, the adoption of stock management, supply, and distribution policies, due to the high investment associated with stocks, is essential. Studies carried out in recent years suggest that the costs associated with stock management in the health sector are between 10% and 18% of healthcare organizations' net revenue [12]. The growth rate of expenditure at the selling price to the public of the total market (includes the NHS market, non-prescription medicines market, and the rest of the market, which includes sub-systems and private insurance) increased 18.5%, between 2003 and 2011 [13].

The reforms in hospital management followed a European trend influenced by the perspective of New Public Management that favored management practices with greater autonomy for managers and greater accountability for their actions, accountability in a clear and public manner, the introduction of incentive mechanisms performance, and the application of management tools [21, 22]. The creation of hospital centres increased efficiency in resource management, avoiding waste, which generated signs of greater profitability and made the health system fairer and of higher quality [14, 15]. With these measures, the contracting of the activity with Hospitals, Hospital Centers, and Local Health Units started to consider efficiency and productivity, the shared management of resources between the different hospitals, to use all the installed capacity in each hospital, fulfilling minimum waiting times, improve user access through the Integrated Access Management System, and an articulated relationship between hospitals and primary care. The provision of healthcare services has been the central concern of healthcare organizations for several years. The increase in demand for health services, together with economic realities, highlights concerns about the sustainability of the public health system [15], making it necessary to seek maximum efficiency in its management [16]. Patient care is supported by a wide range of activities, including inventory management, purchasing, and distribution of supplies to the point of care. These activities, known to many as healthcare logistics or supply chain management, aim to ensure the delivery of the right products where and when needed, with the quality and quantity necessary to provide services, avoiding interruptions in supply [17].

2.1 Stock Management Policy: Stochastic Model

Stochastic models apply when demand and/or supply behave randomly, uncertainly. This uncertainty increases the complexity of stock management, as it is now necessary to deal with the possibility of stock-outs. In order to deal with this random behavior, both on the side of the demand and on the supply side, it is necessary to establish a safety stock to absorb greater variations in the registered average values [4, 18]. In the model of Fig. 1, the quantity to be ordered is fixed (EOQ), but the period between orders is variable (depends on the pace of demand in the period between orders). As demand and delivery times vary, there is a possibility of breakage. If the order cycle is divided into two parts (when the quantity in stock is greater than the order point, and when the quantity in stock is less than the order point), the possibility of breakage exists only in the second part of the cycle, which corresponds to the supplier's delivery time. There will be disruption if the demand during the supplier's delivery period exceeds the order point.

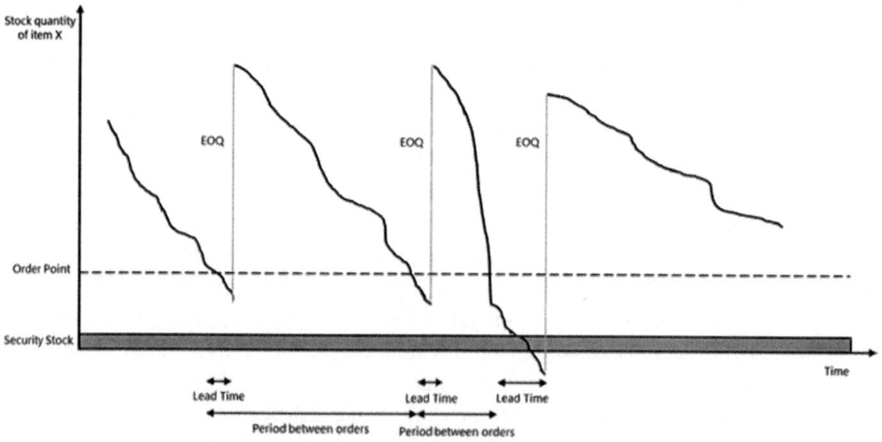

Fig. 1 Graphical representation of the functioning of the continuous review model [6]

3 Methodology

In scientific research, there is always start from a problem whose statement has already the perspective in which the writer stands theoretically and epistemologically. It translates the way of seeing the question to be examined; that is, the enunciation of the problem to be investigated already shows the theoretical view on the theme under examination and it is this look that will direct our investigation and ways of approximation of the survey and analysis of the data obtained. So, it is not the use of quantitative or qualitative data that differentiates forms of approach questions in investigation, but rather the perspective that guides the investigation [19]. The data was obtained directly from the ERP and was subsequently treated. According to the software supplier website, it is a solution that covers areas as relevant as complementary means of diagnosis and therapy, pharmacy and logistics that optimize pharmaceutical and supply services, ensuring response to the area clinic and a correct work base.

For the hospital's consumption analysis, data were collected for the last 5 years. These data are grouped by month to facilitate comparison. A total of 1346 different compounds were collected, resulting in 40,692 consumption data for all medicines in the pharmaceutical warehouse were collected. For the predictions, two different models were used, simple linear regression and Exponential Smoothing with Seasonality. These methods were chosen for their ease of implementation and for the results they normally present in this context. Regarding the treatment and analysis of the data, Microsoft Excel 16 was used, using the Solver tool.

As demand during the delivery period is a random variable, it is necessary to identify it. According to the professionals, deliveries are made between 24 and 48 h, with occasional non-relevant variations. Thus, the Average Lead Time will be 36 h, with a standard deviation of ±12 h.

3.1 Simple Linear Regression

Linear regression finds the line that best represents the relationship between two variables. It is the simplest way to characterize a bivariate sample with quantitative data. When there is only one variable X, the model is designated by simple linear regression. When there is more than one variable X, the model is called multiple linear regression. The simple linear regression model analyzes the relationship between two variables of quantitative nature X and Y, whose tendency is approximately represented by a straight line. Unless the linear correlation coefficient R is 1 or -1, all Y predictions from X's are average forecasts. To make Y explicit in terms of X, other factors also influence the variable dependent and are not specified in the model. These factors are included in the random variable ε_i.

The simple linear regression model is expressed by

$$Y_i = b.X_i + a + \varepsilon_i \tag{1}$$

Where:

Y represents the dependent or response variable
X represents the independent or predictor variable
b represents the slope of the line or slope of the line or change (increase or decrease) in the mean value of Y associated with a unit increase of X
a represents the ordinate at the origin, constant, or intercept the line with the Y – axis, or the average Y value when X is zero
ε_i represents the residual random variable that describes the effects of Y_i not explained by X_i

Additionally, R represents the correlation coefficient and R^2 represents the proportion of variation explained by the model.

3.2 *Exponential Smoothing with Seasonality (Holt)*

The simple exponential smoothing model is suitable for locally stationary series, but is unsuitable for series with other characteristics, namely, when trend and seasonality exist. This type of series shows a consistent increase or decrease all the time in the series. Adjustment to the trend is done using a component of exponential smoothing (S_t) and trend (T_t) for period t. The robustness and accuracy of exponential smoothing methods such as Holt – Winters have led to widespread use in applications where many series necessitate an automated procedure, such as inventory control [20].

$$\textbf{Level } S_t = \pm\left(A_t - I_{t-L}\right) + \left(1 - \pm\right)\left(S_{t-1} + T_{t-1}\right) \tag{2}$$

$$\textbf{Trend } T_t = {}^{2}\left(S_t - S_{t-1}\right) + \left(1 - {}^{2}\right)T_{t-1} \tag{3}$$

$$\textbf{Seasonality } I_t = {}^{3}\left(A_t - S_t\right) + \left(1 - {}^{3}\right)I_{t-L} \tag{4}$$

$$\textbf{Forecast for period } t+1 \text{ and } t+m\; F_{t+1} = S_t + T_t + I_{t-L+1} \quad \text{then} \quad F_{t+m} = S_t + mT_t + I_{t-L+m} \tag{5}$$

S_t is the value of the smoothed forecast for period t
T_t is the estimate of the trend value for period t
I_t is the seasonality value for period t
A_t is the real value observed in period t
α is the exponential smoothing constant, $0 \leq \alpha \leq 1$

β is the trend smoothing constant, $0 \leq \beta \leq 1$
γ is the seasonality smoothing constant, $0 \leq \gamma \leq 1$
F_{t+1} is the forecast value in period $t + 1$
F_{t+m} is the forecast value in the period $t + m$
L spacing of seasonality – number of periods in the seasonal cycle (must be constant in the time and the variation must be greater than the random variations or "noise")

In this study, the alpha, beta, and gamma values are calculated using an algorithm that allows minimizing the mean absolute error of the forecast. This method is preferred when seasonal variations are constant throughout the series. The seasonal component is expressed in absolute terms in the scale of observed series and the level equation, adjusting its seasonality by subtracting the seasonal component [20].

3.3 Errors

The error associated with the forecast is the difference between the value observed in the series time and the forecast. Forecasting errors can be used to calculate forecast accuracy. The mean of the sum of the squared errors is referred to as the mean square error (MSE), (Eq. 6) and it is the most popular measure of synthesis of the forecast errors used in the evaluation of the performance of each method. Although it was considered the MSE is the best method for this study for two main reasons, its wide usage and the series are evaluated under different conditions, consequently having the same measurement level as the underlying series [21].

$$\mathrm{MSE} = \frac{1}{n}\sum_{i=1}^{n}\left(Y_i - \hat{Y}_i\right)^2 \tag{6}$$

3.4 Developed Model

The algorithm implemented allows the calculation of the Economic Order Quantity, Security Stock, Order Point, and Stock Rupture for all items present in the inventory. Figure 2 analyzes the historical consumption data, forecasting the consumption of the following month according to the methods presented above. Later, the method with the least error is proposed to the user. This proposal must be validated using the evolutionary graph of consumption, given that there may be factors such as seasonality or trend that may not have been considered. To analyze the impact of the service level on stocks and their costs, the algorithm was run in three different scenarios. In each of them, the level of service was 95%. Figure 2 shows the implementation pseudocode.

```
Initialize with reading of historical product demand with first ID
  Repeat
      For all products i
        Calculate
            Demand and absolute error
                Simple Linear Regression
                Exponential Smoothing with Trend and Additive Seasonality (Holt-Winters)
                    Calculate α, β and γ trough Excel Solver minimizing error
            Economic Order Quantity
            Security Stock
            Order Point
            Stock rupture
        Choose method with lowest absolute error
        Update cells with this information
      EndFor
Until Stopping criteria
Show results to user
```

Fig. 2 Pseudocode for the application

4 Results

4.1 Hospital Stock

To calculate the intended outputs, some considerations had to be taken, as shown in Table 1. These considerations are closely linked to the fact that there is no concrete and precise information that can be used in the context of the study. Thus, values described in the literature were common in the healthcare sector. Order cost considered two support administrators, part time of the pharmacy director, depreciation of the equipment associated with the administrative and management, and depreciation of the computer program divided by the average number of annual orders. Additionally, the breakage cost was not estimated.

In terms of general data analysis for the 1346 runs made by the algorithm, some interesting data were found. In the Table are the data referring to the total costs of annual stock in the hospital, considering the historical data and the estimates made. In average, 85,032 items are consumed monthly, with an average cost of €28.63. The total cost of stocks is around €1,698,064.0, with a 3% share referring to amounts not directly related to the acquisition, as shown in Table 2.

Regarding monthly expenses with pharmaceutical and laboratory products, these values tend to be stable, around €150,000 monthly, with a growing trend in the last year. The values of post-Covid spending tend to be lower than the homonyms of previous years, values justified by the behavioral changes of patients. Annual expenditure amounts to around €1.5M, with a tendency to increase in the following years. As the data obtained referred only to the first five months of 2020, the associated expenditure is considerably lower. Nevertheless, expenditures in the first five months of the year were like the same period in the previous year, even considering the period of pandemic onset, as shown in Figs. 3 and 4.

The algorithm returned most of the products as modellable from simple linear regression. Despite these results, double attention is needed when analyzed one by one, for the reasons already mentioned, namely, the current inability to analyze the

Table 1 Assumptions taken to calculate the desired values

Variable		Value
Total cost of ownership		15%
	Storage	4%
	Material handling	1%
	Cost of fixed capital	7%
	Personnel costs	2%
	Costs of losses, obsolescence	1%
Average lead time		1.5 days
Standard deviation of lead time		0.5 days
Order cost		10€

Table 2 Annual stock cost of all products

Average monthly consumption in the last 12 months	Average value per unit	Annual acquisition cost	Annual order cost	Annual stock cost	Total cost
85 032,62	28.63€	1,641,768.3€	31,587,8€	24,707.9€	1,698,064.0€

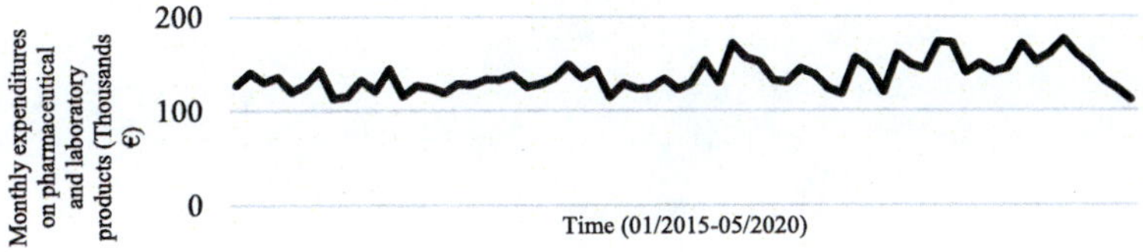

Fig. 3 Evolution of monthly expenditures on pharmaceutical and laboratory products (thousands €)

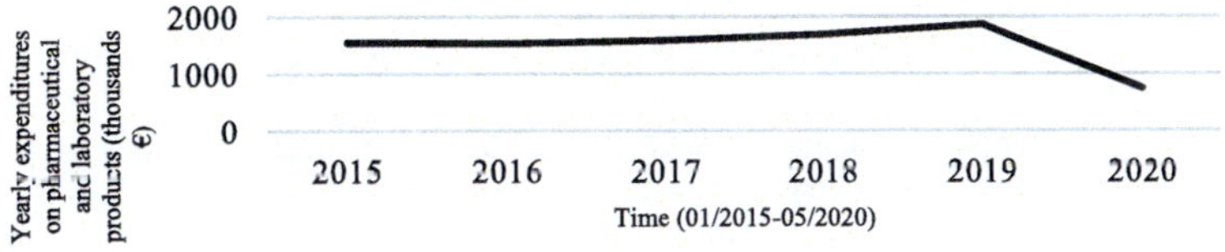

Fig. 4 Evolution of yearly expenditures on pharmaceutical and laboratory products (thousands €)

graphs to detect whether there is a trend, seasonality, or irregularity. Table 3 shows frequency of the best MSE.

Based on the assumptions and using the results obtained from the best methods identified, considering a service level of 95%, comparing with the current stocks indicated in the ERP, it was possible to conclude the following:

- 41% of products have a stock higher than what should be the maximum stock (safety stock + economic order quantity). This figure represents around €147,908.87 of fixed assets (Fig. 5)
- 11% of the products were, at the time, at risk of being out of stock, as shown in Fig. 6.

Table 3 Relationship between the method and the frequency of the best MSE

Method	Frequency of the best MSE	Percentage
Simple linear regression	1322	98.3%
Exponential smoothing with trend and additive seasonality	24	1.7%

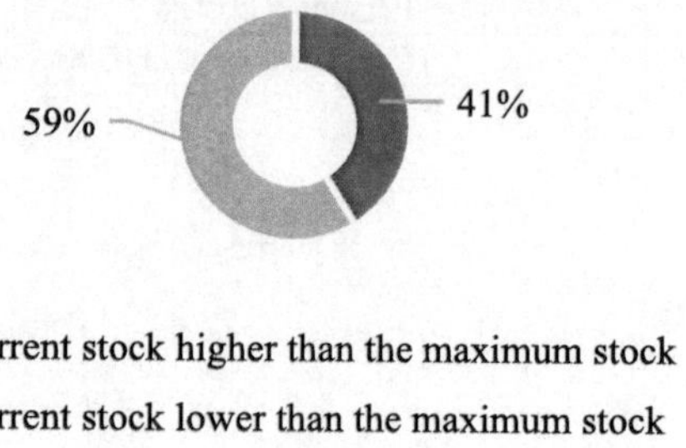

Fig. 5 Percentage of products with current stock above or below the maximum stock

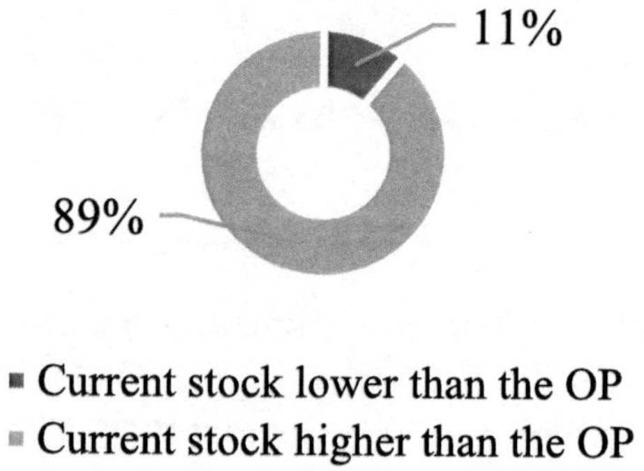

Fig. 6 Percentage of products with current stock above or below the maximum stock

5 Conclusions

The pharmaceutical management of medicines and other products assumes a central role in healthcare services. This plays a relevant and strategic role, mainly due to the high economic burden it entails, keeping stocks of medicines in the same proportion of their consumption, avoiding excessive stocks or stock – outs, is one of the greatest challenges for hospital pharmacists. This challenge is mainly due to the significant fluctuations and high degrees of uncertainty inherent in stock, but also with the aim of reducing costs, including storage costs. In this process, in addition to the financial aspect, quality assurance should also not be discouraged, to ensure the pharmacotherapeutic needs of patients. Health Information Systems allow cooperation, knowledge, and information sharing, as well as the development of service provision activities in the areas of systems and technologies information and communication [22]. They play an important role in the reform of the health system, having as main objectives the improvement of accessibility, efficiency, quality and continuity of care, and increased satisfaction of professionals and citizens. Forecasts and stock management indicators were created for the 1346 different entries present in the system and with consumption recorded in the last five years. These indicators were analyzed individually according to the MSE, the smallest being chosen within the methods used in the forecast. Many products with forecasts based on linear regression prevailed, although there is the possibility of decreasing this value with more adjusted methods if we consider the graphical visualization of historical consumption. In comparative terms of the results obtained with the in – house products, it was concluded that 41% of the products are above what would be the maximum stock, showing that purchases are being made with quantities greater than adequate, representing €147,908.87 of fixed assets. Additionally, 11% of the products are below the order point, being at increased risk of breakage. In terms of limitations, this study has some that can be listed. In a first point, the non – use of KPIs to monitor suppliers and stocks. The fact that the pharmaceutical services do not do this monitoring does not allow a comparison of the pre – and post – implementation status. The automation of the choice of the best predictive methods, being only for the smallest MSE, may not include trend or seasonal lines, being necessary human control in the visual analysis of the graphs of the historical consumption of certain products. Additionally, one disadvantage of exponential smoothing methods is that they are not outlier robust. An observation has an unbounded influence on each subsequent forecast. The selection of the smoothing parameters is also affected since these are estimated by minimizing a sum of squared forecasting errors.

In general, there are several lines of action for the future. Improvement of the algorithm, so that it is more agile and quick in showing results, the inclusion of new predictive methods or evolution of the existing ones to suit the most common time series in pharmacy products, and the inclusion of artificial intelligence for automated analysis of the trend and seasonality lines of these same timelines.

Acknowledgments The authors are grateful to the FCT Portugal for financial support by national funds FCT/MCTES to UNIAG, under Project no. UIDB/04752/2020.

References

1. Castro, C., Pereira, T., Sá, J. C., & Santos, G. (2020). Logistics reorganization and management of the ambulatory pharmacy of a local health unit in Portugal. *Evaluation and Program Planning, 80*, 1–15.
2. Amad Saeed, M., & Kersten, W. (2017). Supply chain sustainability performance indicators – a content analysis based on published standards and guidelines. *Logistics Research, 10*(12), 1–19.
3. Neves, R. P. L. (2009). *Projecto de melhoria da logística hospitalar do Hospital Infante D. Pedro* (p. 71).
4. Rais, A., Alvelos, F., Figueiredo, J., & Nobre, A. (2018). Optimization of logistics services in hospitals. *International Transactions in Operational Research, 25*(1), 111–132.

5. Estêvão, D. (2013). *Gestão Racional da Aquisição de Medicamentos e Outros Produtos Farmacêuticos na Farmácia* (p. 2).
6. Fernandes, C., de Carvalho, J. C. & Guedes (2017). *Logística e Gestão da Cadeia de Abastecimento.*
7. Teixeira, D. R. V. (2012). *Armazém Avançado em Contexto Hospitalar: Método de Reposição por Níveis* (p. 93).
8. Lozano-Diez, J. A., Marmolejo-Saucedo, J. A., & Rodriguez-Aguilar, R. (2020). Designing a resilient supply chain: An approach to reduce drug shortages in epidemic outbreaks. *EAI Endorsed Transactions on Pervasive Health and Technology, 6*(21), 1–12.
9. Volland, J., Fügener, A., Schoenfelder, J., & Brunner, J. O. (2017). Material logistics in hospitals: A literature review. *Omega (United Kingdom), 69*, 82–101.
10. Kefale, A. T., & Shebo, H. H. (2019). Availability of essential medicines and pharmaceutical inventory management practice at health centers of Adama town, Ethiopia. *BMC Health Services Research, 19*(1), 1–7.
11. Moons, K., Waeyenbergh, G., & Pintelon, L. (2019). Measuring the logistics performance of internal hospital supply chains – A literature study. *Omega (United Kingdom), 82*, 205–217.
12. de Oliveira, F. S. P. (2013). *Centralização versus descentralização das compras na logística hospitalar* (p. 116).
13. Furtado, C., & Oliveira, R. (2011). *Análise da Evolução do Mercado Total de Medicamentos entre 2003 e 2010*. Infarmed, pp. 2–19.
14. Nunes, A. M., & De Matos, A. A. (2017). *Revista Eletrônica Gestão & Saúde ISSN: 1982-4785 Nunes AM & Matos AA* (Vol. 08, pp. 284–301).
15. Nunes, A. M. (2017). Gestão Hospitalar: Quatorze Anos de Empresarização em Portugal. *Revista de Gestão em Sistemas de Saúde, 6*(2), 154–162.
16. Moreira, I. M. D. (2011). *Comparação de resultados: Prestação de Cuidados de Saúde Hospitalares Públicos e Privados*. Universidade do Porto.
17. Stank, T. P., Pellathy, D. A., In, J., Mollenkopf, D. A., & Bell, J. E. (2017). New frontiers in logistics research: Theorizing at the middle range. *Journal of Business Logistics, 38*(1), 6–17.
18. de Carvalho, J. C., & Ramos, T. (2016). *Logistica na Saúde.*
19. Gatti, B. A. (2012). Abordagens quantitativas e a pesquisa educacional. In *Seminários – Instituto de Matemática e Estatística* (pp. 1–15).
20. Bastos, J. A. (2019). Forecasting the capacity of mobile networks. *Telecommunication Systems, 72*(2), 231–242.
21. van der Laan, E., van Dalen, J., Rohrmoser, M., & Simpson, R. (2016). Demand forecasting and order planning for humanitarian logistics: An empirical assessment. *Journal of Operations Management, 45*, 114–122.
22. Sequeiros, J. (2020). *Analysis of the supply chain management in a Portuguese public hospital: A case study*. Instituto Superior de Engenharia do Porto.

Towards Digital Transformation: A Case Study to Identify and Mitigate COVID-19 in the Retail Industry

Marisa Lopes, João Reis, Nuno Melão, and Joana Costa

Abstract The purpose of this research is to demonstrate how retailers have used the digital transformation to mitigate the negative effects of COVID-19. As this research aims to understand a real-life phenomenon for which there is very limited knowledge, we created the opportunity to empirically explore the digital transformation in the retail industry during COVID-19 pandemic. In general terms, the research follows a qualitative, descriptive, and exploratory case study design. The results have shown that retailers should focus on technological innovations, adapt their business models, manage their distribution channels, and strengthen their customer-centric strategy. Moreover, it is necessary to emphasize that while smart retail is gradually standing out in the sector, there are also some factors that have not been overcome, such as lack of digital culture, training, and digital leadership. Despite these identified difficulties, the adoption of a digital strategy will allow a differentiating, safe and secure shopping experience, which today is one of the decisive factors for the survival of companies. The COVID-19 pandemic had social and economic effects in all industries – retail was no exception. In turn, the digital technologies already used by companies began to contribute to retailers being able to respond more quickly to customer needs, having been fundamental in fighting the COVID-19 pandemic. To the best of the authors' knowledge, this research is one of the first to explore this topic, bringing new contributions to theory and managerial practice.

M. Lopes
Higher Institute of Management and Administration of Santarém, ISLA-Santarém, Santarém, Portugal

J. Reis (✉)
Higher Institute of Management and Administration of Santarém, ISLA-Santarém, Santarém, Portugal

Industrial Engineering and Management, Lusofona University, EIGeS, Lisbon, Portugal
e-mail: joao.reis@ulusofona.pt

N. Melão
CISeD–Research Center in Digital Services, Polytechnic Institute of Viseu, Viseu, Portugal

J. Costa
Department of Economics, Management, Industrial Engineering and Tourism, GOVCOPP, Campus Universitário de Santiago, University of Aveiro, Aveiro, Portugal

J. C. de Oliveira Matias et al. (eds.), *Quality Innovation and Sustainability*, Springer Proceedings in Business and Economics,
https://doi.org/10.1007/978-3-031-12914-8_29

Keywords COVID-19 · Digital transformation · Case study research · Retail

1 Introduction

In mid-March 2020, the World Health Organization (WHO) classified the new Corona Virus Disease (COVID-19) as a pandemic due to the alarming number of cases and infections. The infectious diseases caused by the Severe Acute Respiratory Syndrome Coronavirus 2 (SARS-CoV-2) quickly became a global public health challenge [1, 2]. As it spread, the disease revealed its damaging effects on global economies and markets [3–5]. In this regard, countries have enacted a series of mandatory control measures for the population, in order to prevent further dissemination. In turn, the industrial sector was also forced to interrupt production lines, reduce human resources, or even close their physical facilities [6]. While people were forced into confinement and deprivation of liberty, the retail sector was one of the few sectors that had to secure products and goods for citizens' livelihood [7].

In an early stage of spreading SARS-CoV-2, which caused an environment of fear and uncertainty, consumers shifted their demand to other types of goods (e.g., non-perishable goods). In light of the above, retailers struggled to obtain all the necessary resources due to the unusual quantities of products that were purchased [8–10]. Retailers also started to adopt new measures, such as (1) additional forms of payment to avoid physical contact; (2) teleworking, to ensure indoor social distance; (3) the use of new digital channels, to allow access to organizations from home [11–14], among many other aspects. The employees in the retail sector also found new risk factors, such as overtime pressure and increased workload; use of additional personal protective equipment and frequent stops for hygiene [6, 13, 15]. In this regard, retailers and logisticians were forced to adopt additional occupational health and safety measures to avoid environments of stress, danger, and burnout [6, 16]. It was in this context that digital and disruptive technologies came to contribute to a better adaptation of companies to the phenomenon of COVID-19.

In the literature, there are some studies that have discussed this topic from the consumer perspective. They revealed that the pandemic caused new consumption habits, a change in the demand for goods, market behavior, and the economic effects of these changes on companies [9, 17, 18]. Furthermore, notable researchers such as Kraus et al. [19] also analyzed the thematic evolution of TD research in the areas of business and management, bringing new light to the subject. In this regard, Kraus et al. [19] identified and reviewed articles published in journals of the Chartered Association of Business and proposed a synergistic framework that linked existing research in DT to the areas of business and management. However, the perspective of professionals who operate in retail and logistics areas is somewhat forgotten in the existing literature. The scarcity of information, largely due to the novelty of the topic, has led the academic community to try to share with professionals the responsibility for finding solutions that will mitigate the effects of COVID-19 in the retail sector. To fill a gap in the literature that is closely associated with the lack of information in the area, we developed the following research question: How is the digital transformation contributing to mitigate the effects of COVID-19 on the retail industry?

Thus, this article is twofold, since it aims to understand and describe the effects of COVID-19 in the retail sector and, on the other hand, propose measures to mitigate the impacts of the disease through a digital transformation strategy. The article is organized as follows: Section 2 describes the methodological process and Sect. 3 reports the results of the research and includes a brief discussion. The final section presents the theoretical and managerial contributions of the research, as well as the research limitations and suggestions for future studies.

2 Literature Review

2.1 Criteria for Articles Selection

This section highlights the results of a literature review that came from articles collection from Elsevier's Scopus on February 14, 2022 (Appendix A). We used Scopus since this database is considered as one of the largest databases that contains peer-reviewed scientific journals. For the research domain of natural and social sciences, Scopus is more adequate when compared with the Web of Science (WoS) or the Google Scholar [20, 21]. We did not use Google Scholar search engine because it includes documents that have not been peer-reviewed and, therefore, tend to be seen as of questionable quality [20].

To select articles for the review we used the Preferred Reporting Items for Systematic Reviews and Meta-Analyzes (PRISMA). The PRISMA statement consists of a checklist of 27 items and a flowchart with four phases (Identification, Screening, Eligibility and Included) [22] and is often used to improve reporting of systematic literature reviews and meta-analyzes [23]. Thus, this method allowed researchers to select the inclusion and exclusion criteria, in order to guarantee a selection of the documents with the greatest criteria.

Our selection criteria started by choosing the terms "Retail" in the topic (title, abstract, and keywords) and "Digital Transformation" or "Artificial Intelligence" or "Internet of Things" in the article title. Regarding the type of document, we only included journal articles since they are considered to be of superior quality. To avoid misinterpretation, only selected documents in English. There was no time restriction, as the theme of COVID-19 is very recent. In addition to the language and type of documents, we also included two conference papers, as these were a preliminary part of this article. Although we had access to all other documents and met all the previous criteria, we resorted to the elimination of two of them, since the topics did not correspond to the research topic. From a total of 240 manuscripts, and considering the PRISMA protocol, we reached 80 documents, as shown in Table 1.

Table 1 Four phases and criteria of the PRISMA protocol

SCOPUS	*n*
Identification	
"Retail" (Title, Abstract, Keywords) AND "Digital Transformation" OR "Artificial Intelligence" OR "Internet of Things" (Title)	240
Screening	
Language (English)	230
Journal	85
Articles	80
Eligibility	
Full-text articles assessed for eligibility	78
Included	80

Within the scope of the literature review, two types of analysis were performed. First, a descriptive statistic was carried out that took into account the graphical data obtained directly by Scopus, such as the distribution by country, thematic area, and year. Second, the content analysis of the 80 manuscripts was carried out, which made it possible to identify the extent to which the digital transformation allowed to mitigate the effects of COVID-19 in the retail industry.

2.2 *Brief Analysis of the State-of-the-Art*

The following results are distributed by countries, subject area, and year. Regarding the results by countries: The United States, India, and the United Kingdom stand out, as shown in Fig. 1. From our analysis, the United States stands out, for having the largest retailers in the world (e.g., Amazon), as well as for the issue of increased sales of e-commerce generated by COVID-19. On the other hand, India and the United Kingdom are countries that are investing in digital transformation of retail infrastructures in order to remain competitive in the international markets [24–26]. Although both countries have a great ambition to develop digital technologies in their economies, they are still not satisfied with their achievements; therefore, these countries have sought to maintain a high level of technological advantage.

To get a holistic view, we considered a more comprehensive value by comprising macro analysis. In this regard, it was possible to ascertain that the European Union (EU) stands out from the remaining 39 countries, either because of the investment that has been promoted by the European Commission or due to the investment of the 45 EU countries identified in the Scopus database. Regarding the distribution per type of document (Fig. 2), there is high distribution of the results, as digital transformation can occur in any sector. However, we highlight the computer science domain (23.2%), which has the highest percentage because it is where the most technical concepts are approached, such as Augmented Reality (AR), Artificial Intelligence (AI), Big Data, Cloud Computing, Internet of Things (IoT), Radio Frequency Identification (RFID), among others. It is expected that computer

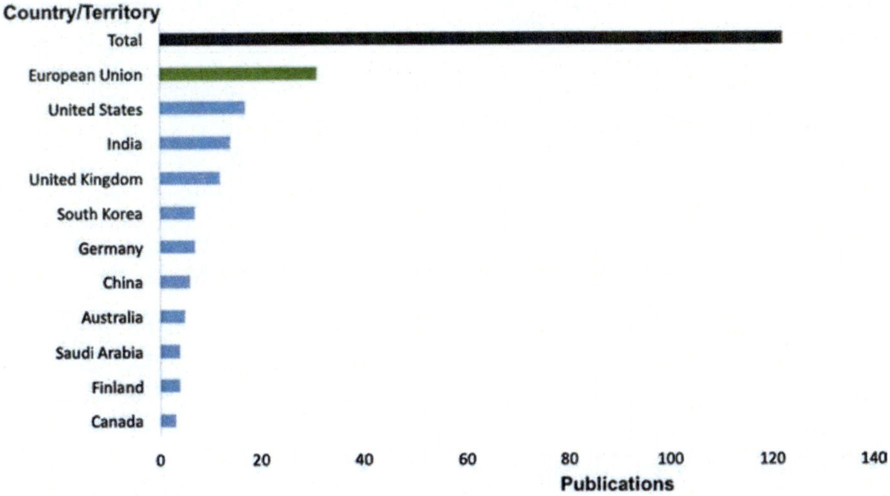

Fig. 1 Distribution per countries (Displayed the top 10 countries)

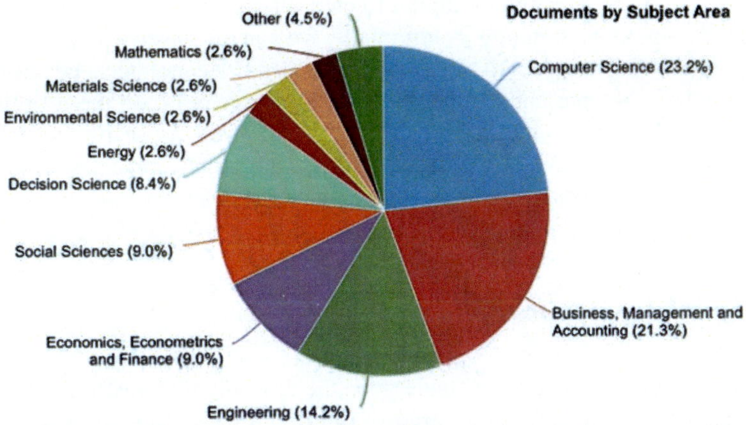

Fig. 2 Distribution per subject area (retrieved from Scopus)

science will continue to grow gradually as a result of the acceleration of the digital transformation process worldwide. Then comes the business and management domain (21.3%), where the retail sector is predominantly discussed and increasingly addressed in relation to digital transformation in companies. These results are in line with other previously published papers [27].

Finally, we present Fig. 3, where we can reveal the distribution of articles published by year. The oldest document dates from 1990 and all the others date from the 2000s. It is only from 2017 that a growing number of publications is noticed, which reveals that the themes "digitization" and "digital transformation" are relatively recent in study and academics.

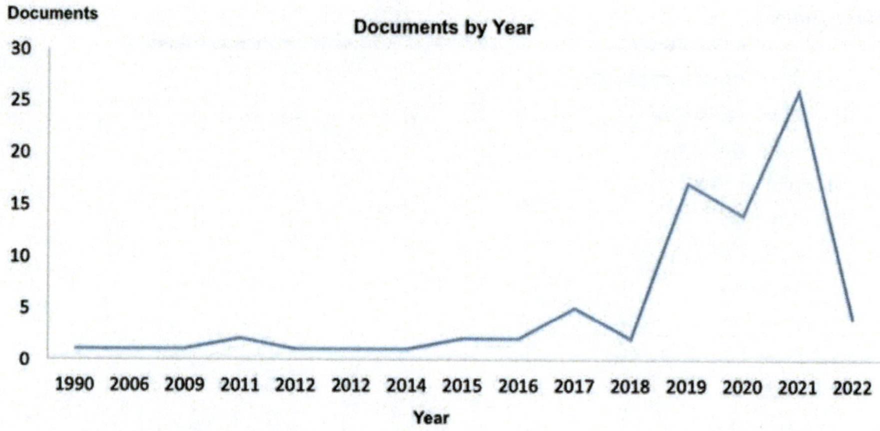

Fig. 3 Distribution per year (retrieved from Scopus)

As of 2019, growth continues to be exponential given that that date goes back to the beginning of the COVID-19 pandemic and that it was identified as the most recent cause of the acceleration of digital transformation in companies, largely due to the increase in online retailing. Additionally, we can also justify that the millennial generation (Y) is considered as the biggest driver of technological advances. In the next sections, we present the qualitative analysis, focused on the 80 selected manuscripts, although it is not limited. If in the state-of-the-art section we tried to present a macro perspective of the digital transformation phenomenon, in the following sections we present the results of the content analysis.

2.3 *Digital Transformation in the Retail Sector*

Retail is one of the sectors that has undergone the most digital transformation in its operations [28]. This change is the result of offering a wide range of items, which resulted in management challenges that in turn were mitigated with digital transformation, but also because retail necessarily has to keep up with the sudden demand of modern economies [29, 30]. Overall, the digital transformation in retail is perceived in the academia as technologies that enable new forms of behavior, interactions, or market experiences, and remodels customer relationships, internal processes, and value propositions [31]. In other words, digital transformation enables new ways to create value while meeting long-standing consumer needs.

Based on our analysis, we identified three relevant episodes of digital transformation in retail. The first episode is associated with the shift from "IT enabled business change" to digital transformation in retail; the second episode is related to the influence of I4.0 on retail (i.e., conceptualization of smart retail); finally, the COVID-19 epidemic and its impact on the relationship with consumers and retail professionals.

Starting with the conceptual shift from "IT-enabled business change" to "digital transformation" and new consumer behaviors, this research advocates that despite the successive new forms that digital transformation may take, it is likely to remain an agenda of change and innovation in the retail sector. That is, if, on the one hand, technological disruption brings new consumption habits, on the other, consumer behavior also contributes to the increase of more advanced technologies, either through purchasing habits that encourage the development of innovative digital channels or by conditions imposed by the external environment (e.g., epidemiological restrictions, lack of time to buy instore, etc.). Since early, the retail sector was forced to incorporate new digital strategies in its business models, in order to respond to the new consumer needs [32]. A good example is the exponential growth in the use of smartphones and mobile payment (m-payment) [33]. This new type of "digital native" consumer has come to the fore [34], since digital technologies enabled better, faster, and more comfortable shopping experiences [34, 35]. Thus, the retail sector also benefits from this shift, as digital consumers allowed retailers to capture more information in order to optimize the supply chain management (pull systems) [36]. An associated landmark worth mentioning was the growth of e-commerce, which has been enhancing digital transformation in the retail sector since the 90s. In that regard, omni-channel operations have been singled out as the future of retail, leading large companies to invest in measures and strategies that result in increased online and offline contact points, strengthened by brand image, increased sales, improved services, integration of channel information, and customer loyalty [37]. For global network retailers, the omni-channel operation is seen as an important solution to meet the needs of customers at any time and place [38].

The I4.0 is another indisputable milestone in the context of digital transformation, as the concepts of IoT, robotics, and AI have radically changed retail operations in addition to human action [39]. Given the growth of I4.0, the concept of smart retail has been expanding. This concept has in its genesis an ecosystem of customer-centered activities and where new business modalities are underlying, such as unattended stores [40]. As the consumer experience is a decisive factor, retailers have invested in offering a combination of quality products and services [41]. This combination is mediated by increasingly intelligent and technological stores. As an example, we can find the autonomous stores that use the latest technology (e.g., vision machine) and where the customer has full autonomy to choose and pay without the help of a store attendant [39]. That is the reason why automated teller machines are becoming a reliable option for retailers, as they offer less costs as a result of replacing human intervention, but also greater customer involvement in the shopping experience, as customers are co-producers of services [42]. In that regard, Amazon Go was created to be one of the world's most advanced checkout-free shopping experience due to the use of computer vision, sensor fusion, and deep learning [39, 43]. According to Amazon [43], the technology "just walk out" automatically detects when products are picked up or returned to the shelves and keeps them in a virtual cart and, as soon as the shopping experience is completed, amazon charges the purchase on the Amazon account. The future of retail points out the autonomous stores model to be increasingly used [32, 42]. Thus, the concept of smart city has also been enabled by the growth of I4.0, insofar as it aims at preparing infrastructures in different areas (i.e., healthcare, energy, etc.) to standardize and use IT systems and networks of sensors more comprehensively. An example from the literature is e-Pharmacy in the United Arab Emirates, where digital data transfer between patients and pharmacies takes place, and whose objective is to optimize the sale of medicines through a system that at the same time has greater efficiency and less margin of error [44].

Finally, the academic community has not neglected the impact of COVID-19 on retail. COVID-19 is impacting both the relationship with customers and the indoor operations of logistics operators. In that regard, digital technologies are once again supporting service delivery or sale of products continuity by increasing the use of existing technologies (e.g., contactless payments) or accelerating new channels and disruptive technologies (e.g., Click & Collect, Grab & Go). The aforementioned acceleration has motivated a series of studies in the area, such as the case of Soto-Acosta [45] who argues that COVID-19 is accelerating the digital transformation of organizations, more specifically, it has fostered sources of value creation for both back-end activities as well as business front-end. As an example, Soto-Acosta [45] states that the potential value comes from the use of digital technologies such as IoT and Blockchain, which offer great opportunities to add transparency and real-time information to supply chains. This is just one example where disruptive technologies can operate together to create value for the organization. In the next section, we present the factors identified in the literature that are necessary for the development of the digital transformation process, while, in the results section, we analyze which factors emerged from the case study.

2.4 Five Managerial Instruments for Retail Digital Transformation

Globalization and rapid changes in consumption patterns, combined with the effect of digitalization, require the retail sector to have flexibility in its business models, innovation and effective management of its capabilities in order to succeed [46]. The fundamental resources for digital transformation are focused on the digitization of processes and the focus on digital technologies [38]. The technologies mentioned so far (e.g., AI, IoT) allow companies to automate processes, reduce costs, customize supply, anticipate demand and supply, and increase competitive power [32, 36, 40, 42, 47]. Beiersdorf is one of the well-known companies that implemented digital technology (SAP) to build digital capabilities in the organization, create business opportunities, and return to customers the products that meet their needs [48]. However, this can lead to the idea that the solution involves only investment in technologies [49], but it will be insufficient if managers neglect the external environment [34].

Today, the biggest challenge that companies face is to achieve an effective combination of physical and digital resources to improve customer service [37]. However, if automation can result in less work, it will, on the other hand, require from retail a more qualified and specialized workforce to accompany technological innovation in processes and activities [50]. For digital transformation, not only digital tools are enough, but also employees with strong technical skills will be needed to be able to get the most out of them [51]. Sund et al. [46] identify as a possible solution for companies the creation of an innovation laboratory responsible for identifying, exploring, and launching new business models that allow centralizing this responsibility in a single department. Today, retailers' operations and decisions remain customer-centric, their needs and requirements must be seen as the engine to change, align, and define the organizational strategy [30, 34, 46, 52]. Omni-channel operations are increasingly reliable for the future of retail trade, by allowing the interconnection of all channels [53]. Retailers should consider digitizing touch points to allow the implementation of techniques in physical retail stores, which previously were only available in online retail [47, 54]. According to Chernova et al. [41], the most important factor for the customer is no longer the price, but the shopping experience. Therefore, the shopping experience will impact the perception of value, and this can be improved through self-service technologies and mobile technologies, as they close the gap between online and offline channels [53].

Digital literacy and organizational culture can mean an important impetus for digital transformation [50]. If companies focus exclusively on themselves, this can be a barrier to the transformation not taking place successfully [55]. It is important that companies are collaborative, empathetic, and motivating in the corporate environment that promotes creativity, team spirit, innovation, and commitment [50]. Leadership will also be central to the digital transformation process, as it will have the responsibility to guide employees to achieve the objective of the digital strategy; it must be participatory and collaborative to promote the involvement and commitment of all with a global objective [28, 56]. In sum, from the data analysis, we identified some of the factors necessary for digital transformation, such as:

1. Physical and digital resources;
2. Employee skills and customer-centric strategy
3. Organizational culture
4. Digital literacy
5. Leadership

In addition to the above factors, we have also identified some barriers, such as scarcity of resources; employees' resistance to change; high cost of implementation of new technologies; security; data privacy; and cyber-attacks [38, 49, 56, 57]. In our understanding, it is necessary to deepen knowledge and develop the above five managerial instruments for retail digital transformation with regard to COVID-19. In the light of the above, we intend to deepen the existing knowledge through empirical confirmation or refutation of the factors identified above (vide Sect. 4).

3 Case Research

This research follows a qualitative, descriptive, and exploratory case study design. This research strategy aims to understand and explain the digital transformation in the retail sector in circumstances of COVID-19. The use of case research is justified by the need to understand in-depth a real-life phenomenon, for which knowledge is very limited [58, 59]. Furthermore, the case study allowed for greater detail and understanding that could not previously be found in the existing literature and that concerned the contribution of digital transformation in mitigating the effects of COVID-19 in the retail sector [60]. The criteria for choosing the unit of analysis is related to questions of convenience [61], since one of the researchers assumes roles of responsibility within the unit of analysis. Multiple case studies were also considered for the analysis. However, given the difficulties of accessing similar logistics companies during the COVID-19 pandemic, we thought it would be more prudent to leave this possibility for future research. The selection of cases used a reduced sample and used an intentional (non-random) selection procedure, as described by Gerring [62].

The research approach was interpretive, as the research starts from the knowledge that reality is a social construction [63, 64]. In that regard, Walsham [64] argues that, within the scope of interpretive research, factors such as authenticity, plausibility, and criticism are very relevant. The authenticity and plausibility of how well the text relates to the professional environment is due to the position the main researcher holds in the organization. However, more critical readers can rightly argue that the author did not have the necessary detachment to carry out the research [65, 66]. In that regard, the coauthors ensured that the entire methodological process was followed to ensure impartiality. In addition, we try to use language accessible enough for junior readers, but with details that might be interesting for seniors. This research used three sources of data collection, namely, semi-structured interviews, official documents, and participant observation [58, 67]. The sources of data collection were important, as they made it possible to triangulate existing information to obtain solid evidence [68]. As for priority, we selected the interview technique as the main source [69]. Thus, the interviews allowed the researcher to ask key questions about the topic in order to collect firsthand data [70]. Official documents also made it possible to provide general and formal information about the organization [71], while participant observation allowed the identification of behaviors and practices in the field [72]. Both official documents and participant observations were considered secondary sources to validate the research results [73, 74]. When corroboration was not possible, the main researcher took responsibility for collecting new information until inconsistencies were eliminated.

For confidentiality reasons, the unit of analysis identification (i.e., a well-known multinational logistics operator located in Portugal-EU) was not revealed, such as the names of the participants, which was also kept anonymous [75]. Overall, the objective of the case research was to understand the effects experienced by employees before and during the COVID-19 pandemic. In particular, the research aims to identify the effects of COVID-19 on the retail sector and the respective digital transformation initiatives with a view to mitigating the spread of SARS-CoV-2.

For a deeper contribution, the selected respondents had a high level of knowledge as they perform different tasks and roles across the organization [76]. Respondents' roles ranged from analysts, warehouse directors, managers, developers, coordinators, and department directors. Prior to the interview, the researchers shared a research protocol and the research objectives. During the interviews, audio recording was authorized, which enabled a more detailed analysis and the identification of the main themes. The survey involved 18 respondents, most of them male (60%), aged between 32 and 49 years. More details are shown in Table 2.

Official documents were made available by the company, so that they could be analyzed and compared with other sources, allowing the validation of the collected information [71]. This information made it possible, for example, to identify internal contingency plans shared with customers, identify the multidisciplinary teams and their responsibilities, as well as the initiatives taken by the company in the context of the pandemic, namely, teleworking.

The participant observation technique allowed the identification of behaviors and attitudes towards the interviewees [77, 78], confirming in loco the practices used in the company. This technique allowed to define topics closer to reality and to delimit questions that emerged during the exploratory phase of the research. While one of the authors of this article works at the logistics company, it was possible to record all the daily interactions with peers (e.g., informal interviews) and measures used by the organization during the COVID-19 pandemic. However, as mentioned earlier, some criticisms of participant observation are known [79, 80], since their observations can be conditioned by their own experiences and ideas [65, 66]. To mitigate this limitation, the researcher used a field diary to record the observations [81], which in turn were reviewed by the second author of this article.

Data analysis was performed using a descriptive and thematic analysis [82]. In a first phase, the descriptive analysis aimed to use a deductive approach [83, 84],

Table 2 Interviewees' details

Department	Respondents	Years working at the company (average)	Age (average)
Administration	1	4	44
Costumer service	3	21	45
Human resources	1	20	49
Improvement and new projects	2	6	37
Information technology systems	1	10	45
Operations	3	5	33
Process reengineering of fast moving consuming goods (FMCG)	1	1	32
Pharma process reengineering	1	0.6	41
Quality and environment, health and safety	2	12	36
Transports	3	5	36

which allowed classifying the scientific articles that were used in the literature review according to their distribution by countries, type of document, thematic area, and evolution over time. This analysis was based on graphs extracted directly from Scopus (cf. 2.2. brief state-of-the-art analysis) which, together with an exhaustive reading of the articles, allowed to justify the trends and interpretations of the statistical results. Additionally, we also carried out an inductive analysis [85], which allowed to obtain an overview of the digital transformation, but also with a focus on the retail sector. In a second phase, the case study data were analyzed using the content analysis technique [75], following the same procedure as Reis et al. [76], it was possible to analyze all the data collected.

For content analysis, we started by reading all the text from the various sources (e.g., interview transcripts, field diary, and official website), integrating all this information into NVIVO 11, a qualitative data analysis software [86]. We then identified most of the phrases and ideas that emerged from the coding analysis [87], crosschecking with the results from the first phase (descriptive and thematic analysis). After completing the previous process, we were able to identify the categories, moving on to the next step, which was to identify patterns in the codes to find relevant topics and generate a map that could provide an overview of the data. Thus, the content analysis technique allowed coding and analyzing a large volume of qualitative data to classify categories and subcategories [88]. This procedure made it possible to identify new patterns/dimensions and emerging ideas. Overall, this procedure followed the guidelines of Gioia et al. [89], allowing conclusions to be drawn with the expected rigor. As new themes emerged, we had to conduct more interviews than we had initially estimated. The interviews stopped when theoretical saturation was reached [90, 91]. That is, when the inclusion of new respondents led to repetition and redundancy, which led us to believe that the time had come to stop. Furthermore, we are convinced that the fact that we interviewed interviewees who were highly informed and able to observe the phenomenon from various points of view gave us the advantage of carrying out data collection with relatively low interviews.

For a better understanding, the methodological process is summarized in Table 3. Table 3 is divided into four columns that deal with the methodology, data collection methods, and data analysis techniques. By presenting information in an organized and transparent manner, we hope to provide readers with a holistic view. It is also relevant to mention that reliability and validity of the case research were achieved through a sequence of procedures [92]. For example, implementing a survey protocol and using multiple sources of data collection. These procedures allowed to find

Table 3 Summary of the methodological process

Methodology	Methods	Sources of data collection	Data analysis techniques
Qualitative	Case study research	Semi-structured interviews Official documents Participant observation	Descriptive and thematic analysis Content analysis

consistent evidence and avoid exclusive and less reliable dependence on a single data collection method.

4 Results

In the results section, we present the empirical results from a case study of a Portuguese logistics operator. In particular, the analysis of how digital transformation in retail has helped to mitigate the effects of COVID-19.

4.1 Digital Transformation: Times of Change

Logistics employees are well aware that digital transformation implies organizational change. The results showed that digital transformation implies the introduction of technological components into the logistic processes, whether they are administrative or operational. In addition, the interviewees revealed that the digital transformation consists of the dematerialization of information and processes, which allows a dynamic, efficient, and active management.

With the logistics operator having a vast portfolio of fast moving consumer goods (FMCG) and pharmaceuticals, it was essential to use technological solutions in order to meet the needs and expectations of customers. While the majority of respondents hold administrative positions, they have contact with customers and partners. Some of the most recent changes identified were the adoption of the following capabilities: (1) POWER Business Intelligence – a tool that allows the analysis of solutions to improve decision making; (2) Enterprise Data – which allows the multinational company to efficiently manage the considerable volume of data from its branches; (3) Data Warehouse – used in the customer service area, assists in conducting research and analysis related to historical data. For instance, this technology makes it possible to measure the levels of productivity and operational efficiency (picking errors, incidents, complaints); (4) Web portal accessible internally (employees) and externally (customers) through credentials – it is also an example of structuring information through the digital channel that facilitates communication, use, and decision; (5) Artificial Intelligence – based on the SPE system (Order Production System), this system prepares orders for pharmaceutical products through a procedure of light direction (pick-to-light and put-to-light); (6) Inventory Maintenance Unit (SKU) – used to search and identify inventory, allowing the separation of large quantities from small inventory maintenance units; (7) the Radio Frequency Identification (RFID) – used in the various stages of the process, from reception, storage, preparation, and shipping, allows easy and effective tracking of goods in real time.

The changes presented above were previously analyzed by multidisciplinary teams and have been implemented gradually since 2007. Each technology took into account not only the decision of the multidisciplinary teams, but also the needs and expectations of all stakeholders and which contributed to the decision of technological implementation. The digital transformation process has made it possible to respond to current COVID-19 challenges. In other words, since the company is responsible for supplying some of the main retailers in Portugal, it has also suffered from the consequences of changing consumption habits through COVID-19. Its prior preparation and incorporation of new digital technologies enabled it to respond in a timely and effective manner to the new needs of the market. The reaction of the companies would not be possible without the ongoing digital transformation process. In the following, we will meet the structural and managerial instruments that allowed the digital transformation process to materialize.

4.2 *Structural and Managerial Instruments for Digital Transformation*

As mentioned above, due to COVID-19, the logistics operator was required to timely respond to the needs of the markets. To this end, this sector needed a prior adopting of new technologies, in order to prepare it to respond to fast and unpredictable phenomena. From our analysis, it was possible to identify some key instruments that were decisive for the success of the digital transformation implementation. Figure 4 shows these five identified instruments for a successful strategy of digital transformation.

One of the instruments that were identified by the respondents were resources (31%), which would have been considered of essential for the successful implementation of new technologies. The subcategories identified within the scope of resources are divided into financial, human, and material resources. First, for the

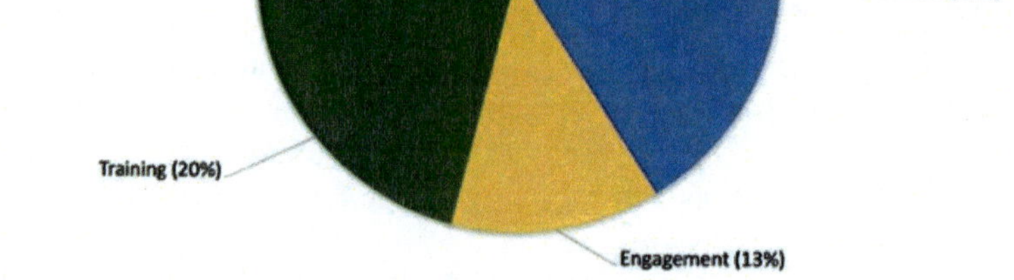

Fig. 4 Instruments for a successful digital transformation strategy

transformation process to take place, significant financial investment is required [93], since disruptive and innovative technologies are often expensive, involving high acquisition, implementation, and maintenance costs. Although expensive, investments in digital resources can bring high profits. For example, Rolls-Royce has implemented sensors in the jet engines it produces to improve their fuel efficiency. With this feature, the company was able to reduce its customers' fuel costs and attract new subscribers to rent jet engines instead of selling them, even creating a new business model [94]. Another example is given by insurance giant Guardian Life, where flexible infrastructure has helped the company to reduce costs, enabling agile operations and accelerating its commitments to startups. The company migrated more than 200 applications to a public cloud, which allowed it to shut down its last data center, the costs associated with running these applications dropped by 20–30% since the migration [94]. However, even investing in digital resources, some companies fail in their attempt to achieve digital transformation. General Electric set out to keep up with the disruption of the market, failed in the transformation it initiated because it focused on the size of the new units rather than the quality of the digital products and services it planned to deliver [95]. Likewise, Ford Motor Company launched a separate entity to incorporate digital technology into new vehicles [95]. Inadequate resource management and funding of the new initiative took a toll on their existing business. In this regard, human resources are also relevant, as it is necessary to delineate a person or team or project that can spend time, knowledge, and work to carry out the transformation. Connecting people can facilitate the process of implementing digital transformation.

A large number of respondents also pointed to organizational culture (26%) as the basis for the company's progression in highly volatile environments. They considered that organizational culture is the "engine of transformation," that is, the company's culture should be guided by values such as openness, initiative for innovation, motivation, teamwork, and dynamism. Under these conditions, the company has the right ingredients to succeed, as there is a clear vision of the mission to be accomplished and everyone is on the right path to achieve it. In the opposite view, the organizational culture resistant to change is a strong barrier to its implementation, since employees understand the digital transformation as something uncertain and of high risk. Clear examples of a lack of vision and organizational culture for digital transformation include the bankruptcy of the Blockbuster movie rental company and the Washington Post, largely a result of these companies' inability to rapidly develop and implement new digital business models [96]. On the other hand, in an effort to support its digital transformation initiative, Publicis Groupe bought a company called Digitas. The acquisition of Digitas was used essentially to drive the total transformation of Publicis Groupe, as Digitas, as a start-up, was used to change and rapid decision-making [97].

No less important, we identified education and training with around 20%, which was also reinforced by the interviewees as an indispensable factor when any type of change occurs, be it social, economic, or technological. Employees must be trained and specialized to improve the use of modern technologies in the company; otherwise, the incorrect use of equipment can impact the company's and the individual's performance, which can result in damage to operational results, personal demotivation, and dissatisfaction with the work. In this regard, it was possible to attend training sessions in the context of the use of new disruptive technologies. Participant observation included the collection of photographic data and feedback from workers on the practical effects of the sessions held by the company. The behavior and reactions of workers made it possible to identify that training is one of the main instruments that contribute to the acceptance of the implementation of new technologies. Although, during the observation, it was possible to affirm some initial resistances. After some time, employees end up realizing that the features of the technologies only came to improve their performance, instead of fearing the elimination of their jobs. According to the McKinsey Global Institute, to emerge stronger from the COVID-19 crisis, companies must start reskilling their workforce now [98]. These virtual practices may decline slightly as economies reopen, but will likely remain above pre-pandemic levels. McKinsey Global Institute gives the example of a pharmaceutical company with more than 10,000 sales representatives that quickly went from an offline model to a 100% remote model [98]. The company now plans to make a 30% online, 70% offline working model permanent, thereby leveraging the newly developed skills of its sales representatives [98].

It also makes sense for people to get engaged (13%) before, during, and after the digital transformation process. Employees must feel part of the process, their opinions must be heard, their work must be monitored and, thus, the group's synergy will bring better results. Disney is a prime example; to capture the hearts and minds of its employees, Disney carefully crafts internal messages so that they are highly engaged [99]. Although the use of technologies may imply greater physical distancing, during participant observation, we found that when employees get involved in the search and implementation of solutions, they feel more available to collaborate. That is, when questioned, they felt that the project was not only the company's but also theirs.

Leadership (10%) will also help in the implementation of a digital transformation strategy, as the leader of the digital transformation projects must be able to clearly convey the objective, trace the way forward, motivate and help his employees to move in that direction. When digital transformation first appeared on the radar, there were no prepared executives with deep experience in digital technologies. Today we have an entire workforce of digitally fluent talent who have worked at big companies like Amazon and Google [99]. Other factors were also pointed out during the formal interviews and direct participation, such as the innovation strategy, market analysis, outsourcing, cost vs. return analysis, and quality control management. However, since these instruments had a less significant impact, we consider only the most relevant for discussion.

4.3 Opportunity and Challenges of Digital Transformation

The data revealed that in the logistics and retail sector, the opportunities for digital transformation outweigh the challenges. The argument used by more than 50% of respondents is that digital transformation increases efficiency, effectiveness, and productivity (Fig. 5). In other words, technology is seen as an instrument that assists in the performance of human work, with a positive impact. It is, therefore, pointed out as a means that makes tasks and processes faster, more fluid, with the use of less time. From the analysis carried out, we found that the respondents' arguments are in line with the literature in other contexts [100, 101] insofar as modern technologies make complex decisions in the analytical-cognitive field with less likelihood of error when compared to humans. However, we also verified that human work is also vital to avoid or minimize possible errors, reduce shortages, or product/service failure in the last mile.

With regard to the cost reduction, digital transformation is a double-edged sword. Thus, if, on the one hand, machines reduce labor costs due to automation, teleworking, etc., on the other hand, the initial cost of technological implementation is high, as it requires significant investments. Nevertheless, it was possible to ascertain that the inclusion of new technologies substantially improved the company's competitiveness vis-à-vis its competitors, by minimizing waste, aiding in decision-making, and transferring jobs to virtual channels during the COVID-19 pandemic.

As for the most relevant challenges, the most common was job loss. Although there is an awareness of the need for continuous improvement, the respondents' perception is that the digital transformation can reduce jobs in the logistics sector. The implementation of new technologies will reduce the importance of the human being, but will also allow greater flexibility (e.g., teleworking). Thus, respondents understand that the digital transformation will leverage the emergence of technical jobs and the enhancement of human skills.

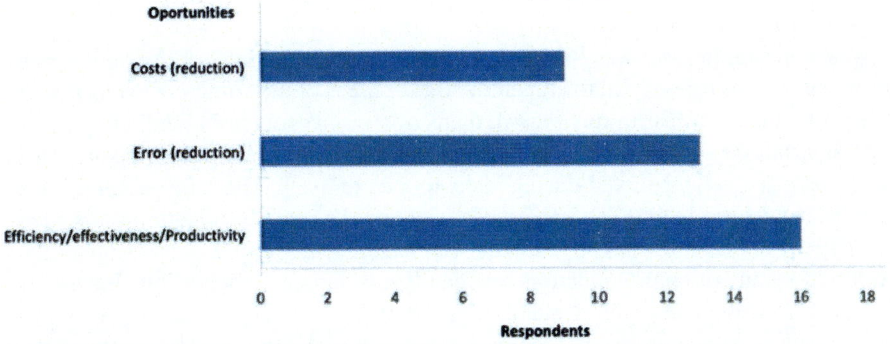

Fig. 5 Top three opportunities for digital transformation

Finally, concepts such as teleworking are twofold, as it was identified as an opportunity and a challenge. That is, respondents identified teleworking as an instrument that offers greater autonomy, greater freedom and focus on work, in addition to reduction of travel costs. In the correspondents' perspective, there is also no division of professional and personal space, which can result in environments of stress, burnout, loss of social bond with the company and colleagues, lack of working conditions, lack of creativity, and isolation. The line that separates the two perspectives is tenuous and, therefore, there must be an understanding on both sides to ensure that the employee has a balance between personal and professional life.

4.4 Mitigating the Effects of COVID-19 Through Disruptive Technologies

In the context of COVID-19, companies faced an unprecedented scenario that affected the management of their operations. In this case study, the unit of analysis studied was no exception and the results corroborate the idea that the pandemic affected the company by accelerating the digital transformation process. In view of the results, we present some contributions on how the digital transformation mitigated the effects of COVID-19.

As it is a logistics operator comprising a portfolio of fast moving consumer goods (FMCG) and pharmaceuticals, the effects of the pandemic directly affected the daily lives of these employees. First because, in the confinement phase, consumption habits have changed and the gross demand from supermarkets and pharmacies has increased the volume of orders for this type of merchandise. With emphasis on basic needs, tobacco, and some generic drugs, consumers preferred online and e-commerce gained expression in this sudden volume of orders. Within the scope of the participant observation, it was possible to verify that the company had to adopt digital tools to support the preparation of orders, namely, Google Sheet. This tool allowed a better management of orders, dimensioning of teams, and agility in the transport of goods.

Teleworking, in turn, was the biggest revolution for the logistics operator, since this measure was transversal to all administrative areas of the company to ensure the safety of all employees. In this regard, there was an investment in media (e.g., laptops, smartphones, tablets) to ensure the normal daily working of the company. This situation was easily observed during visits to the company. The VPN settings have been adapted for all employees to access remotely. Digital conversation tools such as Skype, Zoom, and Microsoft Teams were also used, gaining expression as the preferred means of communication between teams and customers. The creation of Google Sheets made it easier to share information between departments. Some billing processes were improved through the creation of VBA code, to compose the face-to-face regime at the billing conference. The printing of paper guides on some customers has been replaced by digital sending. While the company faced a considerable percentage of teleworkers, additional solutions had to be created for the periodic training process. In this regard, an e-learning platform with mandatory training was created. The company recognizes that this is a scenario affected by the pandemic, but that it will be adopted in the future with good feedback from employees and their managers.

While the pandemic revolutionized the day-to-day management, companies had to adopt new working methods and/or redesign the existing ones. Although the success of digital transformations in the company has been boosted by the pandemic, it was fundamental for cultural change within the organization, whose focus was innovation and continuous improvement. It is thus evident that the digital age is increasingly revolutionizing the organization and its stakeholders, in particular customers who demand greater responsiveness in digital media. The adaptation of the business model to the digital is the result of the desire to follow the market and be up to the needs.

In short, while COVID-19 affected the retail and logistics sector, it also served as a wake-up call to an emerging reality – the digital one. Thus, the modus operandi has been increasingly focused on e-commerce, e-learning, teleworking, innovative organizational culture, and continuous improvement. Which, in turn has been seen as a mitigating force for the effects of the disease, but also a driving force for digital transformation. In the post-COVID-19, the logistics operator under analysis in this article hopes to invest in new technologies, aiming at a greater integration of robotic and artificial intelligence technologies in the logistics process, while intending to maintain the advances made in the digital field.

5 Conclusion

This article aims to fill a gap in the literature in the context of the digital transformation process and how technological advances in the retail sector have mitigated the effects of COVID-19. To do so, we used a case study research that is justified by the need to deal with a timely phenomenon and about which little is known. The discussion section is divided into sections that aim to provide an organized and broad view of the article, as such we focus the key findings into theoretical and managerial contributions, as well as presenting the research limitations and suggestions for future research.

5.1 Theoretical and Managerial Contributions

We have identified several key findings. The first concerns the foundation of a new customer profile and what is the effect of the digital age prior to the COVID-19 pandemic. This profile is due to the development of new digital skills and the adoption of disruptive technological means that enables a greater connection with companies. The second contribution refers to the identification of a gradual transition from I4.0 concepts to the traditional retail sector. In other words, we found that retailers and logistic operators have been looking for technologies that allow them to establish new digital interactions with customers and employees. The third and most relevant finding concerns the COVID-19 epidemic and its impact on the relationship with consumers and retail professionals. In that regard, we identified structural and managerial instruments that allowed the development of advances in the field of digital transformation, as is the case of: (1) obtaining resources; (2) developing an organizational culture that promotes change; (3) training employees; (4) engaging them in processes; as well as (5) leadership that can provide the right guidelines. The instruments identified above are in accordance with the categories we identified during the literature review (cf. Sect 2). However, the novelty of this work, in addition to the identification of structural and managerial instruments, was its empirical validation, classification according to their degree of importance (see Fig. 4), and expanded discussion in light of the current phenomenon of COVID-19. In addition, we also identified the organizational changes that made the mitigation of the COVID-19 effects possible, such as the increase in e-commerce, e-learning, teleworking, innovative organizational culture, and continuous improvement.

This article also aims to encourage other companies in the sector to follow a similar path or to complement the knowledge acquired with new and relevant information. In particular, to bring to the discussion more structural and managerial instruments or expand organizational changes that are capable of efficiently mitigating the COVID-19 effects. It is important to note that COVID-19 had a disruptive effect on companies and reinforced the importance of digital transformation through scenarios that forced both consumers to adopt other means of relationship with companies and companies that had to quickly respond to market demand. While the digital transformation is back at the center of corporate strategies (if it ever ceased to be), academics and professionals must be prepared to describe this new phenomenon, which may arise again in the future and that may once again contribute to the promotion of the digital trend. It is important to mention that this research is far from complete, as we are just beginning to really understand the true impacts of COVID-19 on retail. Thus, a greater number of studies that deepen the theme are needed to ensure a more comprehensive understanding of the topic.

5.2 Limitations and Suggestion for Future Research

This research is not free of limitations. With regard to the systematic review, this research presents a snapshot of the reality, since the Scopus database is constantly being updated. Therefore, it is expected that the results of this research are restricted to the moment in which it was carried out. It is also likely that relevant manuscripts were left out of the analysis due to the application of filters. Although PRISMA allows the incorporation of documents that, in the researcher's perspective, are relevant. This research is limited to the Portuguese reality as qualitative case studies by definition are non-representative due to the limited sample size [62]. In our opinion, the presented results are a good starting point, but they do not have the conditions to guarantee generalization. In other words, the results must be validated in other countries, inside and outside the European context. Future research may also consider conducting a quantitative and more comprehensive post-pandemic study to analyze the effects of COVID-19 on the retail industry. In particular, on those organizations that have chosen not to invest in digital technologies, but also the COVID-19 effects on companies in which they have chosen to do so. The understanding and description of the new coronavirus phenomena is very relevant as it allows companies and citizens to be better prepared for similar future events. It is for this reason that, in addition to the current contribution of this article, we also hope that it will encourage other researchers to analyze the impacts of COVID-19 on logistics and retail companies.

References

1. Rothan, H., & Byrareddy, S. (2020). The epidemiology and pathogenesis of coronavirus disease (COVID-19) outbreak. *Journal of Autoimmunity, 109*, 102433.
2. Pachetti, M., Marini, B., Giudici, F., Benedetti, F., Angeletti, S., Ciccozzi, M., Masciovecchio, C., Ippodrino, R., & Zella, D. (2020). Impact of lockdown on Covid-19 case fatality rate and viral mutations spread in 7 countries in Europe and North America. *Journal of Translational Medicine, 18*(1), 1–7.
3. Andersen, K., Rambaut, A., Lipkin, W., Holmes, E., & Garry, R. (2020). The proximal origin of SARS-CoV 2. *Nature Medicine, 26*(4), 450–452.
4. WHO. (2020). *Coronavirus disease 2019 (COVID-19): Situation report, 82*. World Health Organization. Available online: https://www.who.int/docs/default-source/coronaviruse/situation-reports/20200411-sitrep-82-covid-19.pdf?sfvrsn=74a5d15_2. Accessed on 16 Feb 2021.
5. WHO. (2020). *Coronavirus disease 2019 (COVID-19): Situation report, 51*. World Health Organization. Available online: https://www.who.int/docs/default-source/coronaviruse/situation-reports/20200311-sitrep-51-covid-19.pdf?sfvrsn=1ba62e57_10. Accessed on 16 Feb 2021.
6. Douglas, M., Katikireddi, S., Taulbut, M., McKee, M., & McCartney, G. (2020). Mitigating the wider health effects of covid-19 pandemic response. *BMJ, 369*, 1–6. https://doi.org/10.1136/bmj.m1557
7. Goddard, E. (2020). The impact of COVID-19 on food retail and food service in Canada: Preliminary assessment. *Canadian Journal of Agricultural Economics/Revue canadienne d'agroeconomie*. https://doi.org/10.1111/cjag.12243
8. Laato, S., Islam, A., Farooq, A., & Dhir, A. (2020). Unusual purchasing behavior during the early stages of the COVID-19 pandemic: The stimulus-organism-response approach. *Journal of Retailing and Consumer Services, 57*, 102224.
9. Tanveer, M., Hassan, S., & Bhaumik, A. (2020). COVID-19 quarantine and consumer behavior that change the trends of business sustainability & development. *Academy of Strategic Management Journal, 19*(4), 1–11.
10. Hall, M., Prayag, G., Fieger, P., & Dyason, D. (2020). Beyond panic buying: Consumption displacement and COVID-19. *Journal of Service Management, 32*(1), 113–128.
11. Bartsch, S., Weber, E., Büttgen, M., & Huber, A. (2020). Leadership matters in crisis-induced digital transformation: How to lead service employees effectively during the COVID-19 pandemic. *Journal of Service Management, 32*(1), 71–85.

12. Končar, J., Grubor, A., Marić, R., Vučenović, S., & Vukmirović, G. (2020). Setbacks to IoT implementation in the function of FMCG supply chain sustainability during COVID-19 pandemic. *Sustainability, 12*(18), 7391.
13. Hernandez-Mejia, G., & Hernandez-Vargas, E. (2020). When is SARS-CoV-2 in your shopping list? *Mathematical Biosciences, 328*, 108434.
14. Ntounis, N., Mumford, C., Loroño-Leturiondo, M., Parker, C., & Still, K. (2020). How safe is it to shop? Estimating the amount of space needed to safely social distance in various retail environments. *Safety Science, 132*, 104985.
15. Shahbaz, M., Bilal, M., Moiz, A., Zubair, S., & Iqbal, H. (2020). Food safety and COVID-19: Precautionary measures to limit the spread of coronavirus at food service and retail sector. *Journal of Pure and Applied Microbiology, 14*(1), 749–756.
16. Luceño-Moreno, L., Talavera-Velasco, B., García-Albuerne, Y., & Martín-García, J. (2020). Symptoms of posttraumatic stress, anxiety, depression, levels of resilience and burnout in Spanish health personnel during the COVID-19 pandemic. *International Journal of Environmental Research and Public Health, 17*(15), 5514.
17. Pantano, E., Pizzi, G., Scarpi, D., & Dennis, C. (2020). Competing during a pandemic? Retailers' ups and downs during the COVID-19 outbreak. *Journal of Business Research, 116*, 209–213.
18. Wang, Y., Xu, R., Schwartz, M., Ghosh, D., & Chen, X. (2020). COVID-19 and retail grocery management: Insights from a broad-based consumer survey. *IEEE Engineering Management Review, 48*(3), 202–211.
19. Kraus, S., Durst, S., Ferreira, J., Veiga, P., Kailer, N., & Weinmann, A. (2022). Digital transformation in business and management research: An overview of the current status quo. *International Journal of Information Management, 63*, 102466.
20. Falagas, M., Pitsouni, E., Malietzis, G., & Pappas, G. (2008). Comparison of PubMed, Scopus, web of science, and Google scholar: Strengths and weaknesses. *The FASEB Journal, 22*(2), 338–342. https://doi.org/10.1096/fj.07-9492LSF
21. Mongeon, P., & Paul-Hus, A. (2016). The journal coverage of Web of Science and Scopus: A comparative analysis. *Scientometrics, 106*(1), 213–228.
22. Liberati, A., Altman, D., Tetzlaff, J., Mulrow, C., Gøtzsche, P., Ioannidis, J., Clarke, M., Devereaux, P., Kleijnen, J., & Moher, D. (2009). The PRISMA statement for reporting systematic reviews and meta-analyses of studies that evaluate health care interventions: Explanation and elaboration. *Journal of Clinical Epidemiology, 62*(10), e1–e34.
23. Moher, D., Liberati, A., Tetzlaff, J., & Altman, D. (2009). Preferred reporting items for systematic reviews and meta-analyses: The PRISMA statement. *PLoS Medicine, 6*(7), e1000097.
24. Giri, A., Chatterjee, S., Paul, P., & Chakraborty, S. (2019). Determining the impact of artificial intelligence on 'Developing Marketing Strategies' in Organized Retail Sector of West Bengal, India. *International Journal of Engineering and Advanced Technology, 8*(6), 3031–3036.
25. Jain, S., & Gandhi, A. (2021). Impact of artificial intelligence on impulse buying behaviour of Indian shoppers in fashion retail outlets. *International Journal of Innovation Science, 13*(2), 193–204. https://doi.org/10.1108/IJIS-10-2020-0181
26. Rana, J., Gaur, L., Singh, G., Awan, U., & Rasheed, M. (2021). Reinforcing customer journey through artificial intelligence: A review and research agenda. *International Journal of Emerging Markets*. ahead-of-print. https://doi.org/10.1108/IJOEM-08-2021-1214
27. Reis, J., Amorim, M., Melão, N., & Matos, P. (2018). Digital transformation: A literature review and guidelines for future research. In *World conference on information systems and technologies* (pp. 411–421). Springer.
28. Krasonikolakis, I., Tsarbopoulos, M., & Eng, T. (2020). Are incumbent banks bygones in the face of digital transformation? *Journal of General Management, 46*(1), 60–69. https://doi.org/10.1177/0306307020937883
29. Almaslamani, F., Abuhussein, R., Saleet, H., AbuHilal, L., & Santarisi, N. (2020). Using big data analytics to design an intelligent market basket-case study at Sameh mall. *International Journal of Engineering Research and Technology, 13*, 3444–3455.

30. Ivanova, N. (2021). The future of retail: Innovations and basic trends. In S. Ashmarina, J. Horák, J. Vrbka, & P. Šuleř (Eds.), *Innovative economic symposium* (Vol. 160, pp. 157–163). Springer. https://doi.org/10.1007/978-3-030-60929-0_21
31. Reinartz, W., Wiegand, N., & Imschloss, M. (2019). The impact of digital transformation on the retailing value chain. *International Journal of Research in Marketing, 36*(3), 350–366.
32. Singh, A., & Thirumoorthi, P. (2019). The impact of digital disruption technologies on customer preferences: The case of retail commerce. *International Journal of Recent Technology and Engineering (IJRTE), 8*, 1255–1261.
33. Wohllebe, A., Dirrler, P., & Podruzsik, S. (2020). Mobile apps in retail: Determinants of consumer acceptance – A systematic review. *International Journal of Interactive Mobile Technologies, 14*, 1–12.
34. Ferreira, M., Moreira, F., Pereira, C., & Durão, N. (2020). The digital transformation at organizations – The case of retail sector. In *World Conference on Information Systems and Technologies* (pp. 560–567). Springer. https://doi.org/10.1007/978-3-030-45688-7_56
35. Bulović, V., & Čović, Z. (2020, October 8). *The impact of digital transformation on sustainability in fashion retail*. Proceedings of 2020 IEEE 18th International Symposium on Intelligent Systems and Informatics (SISY), Subotica, IEEE: New York City, 149–154. https://doi.org/10.1109/SISY50555.2020.9217087
36. Suel, E., & Polak, J. (2018). Incorporating online shopping into travel demand modelling: Challenges, progress, and opportunities. *Transport Reviews, 38*(5), 576–601. https://doi.org/10.1080/01441647.2017.1381864
37. Chaparro-Peláez, J., Acquila-Natale, E., Hernández-García, Á., & Iglesias-Pradas, S. (2020). The digital transformation of the retail electricity market in Spain. *Energies, 13*(8), 2085. https://doi.org/10.3390/en13082085
38. Alexandrova, E., & Poddubnaya, M. (2020). Digital technologies development in industry sectors and areas of activity. In *International Conference on Integrated Science* (pp. 112–124). Springer. https://doi.org/10.1007/978-3-030-49264-9_10
39. Hamidi, S., Yusof, M., Shuhidan, S. M., & Kadir, S. (2019). IR4.0: Unmanned store apps. *Indonesian Journal of Electrical Engineering and Computer Science, 17*(3), 1540–1547.
40. Sujata, J., Mukul, P., & Hasandeep, K. (2019). Role of smart communication technologies for smart retailing. *International Journal of Innovative Technology and Exploring Engineering, 8*, 213–218.
41. Chernova, D., Sharafutdinova, N., Nurtdinov, I., Valeeva, Y., & Kuzmina, L. (2020). The transformation of the customer value of retail network services under digitalization. In S. Ashmarina, M. Vochozka, & V. Mantuleko (Eds.), *International Scientific Conference, Proceedings of digital transformation of the economy: Challenges, trends, new opportunities* (pp. 252–260). Springer. https://doi.org/10.1007/978-3-030-27015-5
42. Hauser, M., Günther, S., Flath, C., & Thiesse, F. (2019). Towards digital transformation in fashion retailing: A design-oriented IS research study of automated checkout systems. *Business & Information Systems Engineering, 61*(1), 51–66.
43. Amazon (2021). *Amazon Go. Amazon Go – Amazon.com*, Available online: https://www.amazon.com/b?ie=UTF8&node=16008589011. Accessed on 29 May 2021.
44. Rahaman, S., Mohammed, S., Manchanda, T., & Mahadik, R. (2019). E-Pharm assist: The future approach for dispensing medicines in smart cities. In *2019 International conference on digitization (ICD), Sharjah, United Arab Emirates, Nov 18–19* (pp. 263–267). IEE.
45. Soto-Acosta, P. (2020). COVID-19 pandemic: Shifting digital transformation to a high-speed gear. *Information Systems Management, 37*(4), 260–266.
46. Sund, K., Bogers, M., & Sahramaa, M. (2021). Managing business model exploration in incumbent firms: A case study of innovation labs in European banks. *Journal of Business Research, 128*, 11–19.
47. Souza, C., Szafir-Goldstein, C., & Aagaard, A. (2020). IoT in the context of digital transformation and business model innovation: The case of a traditional Brazilian wholesaler. In

Proceedings of the 2020 Global Internet of Things Summit (GIoTS), Dublin, Ireland, June 3 (pp. 1–6). IEEE.
48. Vom Brocke, J., Fay, M., Böhm, M., & Haltenhof, V. (2017). Creating a market analytics tool that marketers love to use: A case of digital transformation at Beiersdorf. In G. Oswald & M. Kleinemeier (Eds.), *Shaping the digital enterprise* (p. 20). Springer.
49. Bollweg, L., Lackes, R., Siepermann, M., & Weber, P. (2020). Drivers and barriers of the digitalization of local owner operated retail outlets. *Journal of Small Business & Entrepreneurship, 32*(2), 173–201. https://doi.org/10.1007/978-3-319-40967-2_10
50. Brehm, L., & Klein, B. (2017). Applying the research on product-service systems to smart and connected products. In *International conference on business information systems* (pp. 311–319). Springer. https://doi.org/10.1007/978-3-319-52464-1_28
51. Kothari, U., & Seetharaman, A. (2020). Impact of digital transformation on retail banking industry in the UAE. In S. Sharma, Y. Dwivedi, B. Metri, & N. Rana (Eds.), *International working conference on transfer and diffusion of IT, Tiruchirappalli, India, Dec 18–19* (pp. 425–438). Springer. https://doi.org/10.1007/978-3-030-64849-7
52. Wolpert, S., & Roth, A. (2020). Development of a classification framework for technology based retail services: A retailers' perspective. *The International Review of Retail, Distribution and Consumer Research, 30*(5), 498–537. https://doi.org/10.1080/09593969.2020.1768575
53. Barann, B., Betzing, J., Niemann, M., Hoffmeister, B., & Becker, J. (2020). Exploring customers' likeliness to use e-service touchpoints in brick and mortar retail. *Electronic Markets*, 1–23.
54. Barann, B. (2018). *An IS-perspective on omni-channel management: Development of a conceptual framework to determine the impacts of touchpoint digitalization on retail business processes*. Research-in-Progress Papers, Proceedings of ECIS 2018 Proceedings, Portsmouth, UK, 18.
55. Mugge, P., Abbu, H., Michaelis, T., Kwiatkowski, A., & Gudergan, G. (2020). Patterns of digitization: A practical guide to digital transformation. *Research-Technology Management, 63*(2), 27–35. https://doi.org/10.1080/08956308.2020.1707003
56. Van Dyk, R., & Van Belle, J. (2019). Drivers and challenges for digital transformation in the South African retail industry. In E. Ziemba (Ed.), *Information technology for management: Current research and future directions* (pp. 42–62). Springer. https://doi.org/10.1007/978-3-030-43353-6
57. Zaychenko, I., Smirnova, A., Shytova, Y., Mutalieva, B., & Pimenov, N. (2020). Digital logistics transformation: Implementing the internet of things (IoT). In H. Schaumburg, S. Korablev, & U. Laszlo (Eds.), *International conference on technological transformation: A new role for human, machines and management, Warsaw* (pp. 189–200). Springer.
58. Yin, R. (2018). *Case study research and applications: Design and methods*. SAGE Publications.
59. Rashid, Y., Rashid, A., Warraich, M., Sabir, S., & Waseem, A. (2019). Case study method: A step-by-step guide for business researchers. *International Journal of Qualitative Methods, 18*, 1609406919862424.
60. Baškarada, S. (2014). Qualitative case study guidelines. *The Qualitative Report, 19*(40), 1–25.
61. Etikan, I., Musa, S., & Alkassim, R. (2016). Comparison of convenience sampling and purposive sampling. *American Journal of Theoretical and Applied Statistics, 5*(1), 1–4.
62. Gerring, J. (2009). Case selection for case-study analysis: Qualitative and quantitative techniques. In *The Oxford handbook of political methodology*. Oxford University Press. https://doi.org/10.1093/oxfordhb/9780199286546.003.0028
63. Klein, H., & Myers, M. (1999). A set of principles for conducting and evaluating interpretive field studies in information systems. *MIS Quarterly, 23*(1), 67–93.
64. Walsham, G. (2006). Doing interpretive research. *European Journal of Information Systems, 15*(3), 320–330.
65. Merriam, S., & Tisdell, E. (2015). *Qualitative research: A guide to design and implementation*. Wiley.

66. Flick, U., von Kardoff, E., & Steinke, I. (Eds.). (2004). *A companion to qualitative research*. SAGE Publications.
67. Patton, M. (2005). Qualitative research. *Encyclopedia of Statistics in Behavioral Science*. https://doi.org/10.1002/0470013192.bsa514
68. Triangulation, D. (2014). The use of triangulation in qualitative research. *Oncology Nursing Forum, 41*(5), 545–547.
69. Kvale, S. (1994). *Interviews: An introduction to qualitative research interviewing*. SAGE Publications.
70. Gill, P., Stewart, K., Treasure, E., & Chadwick, B. (2008). Methods of data collection in qualitative research: Interviews and focus groups. *British Dental Journal, 204*(6), 291–295. https://doi.org/10.1038/bdj.2008.192
71. Hancock, D., Algozzine, B., & Lim, J. (2021). *Doing case study research: A practical guide for beginning researchers*. Teachers College Press.
72. Fernandes, J., Reis, J., Melão, N., Teixeira, L., & Amorim, M. (2021). The role of industry 4.0 and BPMN in the arise of condition-based and predictive maintenance: A case study in the automotive industry. *Applied Sciences, 11*(8), 3438.
73. Bowen, G. (2009). Document analysis as a qualitative research method. *Qualitative Research Journal, 9*(2), 27–40. https://doi.org/10.3316/QRJ0902027
74. Jorgensen, D. (1989). *Participant observation: A methodology for human studies*. Sage.
75. Blaxter, L. (2010). *How to research* (pp. 99–131). McGraw-Hill Education.
76. Reis, J., Santo, P., & Melão, N. (2020). Impact of artificial intelligence research on politics of the European Union member states: The case study of Portugal. *Sustainability, 12*(17), 6708.
77. Petrova, E., Dewing, J., & Camilleri, M. (2016). Confidentiality in participatory research: Challenges from one study. *Nursing Ethics, 23*(4), 442–454.
78. Cross, R., & Cummings, J. (2004). Tie and network correlates of individual performance in knowledge-intensive work. *Academy of Management Journal, 47*(6), 928–937.
79. Spradley, J. (2016). *Participant observation*. Waveland Press.
80. Mills, A., Durepos, G., & Wiebe, E. (2010). *Encyclopedia of case study research*. SAGE Publications.
81. Given, L. (Ed.). (2008). *The Sage encyclopedia of qualitative research methods*. SAGE Publications.
82. Paschou, T., Rapaccini, M., Adrodegari, F., & Saccani, N. (2020). Digital servitization in manufacturing: A systematic literature review and research agenda. *Industrial Marketing Management, 89*, 278–292.
83. Elo, S., & Kyngäs, H. (2008). The qualitative content analysis process. *Journal of Advanced Nursing, 62*(1), 107–115.
84. Azungah, T. (2018). Qualitative research: Deductive and inductive approaches to data analysis. *Qualitative Research Journal, 18*(4), 383–400.
85. Thomas, D. (2006). A general inductive approach for analyzing qualitative evaluation data. *American Journal of Evaluation, 27*(2), 237–246.
86. Maher, C., Hadfield, M., Hutchings, M., & de Eyto, A. (2018). Ensuring rigor in qualitative data analysis: A design research approach to coding combining nvivo with traditional material methods. *International Journal of Qualitative Methods, 17*(1), 1609406918786362.
87. Auerbach, C., & Silverstein, L. (2003). *Qualitative data: An introduction to coding and analysis*. New York University Press.
88. Hsieh, H., & Shannon, S. (2005). Three approaches to qualitative content analysis. *Qualitative Health Research, 15*(9), 1277–1288.
89. Gioia, D., Corley, K., & Hamilton, A. (2013). Seeking qualitative rigor in inductive research: Notes on the Gioia methodology. *Organizational Research Methods, 16*(1), 15–31.
90. Eisenhardt, K. (1989). Building theories from case study research. *Academy of Management Review, 14*(4), 532–550.

91. Saunders, M., & Townsend, K. (2016). Reporting and justifying the number interview participants in organizational and workplace research. *British Journal of Management, 27*(4), 836–852.
92. Golafshani, N. (2003). Understanding reliability and validity in qualitative research. *The Qualitative Report, 8*(4), 597–607.
93. Tekic, Z., & Koroteev, D. (2019). From disruptively digital to proudly analog: A holistic typology of digital transformation strategies. *Business Horizons, 62*(6), 683–693.
94. Gurumurthy, R., Schatsky, D., & Camhi, J. *Uncovering the connection between digital maturity and financial performance*. Deloitte Development LLC. Available online: https://www2.deloitte.com/us/en/insights/topics/digital-transformation/digital-transformation-survey.html. Accessed on 19 Feb 2022.
95. Margiono, A. (2021). Digital transformation: Setting the pace. *Journal of Business Strategy, 42*(5), 315–322. https://doi.org/10.1108/JBS-11-2019-0215
96. Hess, T., Matt, C., Benlian, A., & Wiesböck, F. (2016). Options for formulating a digital transformation strategy. *MIS Quarterly Executive, 15*(2), 123–140.
97. Loonam, J., Eaves, S., Kumar, V., & Parry, G. (2018). Towards digital transformation: Lessons learned from traditional organizations. *Strategic Change, 27*(2), 101–109.
98. McKinsey. (2022). *To emerge stronger from the COVID-19 crisis, companies should start reskilling their workforces now*. Available online: https://www.mckinsey.com/business-functions/people-and-organizational-performance/our-insights/to-emerge-stronger-from-the-covid-19-crisis-companies-should-start-reskilling-their-workforces-now. Accessed on 19 Feb 2022.
99. Kane, G., Palmer, D., Phillips, A., Kiron, D., & Buckley, N. (2015). Strategy, not technology, drives digital transformation. *MIT Sloan Management Review and Deloitte University Press*, 14(1–25).
100. Reis, J., Melão, N., Salvadorinho, J., Soares, B., & Rosete, A. (2020). Service robots in the hospitality industry: The case of Henn-na hotel, Japan. *Technology in Society, 63*, 101423.
101. Reis, J., Cohen, Y., Melão, N., Costa, J., & Jorge, D. (2021). High-tech defense industries: Developing autonomous intelligent systems. *Applied Sciences, 11*(11), 4920.

Technician Routing and Scheduling Problem: A Case Study

César Nunes and Manuel P. Lopes

Abstract The problem of routing and scheduling of technicians is a problem that technical assistance and maintenance companies face nowadays, market competitiveness requires quick response, service diversification, and customer satisfaction. The relationship between competitiveness and profitability of companies involves the effective management of their resources. The work developed addresses a real problem of a major Portuguese company providing technical assistance to the home, a varied set of services (need for specific skills and execution times) must be scheduled for a set of technicians with heterogeneous skills and geographical locations (start and end of the route) based on their different places of residence. The results show a considerable increase in the efficiency levels of the solution obtained when compared to the company's current solution and reveals that the lack of homogeneity of skills among technicians and the variation in service flows are factors that should be considered in the operational management of resources and the contracting of work, and that the increase in working hours can also contribute to improving the efficiency of the process.

Keywords Technician routing and scheduling problem · Technician and task scheduling problem

---

C. Nunes
IPP-ISEP – School of Engineering of the Polytechnic of Porto, Porto, Portugal

M. P. Lopes (✉)
IPP-ISEP – School of Engineering of the Polytechnic of Porto, Porto, Portugal

INESC-TEC, Porto, Portugal
e-mail: mpl@isep.ipp.pt

J. C. de Oliveira Matias et al. (eds.), *Quality Innovation and Sustainability*, Springer Proceedings in Business and Economics,
https://doi.org/10.1007/978-3-031-12914-8_30

1 Introduction

The problem of routing and scheduling of technicians (TRSP) is a problem that technical assistance and maintenance companies face nowadays, market competitiveness requires quick response, service diversification, and customer satisfaction. TRSP in its static definition treats a limited team of technicians who must meet a set of services, it can be seen as an extension of VRPTW where technicians play the role of vehicles. In TRSP, each technician has a set of skills, tools, and spare parts, while services require a subset of each. The goal is to determine a set of routes so that each service is performed exactly once in its time window by a technician with the required skills, tools, and spare parts. TRSP naturally has a wide range of applications from telecommunications, healthcare, and maintenance services. Technicians generally start their daily route from their home with a set of tools and parts that enable them to meet an initial set of services; during the day, technicians can replenish tools and parts at a central depot and thus meet more services. Tools are treated as renewable assets and parts as consumables. TRSP presents restrictions on compatibility between technicians and services, although skills are intrinsic attributes, technicians can load different tools and parts [1].

The DTRSP addresses the problem in a scenario in which new requests arise while technicians execute their route; in this type of scenario, two types of decisions must be made in real time: first, whenever the technician finishes a request, it must be decided what will be the next request; second, whenever a new request appears, the algorithm must decide whether it is possible or desirable to accept it or not. Rejection of a new request leads to a cost penalty corresponding to its outsourcing or postponement of the request [1].

2 Literature Review

The Technician and Task Scheduling Problem (TTSP) family consists of five types of problems [2]:

- Technician and Task Scheduling Problem (TTSP)
- Technician routing and scheduling problem (TRSP)
- Service Technician Routing and Scheduling Problem (STRSP)
- Multi-period Field Service Routing Problem (MFSRP)
- Multi-period Technician Routing and Scheduling Problem (MTRSP – Multi-period Technician Routing and Scheduling Problem)

This typology of problems is classified as NP-hard, that is, there are no known algorithms that can solve them in polynomial time, for large instances the exact solution techniques are not applicable to these problems, so the approximation techniques despite not guaranteeing the optimal solution find good quality solutions with computational times considered reasonable [3].

TTSP addresses the allocation of technicians to tasks with skill requirements that must be met [2]. TTSP was the basis of the ROADEF 2007 challenge, a biannual competition proposed by the French Operational Research Society in 2007 being that the problem used a real dataset provided by France Telecom with data instances ranging from 500 to 800 jobs for a set of 150 technicians.

TRSP can be seen as an extension of the vehicle routing problem with time windows (VRPTW) where additional constraints such as skills, tooling, and spare parts information are addressed; however, it does not include the complexity of team formation or precedence [4]. In an extension of the VRP to a TRSP, skill requirements are generated for each job and teams depart from headquarters to visit a set of customers. This version does not consider some constraints present in the 2007 ROADEF challenge problem, such as precedence relationships, priority levels, unavailability of technicians, length of the working day, and outsourcing; however, it does contemplate the complexity of routing [5].

Adaptive large nearest neighbor search (ALNS) heuristics have been used to solve problems such as TTSP and TRSP [5, 6] – they use several destructive and repair operators. A destructive operator removes a part of the current solution, and the repairer reinserts the destroyed tasks back into the solution. Each repair and destruction operation is tracked in order to increase the probability of always selecting the best performing operators and a consequent good quality solution.

The greedy random search procedure (GRASP) was one of the techniques used to solve the TTSP in the ROADEF 2007 contest [7]; this approach was so successful in a way that it won the contest. The solution is obtained through the greedy algorithm that iteratively selects the best decision at each step of the process, while local search is used to improve the solution in a small amount of time. The process is run until the best solution is found.

Local search, another method used to solve the 2007 ROADEF challenge [8], obtained the second place through a process where initially an initial solution is built through a greedy heuristic and then, in an improvement phase, operators are iteratively applied in order to improve the solution.

The intelligent decision heuristic was developed to tackle the TTSP instances for the ROADEF 2007 challenge [2]; it was characterized by building different team configuration scenarios at each stage of the allocation process and operators providing flexibilities in team configurations.

The parallel matheuristic was proposed to solve a TRSP [4] that included many constraints; the process is composed of three phases, a construction algorithm, a parallel ALNS, and a post-optimization mathematical programming phase. The parallel ALNS takes advantage of parallel architectures which results in significant computational acceleration.

The dynamic technician routing and scheduling problem (DTRSP) presents the main constraints of real problems, routing, time windows, skills, tools, and spare parts. It handles the new service requests in real time as the scheduling is implemented, this particularity allows the allocation of urgent services.

DTRSP [1] admits requests over time through a fast reoptimization approach based on a parallel ALNS that produces a new routing plan each time a new request comes in and a multi-plan approach that continuously optimizes a set of routes, these plans are used to make routing decisions.

The DTRSP, given its dynamic aspect, is the one that gathers the largest set of characteristics that could be linked to our problem.

3 Problem Description

This study addresses the real problem of a company that provides, among other services, technical assistance to home appliances. The typology of services varies between the type of equipment (large domestic, brown line, air conditioning, and water heating equipment) and the type of service provided (repair, maintenance, inspection, and installation). The services can arrive through the customer's telephone contact or through digital platforms; in the first case, the scheduling is carried out during the same telephone call; in the second case, the service must be scheduled with the customer within a maximum period of 24 h after its receipt. A strategy implemented in the company is that all services are always scheduled with the customer, the scheduling may be conditioned by customer availability; however, if there is a time window, it is provided by the operator to the customer. Scheduling with a time window can be seen as a strategy to increase the quality of service to the customer; however, this strategy can in part be replaced by a contact prior to the visit of the technician to the customer. In the real case of the company, regardless of the existence of a time window, for all assistances, the technician makes telephone contact with the customer before starting the visit.

The considerable number of technicians, the diversification of the typology of services, and the wide range of equipment on the market lead to heterogeneity in the skills of the technicians, so through their training, certification, and tools, skills are assigned to the technicians. Due to the specificity of each service, different times are needed for its realization; these times are essential to define the daily route of a technician because they must be added to the time used in travel.

The problems of the TTSP family present in general a set of additional constraints similar to the problem under study; however, there is a set of particularities of the problem under study to be considered:

- All technicians leave at the beginning of the working day from their residence and return at the end of the working day to their residence. The total time of a working day must contemplate these two journeys.
- All technicians have all the necessary tools in their vehicles to carry out the associated tasks.
- The supply of spare parts is not contemplated in the route since, in general, the parts are sent by carrier to the technician.
- The total time of the route for a day's work should include the time necessary to carry out the service; a day's work should be the sum of the journeys plus the sum of the times necessary for the interventions.

- The formation of teams is not contemplated, the services that require more than one technician are associated with one technician with a helper.
- The time windows define the arrival time of the technician to the service with a two-hour gap.
- The times to carry out the services vary according to their specificity.
- After being scheduled with the customer, the day and time (if applicable), can no longer be changed.
- The time frame for scheduling is variable depending on customer availability or quantity of service to be carried out.
- The objective is for all technicians to have a full agenda before starting the working day; however, when this is not possible the scheduling of service for the technician may be carried out on the day itself.
- The speed of response to a service may detract from the cost of a route.
- Service outsourcing is not contemplated.
- There are no precedence relationships between jobs.
- The lunch break is defined by the technician; the time windows applied allow the technician to manage this.
- There are no priority levels established for the services.

4 Solution Approach

Considering the specific characteristics of the problem under study and the need to obtain good solutions in a very short computational time, which allow the fast construction of operational decision support scenarios, the option goes for constructive heuristics commonly used in solution construction processes in meta-heuristic approaches. Based on the literature review, two constructive heuristics were considered: the cheapest insertion heuristic and the cheapest insertion heuristic with regret. Regret heuristics not only consider the minimum insertion cost, but they also consider the second lowest cost, the third lowest cost, and so on. In generality the heuristic (*Regret-n*) computes the part with the largest cost difference between the lowest cost insertions and the $n - 1$ next lowest cost insertions [9]. For instances of 250 or 500 services, the least cost insertion heuristic is able to produce lower cost solutions compared to the regret heuristic [10].

The goal of the algorithm is to fill the first work day for all technicians. In the insertion phase, it tests the first service without scheduling by going through the schedules of all technicians with the necessary skills for its realization until it can allocate it in the route of a technician at the lowest cost. If the algorithm cannot insert the service on the first day, it remains unscheduled until the algorithm moves to the second day and so on consecutively until it is possible to insert it.

The working day of a technician consists of 8 h (480 min), the route of the technician calculated in minutes includes the time of travel from home to the first service, the travel time between services, and the return time from the last service to his residence. Travel times are obtained using the service geographic coordinates and the Geodist function [11], with an average speed of 60 km per hour and adjustment of 15% compensation for differences between linear and road distances. The breaks in the working day are at the discretion of the technician and are not foreseen in the algorithm.

Scheduling is a continuous process throughout the day; however, the process of creating new services is generally a process that takes place at certain periods in the day.

The service data matrix contains the service geographic coordinates, service coding, and status. The definition of the service status is characterized by the numbers zero (service still to be scheduled), one (service allocated by the algorithm), and two (service already scheduled with restriction on the technician, day, and time window), for services characterized with status two, the day, technician, and time window cannot be changed.

The time windows are a restriction of the problem, the existence of services already scheduled for a particular technician, day, or within a certain time window results from cases where there is a time window at the customer's request or by commercial imposition in which the appointments must be made on a particular day within a specific time window. The beginning of the time window is not a fixed time of arrival of the technician to the service, for example, for the time window 10–12 h, it is indicated to the customer in the schedule that the technician has an estimated time of arrival between 10 and 12 h.

The times of completion for each type of service are obtained through the average time used by all technicians with that skill in the last 6 months, these are approximate values of reality; however, there are factors not considered that may lead to deviations in these values:

- The time used for the same repair varies from technician to technician, this factor comes from the existence of different diagnosis/repair methods, level of knowledge and experience among technicians.
- Within the same type of service, the service time varies between equipment and type of fault, there are faults that can be resolved quickly and faults that are more complex.

The times established for a technician's workday may be conditioned by the variation in intervention time, but also by other external factors such as customer delays, customer absences, and traffic situations that condition travel times. These factors make it unfeasible to apply fixed timetables for the arrival of the technician to the customer, thus, for cases where a time window is applied, this is provided to the customer in the form of a period of time.

The verification of the time window is contemplated in the algorithm during the insertion process, this verification is performed:

- To the left of the insertion: the sum of all service times and travel times must be less than the start time of the time window,
- To the right: the sum of all service times and travel times must be less than the end time of the time window

The algorithm was developed in the Excel platform using the VBA language. The algorithm pseudocode is shown in Fig. 1.

The tests were performed with a set of 350 services referring to the year 2020 and 2021; from 2020 are the services of the end of the year that were not scheduled (unreachable customers) and those that were scheduled for 2021; regarding the services of 2021 are the processes created in the first two working days of 2021. Table 1 shows the comparison of the results between the current solution and the algorithm's result for the first day.

The result of the algorithm is an increase of 18 more services in relation to the current solution, which translates into a reduction in travel time.

For the analysis of the algorithm performance, three performance indicators (KPI) were created.

```
Algorithm
Input: Matrix of services
Output: Scheduling of services
Begin
iday ← 0
while termination criteria not met do
  iday ← iday+1
  mcs ← 0
  min_cost ← M
  for each service do
    for each tech do
      if tech is skilled for the service then
        if service duration < route time available then
          travel_time ← Call Geodistance
          ins_cost ← Call Cheapest Insertion Heuristic
          if ins_cost < min_cost Then
            if travel_time +service duration <= route time available then
              if time windows of next firm services are not violated then
                min_cost ← ins_cost
                mcs ← service
              end if
            end if
          end if
        end if
      end if
    end for
  end for
  if mcs > 0 then
    Call Insert_Service
  end if
loop
End
```

Fig. 1 Algorithm pseudocode

Table 1 Comparison between solutions

	Actual solution	Algorithm
Number of scheduled services (*SA*)	126	146
Number of technicians used (*TU*)	25	25
Travel time (*TD*) (min)	3956	3182
Service time (*TS*) (min)	6900	7925
Idle time (*IT*)	1144	893

Table 2 Performance analysis for the first day

	Actual solution (%)	Algorithm (%)
RT	63.6	71.4
EA	57.5	66.0
EG	90.5	92.6

1. The routing efficiency (*RT*):

$$RT = \left(\frac{TS}{TD + TS} \right) * 100 \tag{1}$$

2. The algorithm efficiency (*EA*):

$$EA = \frac{TS}{TT} * 100 \tag{2}$$

3. The global efficiency (*EG*):

$$EG = \left(1 - \frac{TI}{TT} \right) * 100 \tag{3}$$

where TT is the total availability time of all technicians for one working day.

Table 2 presents the results obtained in the performance analysis for the first day. The result of the algorithm for the first day shows better results in all performance indicators relative to the current solution.

The constraints associated with the problem can condition the results of the algorithm, a sensitivity analysis to its constraints is required. The tests are performed with the instances and scheduling sequences for the results obtained on the first day.

For this analysis four different scenarios are tested:

- Scenario 1 – Large Service Time Windows (TW)
  The first scenario analyzes the use of larger service time windows, the periods of the day (morning/afternoon), i.e., the existing time slots in the morning shift from 2 to 4 h in the morning and the existing time slots in the afternoon shift from 2 to 4 h in the afternoon.

Table 3 Results of the four tested scenarios

	Actual solution	Algorithm	Large TW	Small TW	HS	DH
Number of scheduled services (*SA*)	126	146	149	144	154	150
Number of technicians used (*TU*)	25	25	25	25	25	25
Travel time (*TD*) (min)	3956	3182	3187	3203	2899	3087
Service time (*TS*) (min)	6900	7925	8075	7825	8325	8100
Idle time (*IT*)	1144	893	738	972	776	813
RT	63.6%	71.4%	71.7%	71.0%	74.2%	72.4%
EA	57.5%	66.0%	67.3%	65.2%	65.2%	67.5%
EG	90.5%	92.6%	93.9%	91.9%	93.5%	93.2%

- Scenario 2 – Small Service Time Windows
  The second scenario uses shorter than existing service time windows, one-hour windows.
- Scenario 3 – Homogeneous Skills (HS)
  The restrictions associated to technicians' skills may condition the allocation of service to technicians holding a lower number of skills, the third scenario considers that technicians hold homogeneous skills.
- Scenario 4 – Departure from Headquarters (DH)
  The distribution of technicians associated with the location of services can affect the performance of the algorithm; the fourth scenario considers that technicians start and finish their routes at the company's headquarters.

Table 3 presents the results of the four tested scenarios.

The routing efficiency (RT) in relation to the algorithm's result increases slightly for the large time windows scenario and decreases slightly for the small time windows scenario. There is a 1% increase when all technicians start at the company's headquarters, the most relevant increase occurs in the scenario where the technicians' skills are homogeneous, 2.5% with respect to the algorithm's solution and 10.6% with respect to the current solution.

The efficiency of the algorithm (EA) increases with respect to the algorithm's solution for the scenario of large working hours, departure from headquarters and homogeneous skills (greatest increase, 3.4%). It decreases slightly for the small time windows scenario.

Overall efficiency (EG) decreases in relation to the result of the algorithm for small time windows and increases for the scenario of large time windows, homogeneous valences and departure from headquarters. In this indicator, for the first time, the large time windows scenario presents a better result than the homogeneous valences scenario.

5 Conclusions

The work developed addresses a real problem of a major Portuguese company providing technical assistance to the home, a varied set of services (need for specific skills and execution times) must be scheduled for a set of technicians with heterogeneous skills and geographical locations (start and end of the route) based on their different places of residence. An algorithm to solve the DTRSP, based on the cheapest insertion heuristic was developed and implemented in the Excel platform using VBA. For the analysis of the algorithm performance, three performance indicators (KPI) were created and computational results show an overall increase in all of them, comparing the efficiency levels of the solution obtained to the company's current solution. The results also reveal that the lack of homogeneity of skills among technicians and the variation in service flows are factors that should be considered in the operational management of resources and the contracting of work, and that the increase in working hours can also contribute to improving the efficiency of the process.

For future work, we propose the comparison of the results of this approach with exact and metaheuristics approaches and perform more extensive computational tests.

References

1. Pillac, V., Guéret, C., & Medaglia, A. (2012). *On the dynamic technician routing and scheduling problem*. Diss. Ecole des Mines de Nantes.
2. Khalfay, A., Crispin, A., & Crockett, K. (2017). A review of technician and task scheduling problems, datasets and solution approaches. In *2017 intelligent systems conference (IntelliSys)* (pp. 288–296). IEEE.
3. Ernst, A. T., Jiang, H., Krishnamoorthy, M., & Sier, D. (2004). Staff scheduling and rostering: A review of applications, methods and models. *European Journal of Operational Research, 153*(1), 3–27.
4. Pillac, V., Gueret, C., & Medaglia, A. L. (2013). A parallel matheuristic for the technician routing and scheduling problem. *Optimization Letters, 7*(7), 1525–1535.
5. Kovacs, A. A., Parragh, S. N., Doerner, K. F., & Hartl, R. F. (2012). Adaptive large neighborhood search for service technician routing and scheduling problems. *Journal of Scheduling, 15*(5), 579–600.
6. Cordeau, J. F., Laporte, G., Pasin, F., & Ropke, S. (2010). Scheduling technicians and tasks in a telecommunications company. *Journal of Scheduling, 13*(4), 393–409.
7. Hashimoto, H., Boussier, S., Vasquez, M., & Wilbaut, C. (2011). A GRASP-based approach for technicians and interventions scheduling for telecommunications. *Annals of Operations Research, 183*(1), 143–161.
8. Estellon, B., Gardi, F., & Nouioua, K. (2009). High-performance local search for task scheduling with human resource allocation. In *International workshop on engineering stochastic local search algorithms* (pp. 1–15). Springer.
9. Pisinger, D., & Ropke, S. (2007). A general heuristic for vehicle routing problems. *Computers & Operations Research, 34*(8), 2403–2435.
10. Lutz, R. (2015). *Adaptive large neighborhood search*. Bachelor thesis at Ulm University.
11. TBonfi. GitHub. Obtained in 2021, from GitHub: https://github.com/TBonfi/Haversine-en-excel/blob/master/Haversine-Distancia.bas

Application of Fuzzy Methodologies in Navy Systems Maintenance

Suzana Lampreia, Teresa Morgado, Helena Navas, and Inês Mestre

Abstract In an environment of scarce resources, enhancing the ship's performance by minimizing equipment intervention actions and maintaining safety standards is an actual challenge. Ships, not yet autonomous, are maritime means that transport personnel and systems. Their good condition is a permit for the safety of personnel and material, avoiding damage to the ship itself and possible occurrences of pollution at sea or damage to other systems – seafarers and people outside the east. Companies, organizations, and the scientific community with an interest in ship maintenance management have been developing advanced systems for monitoring ship equipment; to prevent malfunctions and have on-site knowledge of the state of the equipment. These systems use condition control techniques and data processing through algorithms, statistical systems, and other methodologies. The methodology that will be developed and applied in this investigation is Fuzzy. The equipment chosen for the case study is a fire pump from an ocean patrol vessel, which is a vital

S. Lampreia
CINAV-Escola Naval/Naval Research Centre – Portuguese Naval Academy, Almada, Portugal

T. Morgado (✉)
CINAV-Escola Naval/Naval Research Centre – Portuguese Naval Academy, Almada, Portugal

UNIDEMI, Department of Mechanical and Industrial Engineering, NOVA School of Science and Technology, Universidade NOVA de Lisboa, Lisbon, Portugal

ISEL – College Engineering Institute of Lisbon, Lisboa, Portugal
e-mail: teresa.morgado@isel.pt

H. Navas
UNIDEMI, Department of Mechanical and Industrial Engineering, NOVA School of Science and Technology, Universidade NOVA de Lisboa, Lisbon, Portugal

Department of Mechanical and Industrial Engineering, NOVA School of Science and Technology, Universidade NOVA de Lisboa, Lisbon, Portugal

I. Mestre
Department of Mechanical and Industrial Engineering, NOVA School of Science and Technology, Universidade NOVA de Lisboa, Lisbon, Portugal

J. C. de Oliveira Matias et al. (eds.), *Quality Innovation and Sustainability*, Springer Proceedings in Business and Economics,
https://doi.org/10.1007/978-3-031-12914-8_31

equipment on board, and which is part of the equipment selected by the Technical Management of the Organization understudy to monitor operating hours and operational status.

Keywords Fuzzy · Maintenance · Risk · Ship · Equipment

1 Introduction

The importance of maintenance as a production support activity has become unquestionable due to its vast contribution in terms of operability and improvement of physical assets.

In this context, the concept of maintenance management was developed, defined by *Instituto Português da Qualidade* (IPQ) [1] as "all management activities that determine the objectives, strategy and responsibilities regarding maintenance and that implement them by various means such as planning, controlling and supervising maintenance and improving methods in the organization, including economic aspects."

Maintenance planning concerns all activities related to developing of a regularly scheduled work program to ensure the satisfactory operation of the equipment and avoid serious problems [2].

On the other hand, maintenance control and supervision are related to the various aspects that must be monitored to ensure the correct implementation of maintenance management [3]. According to Morvay and Gvozdenac [4], any management system comprises an organizational structure supported by three main aspects: procedures, people, and technology to control and optimize the use of resources and ensure the fulfillment of the company's objectives.

An information system can support the maintenance planning, execution processes to ensure the management of work lists, inventory, procedures and technical specifications, scheduling, and resources management. It can also conduct to transmission of interventions requests, monitoring, and spare parts reporting and cost control management. In this logic, the role of technology appears as a need for an information system closely connected to all relevant company assets and as a valuable tool for operators to ensure more reliable interventions [5].

However, no management is easy, and the maintenance management of maritime assets is complex due to the uncertainties and constraints involved, such as weather conditions [6].

If objectives like reducing the number of accidents, developing a flexible and multifaceted organization, improving production efficiency and, for example, creating a more accessible and perceptible system for all the people involved are achieved, then it can be said that the fundamental reasons for maintenance management are also being fulfilled, maximizing profit and offering a competitive advantage to the organization. This varied set of arguments has led to the choice and implementation of the "ideal model" no longer just a vague thought to become a recurring topic of research and a fundamental issue to achieve the effectiveness and efficiency of maintenance management [7].

In the context of maintenance management, with the objective of minimizing unexpected breakdowns, the Fuzzy methodology was chosen to develop a decision support model in the context of the maintenance of a ship fire pump.

2 The Fuzzy Methodology in Maintenance

Fuzzy logic came to give some answers regarding data that are neither false nor entirely true. This logic can develop a system for decision-making based on inputs of linguistic variables and, therefore, often vague and subjective from a mathematical point of view [8]. This methodology is used to express the absence of a clear boundary between sets of information [9]. According to Ierace and Cavalieri [8], this technique is suitable for maintenance since the authors consider that, for the most part, the general objectives of maintenance are intangible and depend on the worker's experience. However, according to the same authors, this method is sometimes too subjective, making it difficult to assess the importance of each objective, and the ability of each maintenance policy to achieve them.

2.1 Fuzzy Process

This methodology is suitable for working with imprecise human reasoning because it considers experience and knowledge. It also provides a mathematical framework to model the uncertainty of human cognitive processes that can be controlled by a computer [10].

To accomplish the Fuzzy methodology appliance, various steps and phases inherent to Fuzzy Logic process were considered [11]. The phases of this study and the Fuzzy steps are identified in Fig. 1 [11, 12].

Fuzzification is the process that converts input data into fuzzy values using membership functions. At this stage, the degrees of belonging of the elements are determined. In addition, as an essential process step, the specialist's contribution to the phenomenon under study that is intended to be modelled is usually justified [13]. On the other hand, the inference is the process that aims to enable a system action by evaluating the compatibility of the inputs with the conditions stipulated by the rule base. The number of fuzzy output sets must equal the number of stipulated rules [14].

Four types of inference methods are commonly discussed in the literature: the Mamdani model and Larsen's model. In these models, the antecedent and consequent are fuzzy propositions. The Tsukamoto model differs from previous models. The consequent is represented by a monotonic membership function. The Takagi-Sugeno model, in which the difference lies in the consequent being a

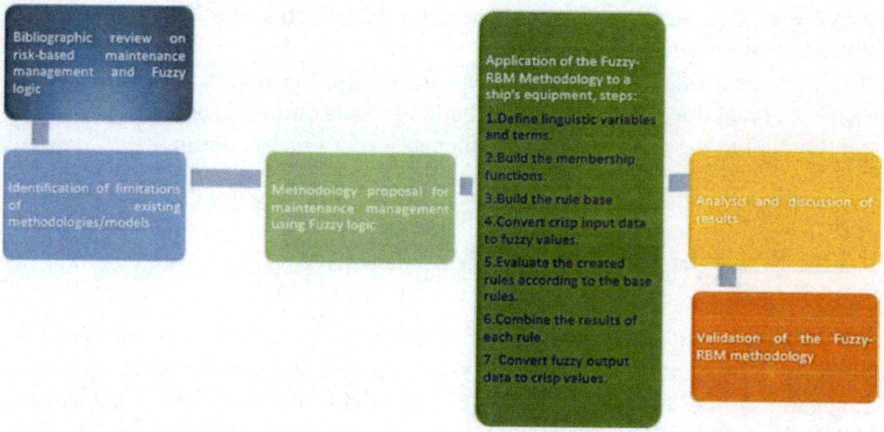

Fig. 1 Phases of study and steps in a fuzzy system. (Source: adapted from [7, 11, 12])

polynomial function [15]. It should be noted that the four methods differ only in the consequent.

Defuzzification is the step in which the system's output values are calculated based on the inference process and according to the belonging functions of the linguistic variables. The number of rules stipulates the number of outputs; if, for example, five rules are determined, then there will be five output values. However, there can only be a single final response value determined on a case-by-case basis since it has to be the one that best reproduces the region obtained through the output values [16]. There are several defuzzification methods, such as the Center of Gravity (COG), Center of Area (COA), and Mean of Maximums (MOM), among others. Only because it is the most used and referred to by various authors, in the present investigation, the authors chose to use the COG.

COG finds the point where a vertical line cuts the fuzzy set into two equal parts. The method finds a center of gravity of a set A in an interval ab. Mathematically, the COG can be calculated in at least two ways; the first, presented in Eq. 1 [12].

$$\mathrm{COG} = \frac{\int_a^b \mu_{\mathrm{A}}(x)\,x\mathrm{d}x}{\int_a^b \mu_{\mathrm{A}}(x)\,x\mathrm{d}x} \tag{1}$$

The second way is presented in Eq. 2; the COG is calculated through an estimate obtained through a sample of points [10].

$$\mathrm{COG} = \frac{\sum_{x=a}^{b} \mu.(x)\,x}{\sum_{x=a}^{b} \mu.(x)} \tag{2}$$

The symbols and their definitions in Eqs. 1 and 2 are as follows:

a and b represent an ab interval
x is the variable
μ_A is the degree of membership in relation to the fuzzy set A

Following the example of age, the linguistic variable is age, which can take linguistic terms such as young, middle-aged, or old, this phenomenon happens because usually, human beings express themselves in this way, in words instead of numbers [17]. Therefore, these linguistic terms can be described through fuzzy sets, which in turn are represented through membership functions that will be presented in the case study of this article.

2.2 *Conjuntos Fuzzy*

A fuzzy set A, belonging to a universe U, is defined through a membership function:

$$\mu_A(x) : U \rightarrow [0,1] \tag{3}$$

This membership function represents a fuzzy set through a set of ordered pairs given by:

$$A = \{(x, \mu_A(x)) | x \in U\} \tag{4}$$

In which there is only one match, a real number from the range [0,1], for each element x. The closer the value of μ_A is to 1, the greater the possibility that the element x belongs to the set A. This degree of belonging of the elements allows the occurrence of gradual transitions between true and false.

The gradual transition is given by:

$$\mu_A(x) : x \rightarrow [0,1], \begin{cases} \mu_A(x) = 0 \\ 0 < \mu_A(x) < 1 \\ \mu_A(x) = 1 \end{cases} \tag{5}$$

The membership functions use numerically translated linguistic terms that allow obtaining output values in the fuzzification and defuzzification stages, respectively. These functions can be of different types such as rectangular, triangular, trapezoidal, and others, the last two being the most frequent in the literature [13]. Triangular and trapezoidal membership functions can be described through the following equations respectively (Eqs. 6 and 7) [18]:

$$\mu_{\mathrm{A}} = \begin{cases} 0 & se \quad x < x_1 \\ \dfrac{x - x_1}{x_2 - x_1} & se \quad x_1 < x \leq x_2 \\ \dfrac{x_3 - x}{x_3 - x_2} & se \quad x_2 < x \leq x_3 \\ 0 & se \quad x > x_3 \end{cases} \tag{6}$$

$$\mu_{\mathrm{A}} = \begin{cases} 0 & se \quad x < x_1 \\ \dfrac{x - x_1}{x_2 - x_1}, & se \quad x_1 < x \leq x_2 \\ 1 & se \quad x_2 < x \leq x_3 \\ \dfrac{x_3 - x}{x_3 - x_2}, & se \quad x_3 < x \leq x_4 \\ 0 & se \quad x > x_4 \end{cases} \tag{7}$$

In the current state of the art, it is possible to find several examples of complex problems in which this logic has contributed to their understanding: as Meng Tay and Peng Lim [19] in its application in the manufacture of semiconductors, and Chung and Kim [20] in the wastewater prioritization.

The application of fuzzy logic has advantages and disadvantages. The advantages are as follows: improves the handling of inaccurate data; facilitates the process of specifying the system's rules; is more intuitive due to the use of words instead of numbers; facilitates the resolution of complex problems; provides faster development of system prototypes [21]. Moreover, the disadvantages: it makes it difficult to analyze aspects such as optimization; the specialist's experience and knowledge limit the accuracy of the fuzzy system; system is influenced by all its parameters, such as the method chosen for fuzzification, or the number of rules [22].

3 Methodology

The proposed methodology for implementing fuzzy logic assessing of the risk of failure of a fire pump on board a ship will contribute to determining the type of maintenance best suited to the reality in the organization and in the maintenance planning of assets. Therefore, the integration of this methodology within the scope of risk-based maintenance will then be demonstrated. The first step for implementing of Fuzzy logic will be the choice of equipment to implement the methodology. In Fig. 2 is presented the Fuzzy methodology for a ship's equipment [23].

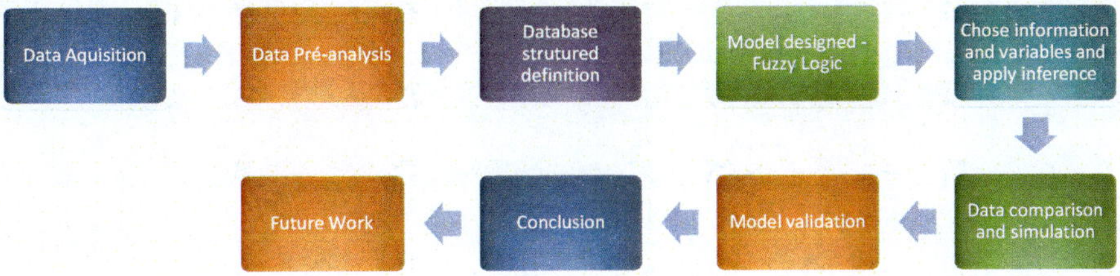

Fig. 2 Fuzzy methodology for a ship's equipment

Fig. 3 Fire electro pump. (Original photo from the corresponding author)

4 Case Study: Fire Electro Pump

In this study, the fuzzy methodology is applied to the electro pump system (Fig. 3).

To build the fuzzy methodology with information about a fire electric pump is essential to mention that the ocean patrol vessel has several fire electric pumps. Electric fire pumps are selected equipment, which means that the number of operating hours is counted for their maintenance management and that it is considered a vital piece of equipment in the ship's operation. The general characteristics of the electric fire pump shown in Fig. 4 are 100 m^3/h of Capacity; Head, 99.6 m; 2950 l/min of speed; 44.7 kW of power absorbed and 55 kW of electric motor power [24].

The basic preventive maintenance procedures for the fire pump are visual inspection, vibration measure, sealing organ substitution, pump external preservation, gauge calibration, and measure of insulation resistance.

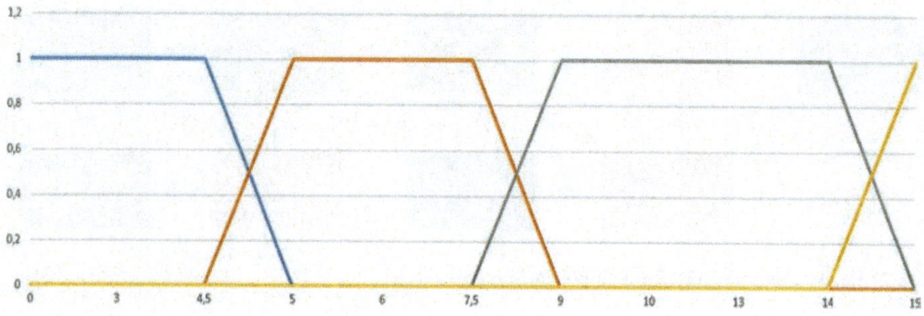

Fig. 4 Trapezoidal membership function

4.1 Fuzzy Parameterization

To perform the parameterization in the application of the fuzzy system, various simulation of criteria and its limits was taken. Finally, the frequency (F) of the failure was considered from remote to very frequent (4–<15), the operational impact (OI) from low to extremely high (0–<5), safety and environment impact (SI) from low to high (1–<5), low to high maintenance cost (MC) (1–<5). The consequence (C) is a result of the sum of OI, SI, and MC that contribute equitably. These criteria and limits were the ones that showed to be more applicable for the equipment performance. This subject is not more detailed because of the conference article dimension limitation.

Table 1 presents the inference rules and Table 2 the risk level classification.

For the defined attributes and the considered situations, the risk calculation is presented in Table 3. And it was based, after various simulation, in Eq. 8:

$$R = F \times C \tag{8}$$

The membership functions used to convert the frequency input values into linguistic terms, and the linguistic terms into output values are translated in Fig. 4.

The attributes levels that served as the basis for the developed methodology were defined based on the maritime maintenance expertise, and after various scenarios, simulation and attribution of various intervals values on considered criteria.

Based on the developed inference in fuzzy Logic, in the study of the ship's fire pump, the obtained results were in a range between non-critical and critical.

There was no overlapping of results in the trapezoidal membership function, which translates into a sufficiently adjusted model. Despite the obtained results, we suggest the need to observe equipment operation under the present rules and maybe adjust it if necessary.

Table 1 Inference rules

Rule no.	Rule	Rule no.	Rule
1	If (**F** is **MF**) and (**C** is **Ba**) then (**R** is **NC**)	11	If (**F** is **PF**) and (**C** is **Ba**) then (**R** is **NC**)
2	If (**F** is **MF**) and (**C** is **Mo**) then (**R** is **PC**)	12	If (**F** is **PF**) and (**C** is **Mo**) then (**R** is **NC**)
3	If (**F** is **MF**) and (**C** is **A**) then (**R** is SC)	13	If (**F** is **PF**) and (**C** is **A**) then (**R** is **PC**)
4	If (**F** is **MF**) and (**C** is **MA**) then (**R** is **C**)	14	If (**F** is **PF**) and (**C** is **MA**) then (**R** is **PC**)
5	If (**F** is **MF**) and (**C** is **EA**) then (**R** is **MC**)	15	If (**F** is **PF**) and (**C** is **EA**) then (**R** is SC)
6	If (**F** is **Fr**) and (**C** is **Ba**) then (**R** is NC)	16	If (**F** is **Re**) and (**C** is **Ba**) then (**R** is **NC**)
7	If (**F** is **Fr**) and (**C** is **Mo**) then (**R** is **PC**)	17	If (**F** is **Re**) and (**C** is **Mo**) then (**R** is **NC**)
8	If (**F** is **Fr**) and (**C** is **A**) then (**R** is SC)	18	If (**F** is **Re**) and (**C** is **A**) then (**R** is **NC**)
9	If (**F** is **Fr**) and (**C** is **MA**) then (**R** is **C**)	19	If (**F** is **Re**) and (**C** is **Ma**) then (**R** is **NC**)
10	If (**F** is **Fr**) and (**C** is **EA**) then (**R** is **C**)	20	If (**F** is **Re**) and (**C** is **Ea**) then (**R** is **PC**)

F frequency, C consequence, R risk, MF high frequency, Fr frequent, PF infrequent, Re remote, Ba low, Mo moderate, A high, MA very high, EA extremely High, NC not critical, PC little critical, SC semi-critical, C critical, MC very critical

Table 2 Risk level classification

Risk level	Value
Very critical	$R > 200$
Critical	$150 < R \leq 200$
Semi-critical	$100 < R \leq 150$
Less critical	$50 < R \leq 100$
Not critical	$R \leq 50$

5 Conclusion

The fuzzy methodology allows the measurement of human cognitive records, for which some criteria must be considered, and these must have several levels to be able to obtain a quantification of results through an algorithm.

Fuzzy logic converts text input data into fuzzy values using membership functions.

The result of the algorithm can be a risk level that allows the perception of the urgency of performing maintenance on equipment.

Table 3 Risk calculus result

Risk	TL R
0	Not critical
12	Not critical
30	Not critical
48	Not critical
128	Semi-critical
36	Not critical
40	Not critical
110	Semi-critical
120	Semi-critical
182	Critical
200	Critical
150	Semi-critical

The fuzzy methodology was applied to a ship's fire pump to serve as a decision support system for its maintenance.

In this decision support system, considering the methodology and method, the trapezoidal membership function did not result in overlapping results. It seems to be partially validated but must be subjected to a practical application under supervision of those responsible for implementing the system.

Applying fuzzy logic in ship equipment may help develop a decision support system in maintenance management of Portuguese Navy ships.

This methodology should be adapted and tested on other equipment.

Acknowledgements The authors from FCT NOVA acknowledge Fundação para a Ciência e a Tecnologia (FCT-MCTES) for its financial support via the project UIDB/00667/2020 (UNIDEMI).

References

1. NP: Portuguese Standard (Norma Portuguesa): NP-EN13306, Maintenance Terminology (Terminologia da Manutenção). IPQ: Instituto Português da Qualidade (2007)
2. Dhillon, B. S. (2002). *Engineering maintenance: A modern approach*. CRC Press.
3. Manzini, R., Regattieri, A., Pham, A., & Ferrari, E. (2010). *Maintenance for industrial systems*. Springer.
4. Morvay, Z., & Gvozdenac, D. (2008). Applied industrial energy and environmental management. Wiley.
5. Silvestri, L., Forcina, A., Introna, V., Santolamazza, A., & Cesarotti, V. (2020). Maintenance transformation through Industry 4.0 technologies: A systematic literature review. *Computers in Industry*.
6. Abbas, M., & Shafiee, M. (2020). An overview of maintenance management strategies for corroded steel structures in extreme marine environments. *Marine Structures, 71*, 102718.
7. Campos, M. L., & Márquez, A. C. (2008). Review, classification and comparative analysis of maintenance management models. *IFAC Proceedings Volumes, 41*(3), 239–244.

8. Ierace, S., & Cavalieri, S. (2008). Maintenance strategy selection: A comparison between fuzzy logic and analytic hierarchy process. *IFAC Proceedings Volumes, 41*(3), 228–233.
9. Maletič, D., Maletič, M., Lovrenčić, V., Al-Najjar, B., & Gomiscek, B. (2014). An application of Analytic Hierarchy Process (AHP) and sensitivity analysis for maintenance policy selection. *Organizacija, 47*, 177–189.
10. Moreno-Cabzali, B., & Fernandez-Crehuet, J. (2020). Application of a fuzzy-logic based model for risk assessment in additive manufacturing R&D projects. *Computers & Industrial Engineering, 145*, 106529.
11. Jaderi, F., Ibrahim, Z. Z., & Zahiri, M. R. (2019). Criticality analysis of petrochemical assets using risk based maintenance and the fuzzy inference system. *Process Safety and Environmental Protection, 121*, 312–325.
12. Kumru, M., & Kumru, P. Y. (2013). Fuzzy FMEA application to improve purchasing process in a public hospital. *Applied Soft Computing, 13*(1), 721–733.
13. Silva, L. (2018). Fuzzy Regression Model (Modelos de Regressão Fuzzy). Monografia curso Estatística do Departamento de Estatística e Matemática Aplicada da Universidade Federal do Ceará.
14. Paixão, A. (2010). Modelo de Sistema de Inferência Difuso para Auxílio da Resposta ao Risco em Projetos. FCT-UNL.
15. Ross, T. (2010). *Logic and fuzzy systems. Em fuzzy logic with engineering applications* (pp. 117–173). Wiley.
16. Cox, E. (1994). *The fuzzy systems handbook: A practitioner's guide to building, using, and maintaining fuzzy systems*. AP Professional.
17. Melo, D. (2009). Implementação do Sistema Fuzzy de Mamdani usando como ferramenta o Visual Basic for Application no Excel. Faculdade de Economia e Finanças – IBMEC.
18. Kumar, G., & Maiti, J. (2012). Modeling risk based maintenance using fuzzy analytic network process. *Journal Expert Systems with Applications, 39*, 9946–9954.
19. Meng Tay, K., & Peng Lim, C. (2006). Fuzzy FMEA with a guided rules reduction system for prioritization of failures. *International Journal of Quality & Reliability Management, 23*(8), 1047–1066.
20. Chung, E.-S., & Kim, Y. (2014). Development of fuzzy multi-criteria approach to prioritize locations of treated wastewater use considering climate change scenarios. *Journal of Environmental Management, 146*, 505–516.
21. Zanette, A., Radanovitsck, E., & Gonçalves, W. (2006). *Fuzzy logic*. UFRGS.
22. Sousa, A. (2014). *Controladores Linguísticos Fuzzy*. Instituto Superior de Engenharia de Lisboa – ISEL.
23. Mestre, I. (2022). *Proposta de Metodologia para a Gestão da Manutenção Baseada no Risco Utilizando a Lógica Fuzzy* (Proposed Methodology For Risk-Based Maintenance Management Using Fuzzy Logic). Master Thesis (Tese de Mestrado). Faculdade de Ciências e Tecnologia da Universidade Nova Lisboa.
24. Portuguese Navy. (2004). *Salt water electropump (fire) operation and maintenance*. Technical Book from Ship Technical Direction.

Index

J. C. de Oliveira Matias et al. (eds.), *Quality Innovation and Sustainability*, Springer Proceedings in Business and Economics,
https://doi.org/10.1007/978-3-031-12914-8

R

S

T

U

W